Planning
and
Urban Change

The Author

Stephen V. Ward is Principal Lecturer in Planning History at Oxford Brookes University. In the early 1970s he was research assistant on the Royal Town Planning Institute History Project, published as G. E. Cherry, *The Evolution of British Town Planning* (1974). Since then he has published many articles on historical topics related to planning and is the author of *The Geography of Interwar Britain: The State and Uneven Development* (1988), editor of *The Garden City: Past, Present and Future* (1992) and co-editor (with Professor John Gold) of *Place Promotion: The Use of Advertising to Sell Towns and Regions.* He is an active member of the International Planning History Society and edits *Planning History*, the Society's bulletin. He is currently working on a major international historical study of place promotion. He is married with three children and lives in Oxford.

Planning and Urban Change

Stephen V. Ward

Paul Chapman Publishing Ltd

To my mother and the memory of my father

Paul Chapman Publishing Ltd
144 Liverpool Road
London
N1 1LA

British Library Cataloguing in Publication Data
Ward, Stephen V.
Planning and Urban Change
I. Title
307.76

ISBN 1 85396-218-X

Typeset by Best-set Typesetter Ltd., Hong Kong
Printed and bound by Cromwell Press Ltd., Broughton Gifford, Great Britain

B C D E F G H 9 8 7 6

CONTENTS

Preface vii

1. Planning and Urban Change 1
   Introduction 1
   The central themes of the book 2
   The structure of the book 8

2. Ideas and the Beginnings of Policy 1890–1914 10
   Introduction 10
   The nineteenth-century city 11
   The origins of town planning ideas 19
   The beginnings of statutory town planning 30
   Overview 38

3. Widening Conceptions and Policy Shifts 1914–39 39
   Introduction 39
   Town extension planning and mass suburbanization 40
   New models for managing urban growth 51
   Origins of planned redevelopment 60
   Origins of balanced regional development 69
   Towards a comprehensive approach 76

4. A New Orthodoxy of Planning 1939–52 80
   Introduction 80
   War and the emergence of a mass consensus for planning 80
   Formulating the post-war planning agenda 1940–2 86
   Partial implementation of the planning agenda 1943–5 92
   Creating the new orthodoxy 1945–7 100
   Completing the new orthodoxy and the first retreats 1947–52 112
   Overview 115

5. Adjustments and New Agendas: I. The Changing Planning System 1952–74 116
Introduction 116
Politics, the 'affluent society' and planning 117
The changing physical planning system 122
Overview 140

6. Adjustments and New Agendas: II. Strategic Policies 1952–74 142
Introduction 142
Planning and redevelopment 143
Planned containment 161
Regional balance 168
Planned decentralization 176
Planning 1952–74: conclusions 186

7. Remaking Planning since 1974: I. The Changing System 188
Introduction 188
The making of a new political economy 189
The changing physical planning system 195
Overview 222

8. Remaking Planning since 1974: II. Specific Policies 224
Introduction 224
Regenerating the urban cores 224
Accepting regional imbalance? 244
Decentralization after the New Towns 248
Planned containment 254
Conclusions 1974–94 258

9. Planning Impacts since 1945 and the Future 260
Introduction 260
Spatial impacts of planning 260
Economic and social impacts 271
The future 276

Bibliography and Sources 280

Index 307

Preface

This book is a product of over two decades studying the development of planning in Britain and its role in changing cities. It began as a student interest when I was fortunate enough to study under Tony Sutcliffe and Gordon Cherry, who have subsequently become the godfathers of planning history, not just in Britain but internationally. Gordon played a particularly important role in sustaining it further by employing me as his research assistant on the Royal Town Planning Institute History Project in 1971–3. Thereafter it became my main teaching interest at South Bank Polytechnic (up to 1979) and Oxford Brookes University. The Planning History Group, now the International Planning History Society, has furthered my knowledge and understanding and brought me into contact with many others interested in the development of planning. All this is an essential underpinning for this book which, although it incorporates some original research, is largely a work of synthesis. It brings together the labours of many scholars, commentators and practitioners in what I hope will prove to be an interesting and informative textbook, accessible and useful for students. Certainly I have greatly enjoyed researching and writing it.

Inevitably many debts are incurred in such a project. First I must record my thanks to the many colleagues at Oxford who contributed in different ways to this study, by lending or giving me material, by advice on sources, by insights drawn from their own research, practice or teaching and, not least, by their informative conversation. Particular thanks are due to Huw Thomas and Patsy Healey (both ex-colleagues), who read an advanced draft of the book and made many suggestions for its improvement (though of course whatever faults remain are my own responsibility). Dennis Hardy and John and Margaret Gold were kind enough to lend me pictures used in their own publications. Rob Woodward has prepared most of the illustrations for this book. Marian Lagrange and the staff at Paul Chapman have been helpful, efficient and encouraging during the period of its writing and production.

Yet, as always, the main debts are personal ones. My own much-loved family has borne a disproportionate part of the burden of the writing of this book. I thank Maggie, Tom, Rosamund and Alice for their immense tolerance of all the distracted holidays, weekends and evenings. In the longer term I also owe an huge debt to my own dear parents, who did so much to encourage my interests in my surroundings and their history. This book is therefore dedicated to my mother but sadly only to the memory of my father, who died a few months before it was completed.

Stephen V. Ward
Oxford

Picture Credits

Bournville Village Trust: 2.2, 3.1.
Controller of Her Majesty's Stationery Office: 4.4, 4.5, 4.6, 5.1, 5.2, 5.3, 6.2, 6.3, 6.11, 7.1, 7.2, 7.7, 8.10.
Consortium Developments Ltd: 6.9.
Corby Borough Council: 7.6.
First Garden City Heritage Museum, Letchworth: 2.6.
Hulton Deutsch Limited: 6.9, 8.1.
Laing Development Company: 6.4.
Liverpool City Housing Department: 2.3, 3.7, 3.8.
Merseyside Development Corporation: 8.2.
Newcastle Chronicle and Journal: 6.7.
Osborn Collection, Welwyn Garden City Library: 3.3, 3.4.
Punch Library: 3.5.
Sheffield City Housing Department: 6.5, 6.6, 6.8, 6.10.
Solo Syndications: 4.2, 4.3.
Thamesdown Borough Council: 6.14.
Town and Country Planning Association: 4.8.
Willan Group Limited: 8.3.
7.5, 7.8, 8.4, 8.5, 8.6, 8.7 and 8.8 were photographed by the author.

Despite all efforts, the copyright on several illustrations remained uncertain. It was not possible to trace the copyright for 2.1 and 3.9; my thanks to Gateshead Central Library and the Rights Department of Victor Gollancz publishers for their painstaking assistance on these. 6.1 was photographed by Brian Wharton, whom it has not been possible to trace, though I am grateful to Oliver Marriott, in whose book *The Property Boom* the picture first appeared.

All other illustrations are from the personal collection of the author, except 3.3 which was lent by John and Margaret Gold. In all cases there were no obvious copyright holders to approach, either because of the age of the material or because the originating agency no longer exists and has no current parallel.

The author and publishers are grateful to all whom gave their permission for the use of copyright material. They apologise if they have inadvertently failed to acknowledge any copyright holder and will be glad to correct any omissions that are drawn to their attention in future reprints or editions.

1

PLANNING AND URBAN CHANGE

INTRODUCTION

Town planning, by its nature, is essentially concerned with shaping the future. This does not mean, however, that town planners are able to ignore the past. In an older urbanized country such as Britain they have, fairly obviously, to work with physical structures and urban arrangements inherited from the past. What is less obvious though is the point that the concerns and ideologies of the town planners themselves are also products of the past. Planners carry with them professional assumptions about the need to regulate and order urban space and about the ways in which they should do this. They also work within a planning system that embodies past political assumptions about the institutional location, purpose and instruments of planning policy. And, not least, they have to live with the consequences of past planning decisions, expressed within the fabric of towns and cities.

All of this is by way of arguing that to understand town planning properly, it is essential to understand how it has developed. This is not to say that planners or indeed society should drive into the future with eyes fixed exclusively on the rear-view mirror. Quite obviously this would be a recipe for disaster, though the analogy aptly reminds us that failing to look behind can also produce disaster, however exhilarating it may be in the short term. Nor is it to say what many planners have certainly thought in more pessimistic moments during the recent years of Thatcherite assaults on their activities, that the past may be the only thing they have to look forward to. Clearly, it is always important to appreciate that town planning as a tradition of thought, policy and action has a breadth, depth and diversity that may not be immediately apparent in the way it is practised today. But however much we might yearn for the Utopian socialism of the early days or the political commitment to the strong and socially concerned planning system that was created in the 1940s, we must also understand the reasons why they were superseded.

The case for the explicitly historical approach of this book is, then, to enable a proper and rounded understanding of town planning as a continuing tradition of thought, policy and action. Other books share this broad historical approach, among them Ashworth (1954), Cherry (1972, 1974b, 1988), Hall (1988, 1992), Ravetz (1980, 1986), Lawless and Brown (1986) and Hague (1984). Each inevitably interprets the story in their own way, stressing different aspects and offering different explanations. The reader should certainly refer to these and the other works referenced in the succeeding chapters of this book to gain fuller understanding. But the present work provides a solid grounding for those training to be town planners or otherwise interested in planning in the 1990s to understand the development of town planning ideas and policies since the late nineteenth century and assess their impacts on urban change. We examine where town planning ideas actually came from, who originated them and why. We also consider how and why governments saw fit to incorporate at least some of the ideas of the town planning movement into state policies, and assess their impacts on the actual processes of urban change. Ideas, policies and impacts are in fact the three continuing themes that run through this book. It is therefore important to establish from the outset what we understand these terms to mean, and briefly rehearse some of the main arguments that we will develop in detail in the following chapters.

THE CENTRAL THEMES OF THE BOOK

Ideas

The genesis of town planning ideas
Before it was anything else, town planning was a series of radical reformist ideas about changing and improving the city which began to take shape from about 1890. The basis of these ideas lay in land reform and, increasingly, housing reform, though with other important dimensions in the enhancement of community and the protection of amenity. The actual term town planning was coined, probably in 1905, to give these ideas a distinct identity and coherence. They were advanced further mainly by the relatively small number of reformers and professionals who rallied behind the new flag of the town planning movement. A few key organizations, most notably the Garden City and Town Planning Association, the National Housing and Town Planning Council and the Town Planning Institute, played central roles in this.

Early conceptual innovation in planning
As the reformist ideas of this new movement were given physical expression in pioneering ventures such as garden cities and suburbs, it acquired a more specifically physical and professional focus. A new professional, design-based repertoire of ideas was assembled, incorporating wider strategic concepts of city extension or comprehensively decentralized 'social cities' and detailed ideas of zoning, site layout, etc. Within a few decades many important new ideas were developed and incorporated within this intellectual tradition of town planning. A strategic model for planned metropolitan decentralization and containment was moulded out of the more radical notions of the social city. Ideas for urban redevelopment were

reinvigorated as the functionalist theories of the modern movement in architecture were extended to entire cities. Before the late 1930s, however, there was usually little immediate prospect of most of these ideas being implemented on a sizeable scale. In fact what was most striking about the process of intellectual innovation over this period was the extent to which it was independent of the rather limited operations of planning policies in practice.

Later conceptual innovation in planning

All this began to change significantly as town planning ideas were comprehensively incorporated into official town planning policies from the 1940s. The essential focus of planning activity now became, as never before, the officially ordered planning system rather than the independent planning movement. Increasingly, and especially after 1960, innovations in planning thought arose more from within the policy process, rather than from the wider town planning movement. The tradition of autonomous intellectual thought and conceptual innovation that had characterized the earlier years now began to atrophy. The wider town planning movement became more concerned with refining and celebrating the contemporary successes of town planning policies (such as the New Towns), rather than with developing new radical models that looked beyond present concerns.

Thus as events and government actions moved very sharply against the established policy conventions of planning in the 1970s and especially the 1980s, the town planning movement has found itself without the autonomous intellectual tradition that would have allowed it to develop alternatives in the manner of earlier generations. Certainly new environmentalist ideas have emerged during these years, but not from the town planning movement. Despite welcome attempts to incorporate such ideas within town planning thought, there is no doubt that this new environmental radicalism has to some extent outflanked the older established town planning tradition. As we will show, one of the main reasons for this lies within our other central concerns with planning policies and impacts.

Policies

What are planning policies?

Put very simply, town planning ideas become policies at the point at which they are incorporated by government into officially endorsed courses of action. The manner of this incorporation varies, depending on the importance of the original idea. Some policies were so fundamental to town planning that they were written directly into the planning system by central government. For example, elements of land market reform were integral to the whole practice of town planning and became intrinsic policies, embodied in the various compensation and betterment provisions of the Acts. Others, however, represent conscious applications of the planning system to pursue particular ends, for example by encouraging town extension rather than containment, or rehabilitation rather than redevelopment, etc. Such conscious policies may also reflect the various scales of planning ideas, with strategic policies such as metropolitan decentralization or containment, and more detailed policies, for example zoning or pedestrianization. The adoption of these conscious policies is inevitably a rather more discretionary process, involving more local decisions.

Town planning policies and party politics

Ideas may have political implications but, while they remain just ideas, this dimension remains fairly passive. Policies, by contrast, are actively political and the course of policy-making in planning cannot therefore be understood without reference to a wider political frame. In this connection it is immediately important to recognize that town planning ideas have generally found a more sympathetic political home within the Liberal and Labour Parties. By its nature town planning as a political project has involved greater state control over private activity, particularly in the use and exploitation of land. Its general political trajectory has been based historically on the assertion of public interest concerns and reduction of the role of private interests in the urban development process. It is therefore readily understandable that first Liberal and later Labour administrations should have set the pace in town planning policy-making over this century.

Conservative governments have generally pursued a more cautious line, usually diminishing earlier Liberal or Labour planning policy initiatives in favour of private development interests. This has been particularly evident in the intrinsic policies on land values that are embodied in planning legislation, on which there is a long history of party political disagreement. Generally Conservatives have wanted to see a bigger proportion of land value increases arising from development remaining with private landowners and developers, while the Liberals and Labour have favoured stronger taxation or public landownership.

In other aspects of planning, party political disagreements have been less significant. Conservative sympathy for the private developers has been tempered by their sensitivity to other important political interests, notably the protection of residential owner-occupiers and other established interests, such as farming. In practice this has encouraged a high degree of political consensus over many planning matters, especially in the 1940–74 period. Elements of this approach have even survived as the traditional post-war consensus on many aspects of state policy broke down in the late 1970s and 1980s. In practice the traditional Conservative desire for planning policies that restricted development 'in their backyards' has probably outweighed the post-1979 Thatcherite Conservative portrayal of it as a drag on the 'enterprise culture', especially in urban fringe areas. But it remains true that the planning policies of the 1980s and 1990s have become far more developer led than they ever were from the 1940s to the 1970s.

Officialdom and policies

Another important element in the continuity of planning policies has been the largely hidden politics of Civil Servants, defending their established departmental interests. In an activity like planning, which has only rarely been a national party political issue of the first order and which raises issues of great technical complexity, Civil Servants and professional experts have had key roles in forming policies. Ultimately they can be overridden by determined ministers, but we will find only a few clear instances where this has definitely happened. By contrast officials have certainly limited the impact of potentially far-reaching planning reforms on some occasions, most notably in 1964–5. Their local equivalents, the municipal town planners, also helped shape the course of policies in their own areas.

Ideas and policies

All of this serves to remind us that the policy dimension introduces many considerations that may have precious little to do with town planning ideas in the purest sense. One of the most important points to understand about the relationship between ideas and policies is that the state generally adopted town planning for reasons rather different to those which motivated the town planning movement to invent it. The recognition of this point can sometimes be difficult, because the advocates of town planning were generally clever enough to present their case in a way that did more than merely preach to the converted. In the pre-1914 period, for example, they addressed their arguments to the concerns of the political élite to defend British imperial and economic dominance and domestic social stability, not themes which were central to the garden city or co-partnership housing ideas.

But there can be little doubt that it was these (and other) wider concerns which propelled the Liberal government towards the first town planning legislation in 1909 rather than the radical reformism and Utopian socialism of many of the inventors of town planning's central ideas. Similarly, external events like the 1930s' Depression and both World Wars, especially the Second, had a dramatic effect on the sympathy of governments for planning ideas that were of rather older origin. Whereas ideas can and frequently do have a purity of purpose, policy-making has been inherently opportunistic.

The lack of congruence between conceptual innovation and policy-making was also reflected in the partial way in which town planning ideas were incorporated in policies. The wider social reformism that animated early planning ideas was quickly narrowed down to environmental policies. The co-operative and voluntarist tradition was soon subordinated to a more government-oriented approach as, for example, co-operatively developed garden suburbs gave way to municipal satellite towns and garden cities to New Towns. Yet such policy-driven changes could also add a creative dynamic as older ideas were adapted to meet new circumstances. This was particularly noticeable in the post-war period as the distinction between ideas and policies became increasingly blurred. By the 1960s, for example, the metropolitan, decentralist garden city/New Town idea was being merged with French growth-pole theory to become a model for regional growth. But these various kinds of slippage between ideas and policies were minor compared to that which occurred when policies were actually implemented.

Impacts

Assessing impacts

Whether town planners like it or not, society essentially evaluates the success or failure of planning by its impacts rather than the intentions embodied in policies, still less in the underlying ideas. Social and political reactions to the perceived impacts of planning policies have helped trigger important changes of direction, for example against town extension policies in the late 1930s and 1940s or against comprehensive inner-city housing redevelopment in the 1970s. It is, however, arguable that much of what was being criticized did not actually result from planning policies, but represented particular modes of implementing planning policies over which planners had little control.

There is, in fact, a major problem in assessing what the impacts of planning have actually been. Planning policies supplement and attempt to order existing processes of urban change, but do not normally replace them. This means that we can never be exactly certain as to what would have happened without planning. Inevitably, therefore, planning impact studies, however well researched, ultimately rest on a degree of intelligent 'counterfactual' guesswork and can never be proven in a completely satisfactory way. As we have suggested though, the need to evaluate planning outcomes is too important to be ignored on grounds of methodological purity.

Impacts, policies and ideas

What we are arguing here is that what are termed the impacts of planning have actually been produced by a much wider and more diverse set of forces than those which have shaped our other two concerns. Both ideas and policies were certainly directed at wider issues, but their characteristics were actively shaped within rather narrower social and political milieux. Thus ideas essentially arose out of the particular social and intellectual setting of a few creative individuals and organizations. Policies were broadly the result of an interaction of these planning ideas with political and institutional processes, but in ways that still reflected the particular concerns and approaches of the policy-makers. Planning impacts, in contrast, arise from the interaction of these ideas and policies with an altogether wider range of economic and social forces, many of them entirely outside the planner's control. Despite commonly held assumptions about their power to shape cities, town planners have actually had rather limited powers over most of this century.

Planning and twentieth-century urban change

Developing this point further, we must immediately note a considerable variation over time in the conscious ability of planners to shape the pattern of urban change. Planning was a significant though somewhat marginal influence before 1939. Then, as wartime brought a state-dominated economy, marginalizing private landed and development interests, town planning briefly became a major force in the 1940s. But when this interventionism gave way, during the 1950s, to the mixed economy, where an active state was retained but within what was now an unequivocally capitalist, market-led economy, town planning also assumed a less directive role in urban change. Increasingly, until the 1970s, town planning sought to harness rather than dominate private interests in development. Decisive change occurred in the 1980s as town planning intentions were more firmly subordinated to a market-dominated process of urban change.

Planning and the market

The net effect of all this fluctuation is that planning has only once, in the 1940s, been in a sufficiently strong position actively to shape urban change fairly independently of land market considerations. Yet this was a period when relatively little urban development took place, compared to what went before and what came after. Most urban development has occurred when town planning was able to exert at best an important minor, at worst a marginal, influence on the process of change. The implementation of plans has relied on actions that have been beyond the planner's control. They have been dependent on investment (and disinvestment) decisions by private developers and land users. Such decisions have been motivated firstly by considerations of profit and only, at best, secondly by the intentions of town planners.

This has been true even of such state-dominated exercises in planning as the early New Towns, where the Development Corporation set up by central government for each town was planner, landowner and developer. Despite such apparent omnipotence, success remained contingent on decisions by employers to locate there. Nor was inner- and central-city planned redevelopment able to transcend the considerations of the market, despite an almost equally powerful array of planning powers. Still less could planning actively shape urban change where it was merely providing blueprints or guidelines for change but remained entirely reliant on private development for their fulfilment. Nor was it just the economic considerations of the market that deflected the outcomes of planning from the original intentions.

Town planning and other state intervention

Another point of great significance is that town planning has never had complete control over other aspects of government intervention in relation to urban change. Town planners actually have very little direct formal influence over road-building decisions and have often had surprisingly little real control over public-sector housing, despite a historically close association between the two functions. More generally there is no automatic assurance that publicly owned land will be managed in ways that are any more consistent with town planning objectives than are those of private landowners. Especially in recent years, municipal estates departments have often needed to look to maximize financial returns on surplus lands and commercial properties, producing outcomes that differ little from those of the market. And town planning influence over the decisions of major public utilities (even when these were under state ownership), health authorities or nationalized industries, etc., has been very limited. All of this introduces other dimensions into plan implementation that the planner has less influence over than might at first appear. At best the town planner will be engaged in a continual process of negotiation with other arms of state intervention, attempting to mould their actions into something consistent with planning intentions.

Planning and social change

Finally we should note that planning impacts also bear the imprint of important social changes that have proved notoriously difficult to predict, let alone consciously incorporate into planning intentions. Since they began making use of population forecasts in the 1930s, town planners have, fairly consistently, had to work with faulty projections. Future population size was underestimated in the 1930s, 1940s and 1950s, then overestimated in the 1960s. And even as the accuracy of overall population size forecasts began to improve in the 1970s, serious mistakes were made in predictions of household size. Such flawed demographic data reverberated throughout town planning, most notably in a very marked failure to allocate sufficient housing land in the post-1945 period.

In turn, faulty population projections were paralleled in weak forecasting of other critical social and economic indicators. Thus predictions of car ownership were also seriously flawed. The net result of these not very well foreseen social changes was that planning outcomes could sometimes diverge markedly from the intentions embodied in major planning policies. For example, metropolitan containment and decentralization policies took on a somewhat different character as greater affluence and mobility allowed far more spontaneously dispersed patterns of living and working than those envisaged in the early post-1945 plans. In contrast to the highly planned outer metropolitan areas envisaged in the 1940s, a much less planned 'outer city' has emerged. And, as we will see, it is not difficult to think of other examples where unforeseen social changes have deflected policy impacts from policy intentions.

The Structure of the Book

This interplay of ideas, policies and impacts is then our central theme, present in every chapter. The book's detailed structure is largely chronological, however, tracing the development of town planning over approximately the last hundred years. In the next chapter we examine the origins of town planning ideas and policies in the formative pre-1914 years, when the term 'town planning' itself came into being. Chapter 3 looks at the widening of planning ideas during the period between the outbreak of the two World Wars. It also assesses the impacts of the first wave of planning policies, encouraging town extension, and considers how planning policies shifted to reflect something of the widening agenda for planning action in the 1930s. Chapter 4 is largely focused on the decisive policy changes of the 1940s, when the circumstances of war created the political momentum for decisive action on the town planning agenda that had been rehearsed between the wars.

The next two chapters look at the years of post-war affluence, from the early 1950s to the early 1970s. Chapter 5 traces the development of the planning system over this important period of consensus mixed-economy planning, highlighting both intrinsic policies and new ideas. Chapter 6 complements this by examining conceptual and policy developments in relation to the four main strategic planning concerns with redevelopment, containment, regional balance and decentralization. This pattern is repeated in Chapters 7 and 8, covering the years since the early 1970s, when the traditional political consensus over planning matters has been substantially undermined. Thus Chapter 7 explores the broader pattern of change in planning thinking and policies, while Chapter 8 addresses the specifics of particular policy initiatives. Finally Chapter 9 considers the overall impacts of post-1945 planning and draws together the main conclusions of the whole book. We begin, however, at the other end of this historical process of urban change, considering how and why the notion of town planning came to be invented and adopted as public policy.

2

IDEAS AND THE BEGINNINGS OF POLICY 1890–1914

INTRODUCTION

The last decade of the nineteenth century and the early years of the twentieth was the critical formative period for the development of town planning thought and policy in Britain and the other major urban industrial countries. It was during these years that the specific reform movements that created the body of thought that underpins modern town planning came to prominence. And it was during these years that the very term 'town planning' was coined, as an umbrella term to encompass the activities of these separate and in some ways rather divergent reform movements. A recognizable and self-conscious town planning movement appeared and began to foster the ideas of town planning, in model schemes of various kinds. It also began to act as a pressure group to lobby politicians with a view to securing the incorporation of these ideas into formal government policies.

In 1909 the first British town planning legislation was passed – the Housing, Town Planning Etc. Act. It was this modest measure which defined what has become the dominant location of town planning activity, within the framework of local government. This was the first step in shifting planning away from the realm of philanthropic and co-operative social reform to the realm of state policy. In the same year too town planning education was begun, a key stage in the professionalization of the new activity of planning that accompanied the transition from idealism to policy. A few years later, in 1914, the body that ultimately grew into the professional body and qualifying association for planning, the Town Planning Institute, was itself formed.

In this chapter we examine and analyse how and why this new activity of town planning appeared. The distinction, introduced in the last chapter, between ideas and policies is developed in this explanation. In particular we will contrast the often radical origins of the ideas that made up town planning thought, based in many cases on Utopian socialist and co-operative traditions, with the more conservative conceptions of reform that led to their incorporation into policies. Both, however, were essentially reactions to the late nineteenth-century city.

THE NINETEENTH-CENTURY CITY

The urban condition

The extent of urban growth

During the nineteenth century, Britain had become an overwhelmingly urban society (Dyos and Wolff, 1973; Carter and Lewis, 1990). Although the rate of urbanization was less rapid than that which has affected many other countries in the twentieth century, we should be under no illusions as to its significance. There has been nothing comparable anywhere in Britain in our lifetime. In 1801 the urban population of England and Wales had been 3 million, just over a third of the whole population. By 1911 the figure was 36.1 million, nearly four-fifths of the total. This growth was integral to the economic transformation of Britain from an agricultural to an industrial economy. Whereas towns and cities had formerly fulfilled essentially trading and mercantile functions, they now became central to the new, more concentrated, mode of capital accumulation. The growing urban populations supplied the labour that was essential to the functioning of this new, more intense mode of economic activity.

Before industrialization, only London, with a population of roughly 575,000 in 1700, would have been regarded as a large city by modern standards. The second largest city in England and Wales, Norwich, had only 29,000. A century later, in 1801, London had risen to 865,000 (almost 10 per cent of the total). A new pattern of regional cities was now emerging, headed by Liverpool and Manchester with 78,000 and 70,000 respectively. London remained the only really large city, however, outweighing all the other urban centres put together. Another century later this had changed. London had continued to grow in absolute and relative importance, with a population of nearly 6.6 million. However, the five major provincial urban concentrations of south-east Lancashire, Merseyside, west midlands, west Yorkshire and Tyneside now together contained 6.8 million. Clydeside added another 1.5 million and there were smaller concentrations in south Wales, the east midlands, south Yorkshire and elsewhere. By 1891 over half the population had been living in towns of over 20,000 population.

Urban spatial structure

The economic and demographic changes associated with urban growth had also remodelled the spatial structure of cities (Johnson and Pooley, 1982). By the late nineteenth century, and before the appearance of town planning, there were already recognizable zones within cities that were highly specialized in particular activities. Central office and shopping districts had emerged. Beyond them, typically, were mixed areas of industry and tightly built working-class housing, usually built in the earlier part of the century. In the midlands and much of the north these were typically built in back-to-back form. In Scotland and a few English towns, multi-storey tenements were usual. Further away from the centre these very high-density areas gave way to more recent and rather less packed and

Figure 2.1 *By the late nineteenth century all large urban areas had densely packed slum areas close to their centres. This photograph, of Gateshead, shows one basic type of slum, the tenement, where many families lived under one roof. Further south, tiny individual houses, often of back-to-back construction, were typical.*

more ordered development of housing, often close to, but less intermingled with, industry. There were also distinct middle-class residential areas. Finally in the urban-fringe areas were large areas of more completely residential development, varying from upper-working-class terraces, to the rather grander detached villas of the better off. By the end of the century the rapid growth of these suburbs, catering for a widening lower middle class was becoming one of the most striking aspects of urban change.

Urban spatial dynamics

Overall this new pattern had arisen relatively spontaneously, reflecting the demands of a changing economy and society (Dennis, 1984). These changing demands were mainly expressed through the operations of an increasingly dynamic urban land market, which began to allocate space for particular activities using the price mechanism. Another closely related and directly formative influence on the structure of the nineteenth-century city was the increasingly sophisticated urban transport system. By the early twentieth century this included railways, electric tramways and first horse and then motor buses. These, especially the first two, were of critical importance in the growth of the new middle-class residential suburbs, allowing them the increasing separation from the typical places of employment on which their inhabitants depended, the offices and shops of the central area.

Urban problems

The immense scale of urban growth and change had not been problem free. Although there had been an abundance of private capital to fuel economic transformation, the vast increase of urban population associated with this created major social problems (Sutcliffe, 1982). There was a prevailing political faith in *laissez-faire* over much of the nineteenth century, allowing market processes to reign supreme in most aspects of economic and social life. Unregulated private enterprise proved quite unable, however, to create or maintain the social investment and services that were the essential cornerstone of the city as an efficient productive unit. The nineteenth-century city was accordingly dogged by serious and recurrent problems of disease and ill-health. Outside London, building took place for much of the century without effective controls over materials or minimum standards. Most cheaper houses were poorly designed and built. Drainage and sanitary provision was usually poor and often abysmal, relying on rural practices of cess pits or earth closets. Water supplies were often inadequate and drawn from tainted sources.

By the last decades of the century the severity of the public-health question was diminishing in significance as governmental action gradually tackled the problems. Other urban problems which private initiative had also failed to address satisfactorily now assumed greater prominence. Education and the creation or ownership of new urban infrastructures of gas, electricity and public transport became more pressing, all issues which had tremendous potential for the overall effectiveness of the city as a productive unit. The issue of the inadequacy of low-cost housing supply, which was to be closely tied to the emergence of town planning, also began to figure increasingly as an urban problem from the 1880s.

Municipal intervention

The Victorian local government system

The main response to urban problems came from the various agencies of local government (Finer, 1950). Urban local government was largely in the hands of the boroughs, established for the main urban centres under the Municipal Corporations Act 1835. Within their areas they increasingly controlled most aspects of local government though the pattern outside was much more chaotic. It was greatly simplified by the creation of the county councils and county boroughs (in effect a new name for many of the municipal corporations) in 1888 and the urban and rural district councils six years later. Although separate single-purpose local bodies survived for education (until 1902) and poor relief (until 1929), these reforms created an enduring framework of comprehensive local government that survived until the 1972 reforms. It was a system which combined unitary authorities, the county boroughs, which undertook the full range of local government functions in the larger urban centres (outside London). Outside the county boroughs a two-tier system was created, with an upper tier of county councils and a lower tier of municipal boroughs, urban and rural districts. A slightly different two-tier system also operated in London, with a very powerful London County Council (LCC) and a lower tier of metropolitan boroughs. The Scottish system exhibited some important differences but for the bigger cities at least it was broadly similar to that in England and Wales.

Local government and urban reform

Much of the real initiative in addressing urban problems rested at local level (Briggs, 1968; Fraser, 1979, 1982). Central government provided an overall enabling framework and exerted general control, but it was local authorities that made the running. Frequent use was made of local legislation, locally promoted parliamentary Acts which gave specific powers to the sponsoring authority. Liverpool, for example, had taken pioneering steps in public health in the 1840s, while Birmingham launched a major drive towards municipal ownership of public utilities in the 1870s under the radical mayoralty of Joseph Chamberlain (Hennock, 1973). By the early twentieth century the LCC and Glasgow were establishing a supremacy in municipal tramways. The political bases of such municipal innovation varied, but it was usual to find progressive local big business and professional interests dominating the most innovative councils. Councils dominated by small business or property interests tended to be much more cautious in their policies. Reflecting the still narrow electoral franchise, Labour was not yet an important political force, though it was beginning to have some influence in London by the 1890s. It was this which fuelled contemporary discussion about 'municipal socialism', describing the increasing municipalization of the social capital of cities that had become well established by 1900. In fact though, this process would have been much better described as municipal capitalism.

Reformist motivations

How then can we explain the reformism of the urban élites? In the formal sense they increasingly sought to legitimate their policies by reference to the concept of the public interest. This was an important legal abstraction resting on a notion of the whole community and supposedly transcending the narrow interests of powerful individuals or social classes. Yet, as town planners found when they subsequently inherited the concept in the twentieth century, it is a very malleable and imprecise construct, capable of widely varying interpretation.

For the urban élites of the nineteenth century, it certainly rested on some degree of altruism, often inspired by religion. Many of the leading business figures, in the industrial cities at least, were active members of non-conformist churches such as Unitarians, Congregationalists or Quakers. This predisposed them to look for ways of doing God's work on earth (rather than viewing urban social problems solely in terms of personal morality). But there can be no doubt that the progressive policy agenda, the 'municipal socialism', of Victorian cities reflected a good deal that was in the general interests of industrial capital during this period. Disease and ill-health damaged industrial productivity. (And did not always respect social class: élite residential areas, such as Edgbaston in Birmingham or Victoria Park in Manchester, were not entirely insulated from the rest of the city.) Nor did the manufacturers approve of the high cost or inefficiency of utilities that characterized the private supply of water, gas or electricity. It all added either directly to production costs or indirectly to wage costs. Similarly cheap and efficient public transport reduced wage costs and gave employers access to a wider labour market.

The other side of this was that all these services were run to earn profits that could be used to keep local taxes down. There was no sense in which labour was being subsidized by the taxes of the better off, which true municipal socialism would have implied. It was noticeable that the business élites who ran urban government in the late nineteenth century adopted a much harder-faced approach in relation to functions like poor relief, which involved transfer payments to the poor, funded from local taxation. Such attitudes also coloured their whole approach to the mounting problem of housing.

The politics of the housing problem

The issue of housing had assumed political prominence during the 1880s, when popular and labour unrest had prompted the appointment of a Royal Commission on the Housing of the Working Classes, which reported in 1885 (Jones, 1971). Generally, real incomes were rising in the later decades of the century, allowing the lower middle classes and a widening section of the working class, particularly skilled workers in regular waged employment, to secure better housing through market mechanisms. The problem was that the position of a large section of the working class remained insecure. This was because many jobs (particularly the less skilled) were subject to cyclical fluctuations (e.g. shipbuilding) or were inherently irregular or organized on a casual basis (e.g. dock work) or simply remained very low paid. And at any one time all the big cities contained huge numbers of underemployed people, many of them attempting to eke out an existence as street traders.

The main working-class response to this was to create trades unions and other increasingly political movements to protect their interests. The middle and upper classes were, however, very reluctant to countenance any wholesale change in the distribution of income. Housing therefore became a more acceptable arena for political action. On the one hand it represented a key part of working-class quality of life, which gave it a real saliency in emergent Labour politics. On the other, it was an area where the urban business élites were prepared to make some concessions, particularly since the private landlords (who provided the bulk of working-class housing at this time) were increasingly being marginalized politically. Moreover, other municipal reformist initiatives had already begun to impinge directly on housing, giving it a tangible reality as an urban and municipal issue.

Housing and municipal action

By 1890 public health and street improvement (that is clearance) initiatives were actively limiting the supply of very cheap housing (Ashworth, 1954). Major schemes of central area rebuilding (such as the creation of Corporation Street in Birmingham) typically involved municipal acquisition and demolition of slum housing, usually replacing it with new streets, shops and offices. There were powers to close particularly unfit housing (e.g. cellar dwellings) in many cities, or even control overcrowding. The introduction of better water supplies, water-borne sewerage and other improvements into older housing also had the effect of pushing up rents. More importantly there were increasing controls on new building, in effect preventing additions to the stock of very cheap housing. London had had some degree of control since the Great Fire in 1666, but from the 1840s the new industrial towns and cities also began to regulate building, initially by local Acts. This regulation became general following the Public Health Act 1875. Most cities banned the very cheap and high-density back-to-back building. Virtually every late nineteenth-century English and Welsh city was now acquiring whole districts of the grid-iron street layouts and characteristic long rows of narrow-fronted 'by-law' houses (Muthesius, 1982; Daunton, 1983). The tenement tradition of Scotland and Tyneside was also more tightly regulated.

In addition, their new powers to create and manage the urban infrastructure were giving municipalities a more positive role in housing development. Most important was the growth of tramways, especially as electricity was adopted from the mid-1890s. The cheapness of the electric trams, in contrast to the higher costs of horse buses and, with some exceptions, railways, offered a tremendous prospect for the dispersal of the tightly packed inner districts, widening social access to suburban living. This was to be a key factor in Edwardian thinking about the housing problem during the first decade of the twentieth century.

The origins of municipal housing

By this stage municipalities themselves had begun to provide housing directly, though on a very modest scale (Wohl, 1974). It was legally possible to build council housing from 1851 but, although municipal lodging houses were built in a few towns, only Liverpool built family accommodation – St Martin's Cottages

Figure 2.2 *Housing built to conform to local by-laws brought marked improvements over the older slums by the last decades of the century. The form of most such housing in England and Wales is shown in this 1930s' photograph of Birmingham. Notice the long continuous rows of narrow-fronted houses, built to make maximum use of the wide (and, for developers, expensive) by-law streets, a sharp contrast with the courts and alleys of earlier slums. The houses themselves were larger with better sanitation and small backyards or gardens, giving increased privacy to individual families.*

(actually a tenement block) in 1869. A few more authorities took action in the 1870s and 1880s using a growing volume of legislation. Usually such initiatives involved the erection of a limited amount of replacement dwellings for central clearance schemes. There was still, however, a strong tendency to prefer private philanthropic initiatives wherever possible (Tarn, 1973). In London especially, it was still usual in the 1880s for bodies like the Peabody Trust or the Improved Industrial Dwellings Company to erect tenement blocks on sites provided by the Metropolitan Board of Works (the precursor of the LCC). These bodies operated on the principle of '5 per cent philanthropy', offering only limited profits of 5 per cent or less and relying on the great and the good of Victorian society to put up the funding.

The Housing of the Working Classes Act 1890

The extent of Victorian philanthropy fell well short of what was required, however, especially in provincial cities. The Housing of the Working Classes Act 1890 marked the beginnings of a comprehensive framework for municipal housing provision and it was certainly more extensively used than previous legislation (Yelling, 1986). Before 1890, rather fewer than 3,000 council houses had been built in the whole country; between 1890 and 1914 the figure was over 20,000, still an extremely small number but a marked increase. The new LCC, for example, built tenemented schemes at Boundary Street in Bethnal Green, Millbank and on several suburban sites, to be discussed more fully below (Beattie, 1980). The pace of building in the already active cities of Liverpool and Glasgow quickened. In fact most cities and many smaller towns built some housing under the Act.

Political limitations on municipal housing

Yet in doing this they quickly moved beyond the limits of their own political agenda (Merrett, 1979; Smith, Whysall and Bevrin, 1986). Municipal housing was an expensive function that, unlike gas, trams or water, could not be expected to earn profits. And although private landlords were often only marginal influences on the business élites who ran most cities, the fact was that municipal housing was competing with a well-established private rental market. Rents were not intended to be subsidized from local tax revenues, but concealed and sometimes overt subsidies were usual. It was easy therefore for councils and electors to become frightened of their growing involvement in municipal housing. Despite a mounting national housing shortage in the early years of the twentieth century, there was paradoxically something of a political backlash against municipal housing during this same period, especially noticeable in London and Birmingham (Nettlefold, 1908; Gibbon and Bell, 1939). This backlash was, as we will see, a

***Figure** 2.3 Higher standards meant higher rents that were beyond the means of most slum dwellers. The increasing need to remove the worst slums and the inadequate supply of cheap housing led many local authorities into direct housing provision in the 1890–1914 period, often using a more carefully designed form of flatted, tenement housing. This example was completed in 1914 by Liverpool, one of the main pioneers of municipal housing.*

crucial element in the mounting belief in town planning as a new policy solution. First, however, we must examine the emergence of the ideas underpinning the town planning movement that appeared in the first decade of the twentieth century.

THE ORIGINS OF TOWN PLANNING IDEAS

Land reform

The land problem

One of the key reformist concerns which underpinned the early development of town planning ideas was land. To an extent that is difficult to appreciate today, many reform interests of the late nineteenth and early twentieth centuries saw land as the most critical dimension in achieving social and political change. As they saw it, there were two main aspects of the land problem (Aalen, 1992). The first arose from the parlous state of rural economy and society, consequent on the decline in agriculture, as imported wheat and refrigerated meat began to replace home produce in the last decades of the nineteenth century. Land reformers argued that the prevailing pattern of land ownership, dominated by large and increasingly absentee owners, was a major contributory factor in the failure to adapt.

Meanwhile the intensification of the operations of the urban land market provided a second focus for their concerns (LLEC, 1914). As well as a general attack on private landlords, there were specific attacks on the growth of land speculation and the related practice of 'sweating'. These were becoming more common as transport improvements were opening up the cheaper land on the urban fringe for building. Meanwhile, however, shrewd speculators, who had contributed nothing to the costs of the improvements, were able to force up land values by buying up land around the new railway stations or tram termini, restricting its availability for development. In turn this nullified the original cheapness of the land, forcing development to occur at a much higher density than otherwise. This would also increase housing costs and thereby contribute to overcrowding and other urban ills.

'Back to the Land'

The whole land reform movement was unified by a strong pastoral, anti-urban impulse (Gould, 1988). The idea of a countryside revived by owner-occupied smallholdings as an alternative to the concentrated cities and derelict tenant farms was a powerful one, expressed in famous slogans such as 'Back to the Land' or 'Three Acres and a Cow'. But although such visions retained great potency in the Celtic periphery, above all Ireland, Britain's (and especially England's) overwhelmingly urbanized character by the late nineteenth century limited the specific appeal of such a peasant-based agrarian Utopia. In a less specific sense, however, the rural imagery of this vision did endure and became a powerful bequest to early town planning.

Land taxation and land nationalization

The advance of specific land reform ideas within the more urbanized sections of British society came to depend on more specific remedies. During the 1880s two bodies were formed that propounded slightly different solutions, both of which were subsequently incorporated within town planning thinking. The English Land Restoration League (and similar Scottish body) pushed the notion of a land tax. In this they were following the ideas of the American economist and increasingly well-known lecturer and writer Henry George, advanced in his important book *Progress and Poverty*, published in 1880 (George, 1911). He argued that the unearned increases in land values that accrued from the efforts of society at large should be appropriated by the state in a single tax. The other organization was the Land Nationalisation Society (Hyder, 1913). Under the leadership of Alfred Russel Wallace, the land nationalizers looked to the gradual elimination of private land ownership. It would be replaced by state ownership, so that communally created increases in land value would automatically be in collective hands, secured as ground rent from tenants.

Both organizations disseminated their ideas with great fervour from the 1880s, sending out fleets of red and yellow horse-drawn vans to spread the word. Both shared the characteristic belief that land reform alone would transform society, a notion which potentially offered a path to reform that avoided any clash between industrialist and worker. It seemed that both classes could find a common cause against private landed interests. Reform was to be sought in rents, not wages. A century later it is easy to see the limitations of these ideas, but we should also recognize their huge significance for the development of town planning. They were crucial in the early years and, incorporated within a wider reformist programme, they have continued to resonate throughout twentieth-century debates on town planning policy.

Housing reform

The origins of housing reform

By about 1900, land reform ideas had also begun to have an important effect on housing reform, itself another important contributor to the emergence of town planning ideas and policy. Throughout the second half of the nineteenth century there had been many important housing reform initiatives, some of which we have already noted (Burnett, 1978). The 5 per cent philanthropy schemes were usually model tenement dwellings, managed in a paternalistic way that was intended to promote a new model existence on the part of tenants. Tenants were to be steered away from social evils such as drink or gambling and towards more physically hygienic and morally uplifting lifestyles. Less draconian methods of management to secure social improvement were developed by Octavia Hill (1883). By the 1880s her emphasis on close tenant supervision to encourage self-help in matters of household management was having an increased influence. Paternalistic management styles, often reflecting at least some of Hill's precepts, were typically incorporated into the new municipal schemes, most of which were also tenement dwellings.

Industrial philanthropy

Several enlightened employers had also been attracted into housing and community development. In common with other forms of company or tied housing, this was usually in less urbanized locations where it was otherwise difficult to retain workers. The quality of a few of these schemes made them wholly outstanding, usually reflecting something more than simple economic self-interest. In common with many of the progressive leaders of late Victorian urban government, almost all such industrial philanthropists were active members of non-conformist churches. The new community of Saltaire, for example, was developed outside Bradford by the Congregationalist mill owner Sir Titus Salt from 1854 (Reynolds, 1983). Extremely important later schemes were those at Port Sunlight, near Birkenhead, developed from 1888 by Congregationalist soap maker W. H. Lever (Hubbard and Shippobottom, 1988) and Bournville, just outside Birmingham, developed from 1894 by the Quaker cocoa manufacturer George Cadbury (BVT, 1955).

Port Sunlight and Bournville

These later schemes and their initiators contributed very directly to the emergence of town planning ideas. Environmentally they were particularly important because they exemplified a new form of working-class housing that was of much lower density than contemporary by-law estates. Bournville, for example, was developed at eight houses per acre (twenty per hectare) compared to twenty (fifty per hectare) or more in adjoining areas. Cottage style housing in short terraced or semi-detached form was provided with large amounts of open space and good community facilities. In practice the rents were rather high and such schemes were certainly beyond the reach of slum dwellers. But rent levels at the lower end overlapped those of by-law housing so that skilled workers could begin to consider living in these kind of environments (Cadbury, 1915). For housing and land reformers the schemes illustrated what was possible when development was organized on non-speculative lines. They showed that the need for municipal housing might be avoided if the land market in the urban fringe could be regulated (as it was voluntarily by Cadbury and Lever) and some viable mechanism of low-cost housing development could be found.

Co-partnership schemes

Housing reformers thought they had found this in the co-partnership approach (Skilleter, 1993). Essentially it involved a co-operative structure, whereby the tenants collectively owned the society that undertook the development. This avoided some of the evils of land speculation and sweating and allowed land value increases after the site had been bought to be enjoyed collectively by the members of the co-partnership societies (LLEC, 1914). The approach had been pioneered in 1888, but was little used until 1901, when Ealing Tenants Society was formed. By this time the law had been strengthened to allow all housing agencies paying less than 5 per cent dividends to borrow most of the capital needed from the Public Works Loan Commissioners. This did not overcome all their financial problems and the continuing need for additional finance tended to give such

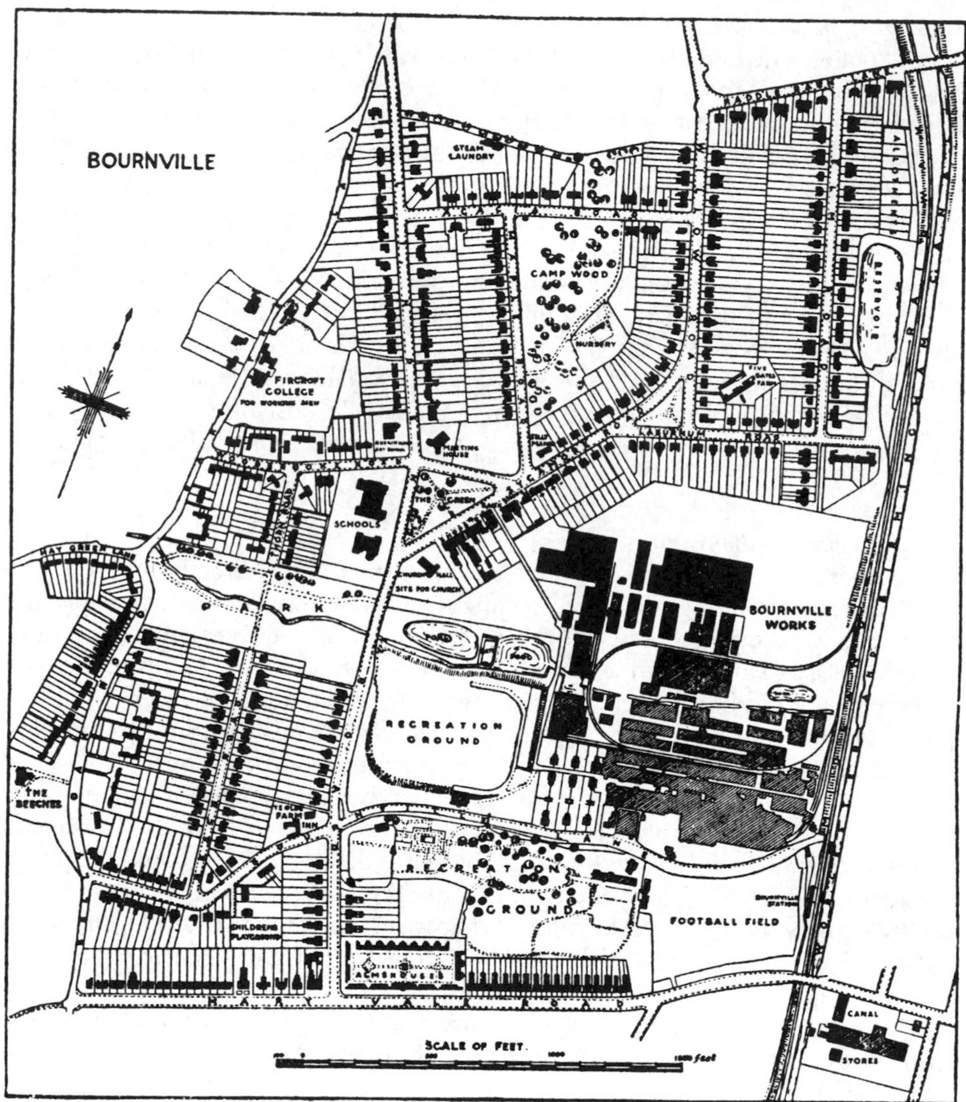

Figure 2.4 *Industrial philanthropists such as Cadbury and Lever developed something closer to the middle-class, rural-inspired ideal of the home as the model for working-class housing. Bournville, shown here in 1914, lay just beyond the 'middle ring' of Birmingham's by-law terraces, yet its cottage houses, large gardens and community facilities built around the Cadbury factory set new standards.*

initiatives a strong lower-middle-class/upper-working-class character. To Edwardian housing reformers, however, it seemed to offer a viable middle way between traditional philanthropy and municipal housing, an ideal way to secure low-cost, good-quality housing in the areas being opened up by the new electric tramways.

The National Housing Reform Council

By this time the whole topic of housing reform was growing in political importance. In 1900 Henry Aldridge, a Land Nationalisation Society lecturer, and William Thompson, a schoolteacher and member of Richmond Town Council in Surrey, formed the National Housing Reform Council (NHRC) (Aldridge, 1915). Initially it was a predominantly working-class organization, but it increasingly assumed a more middle-class character with figures such as Cadbury and Lever in prominent positions. It functioned primarily as a pressure group and quickly established strong parliamentary links. Meanwhile its initial concerns to promote municipal housing and secure central assistance gave way to a more mixed approach that gave great encouragement to co-partnership development and, within a few years, town planning. Its interests in this new area were not, however, unrivalled.

The garden city

Ebenezer Howard

The garden city idea and the movement it spawned were crucial precursors to the town planning movement in Britain. They were the source of many important planning ideas and the means by which existing reformist notions were applied to the solution of urban (and rural) problems. The originator of the idea was a lower-middle class Londoner, Ebenezer Howard (1850–1928), who had tried, and failed at, various things during his life, including homesteading in Nebraska and being an inventor (Beevers, 1988; Fishman, 1977). By the 1890s he was making a modest living as a parliamentary stenographer, producing verbatim shorthand accounts of meetings. More important though was his increasing immersion in the various currents of free thought and social reformism in the last decades of the century. Politically Liberal, Howard became increasingly interested in emergent socialist ideas, though very much of the voluntary and co-operative kind. In effect, he was a non-Marxist Utopian socialist, looking to achieve socialism without the need for class conflict.

From the late 1880s, Howard developed and rehearsed his ideas at great length within reformist circles and his own family, often going well beyond the boredom threshold of his listeners. One of his reformist acquaintances, the emerging playwright George Bernard Shaw, dubbed him 'Ebenezer the Garden City Geyser' (Beevers, 1988, p. 70). To his own family, particularly its younger members, his obsessive commitment was seen as 'an affliction' (*ibid.*, p. 44). Finally, however, in 1898, Howard published a book that introduced to a wider public the ideas which had taken over his life.

This book originally appeared in 1898 under the title *To-morrow: A Peaceful Path to Real Reform*, though it was reissued in 1902, with some modifications, under its better-known title, *Garden Cities of To-morrow* (Howard, 1902). It outlined a Utopian socialist alternative to the evils of existing urban society, specifically the huge urban concentration of London. This vision was to take the form of a social city, a decentralized network of individual garden cities, each of 30,000 population, surrounding a larger central city of 58,000. The garden cities

were to be slumless and smokeless, with good-quality housing, planned development, large amounts of open space and green belts separating one settlement from another. The key to the whole approach was to be the communal ownership of land purchased cheaply at agricultural values, so that the citizens of the garden cities would collectively benefit by the increment in land values consequent on urban development. Many other aspects of the garden city would also be owned and operated collectively, though Howard did not envisage a complete replacement of private capital.

The sources of Howard's ideas

The novelty of Howard's ideas lay in the fact that they were a 'unique combination of proposals' (Howard, 1898, p. 102). It is clear that Howard was heavily influenced by much of the Utopian thinking of the 1890s. Edward Bellamy's 1888 book, *Looking Backward* (Bellamy, 1888), describing a technological socialist future was a profoundly important influence. Important too were the socialist artist William Morris's arcadian visions of a rebuilt London, *News From Nowhere* (1890) (Briggs, 1962) and the various writings of the Russian anarchist Peter Kropotkin (1974, orig. 1899), who also favoured a small-settlement alternative to the big industrial city. Howard also drew heavily on the thinking of the land reform movement reflecting both their strong pastoral impulse and belief in the centrality of land and rent in achieving social change. Howard himself was closely involved with the Land Nationalisation Society and his earliest adherents in the garden city movement came from that source.

He was also influenced by various housing reform ideas, especially the environmental ideals being pioneered at Bournville and Port Sunlight (Creese, 1966) and the organizational innovations of co-partnership. Howard also explicitly acknowledged the importance of the ideas of the economist Alfred Marshall, who in the 1880s had proposed the creation of new industrial colonies, away from the big cities, where better working-class environments could be developed. Apart from Bournville and Port Sunlight, such hopes had remained unfulfilled at the time, but Howard saw his social city as a means of giving specific form to the industrial decentralization that was becoming apparent in the 1890s.

A garden city movement

What differentiated Howard's ideas from many other exercises in Utopian dreaming was that they quickly formed the basis for a wider movement (Hardy, 1991a). In 1899 the Garden City Association was formed to promote the garden city idea and in 1903 the first garden city at Letchworth in Hertfordshire was begun. Thanks to an effective propagandist campaign, many other countries also created their own garden city associations (Ward, 1992b; Buder, 1990). These successes coincided with the beginnings of a marked shift in the leadership and general tone by moving it from Utopian socialism to more respectable bourgeois reformism (Ward, 1992a). Howard had found it very difficult to secure support from the emergent labour movement, where there was scepticism from both the working-class and intellectual wings. By contrast liberal reform interests saw great

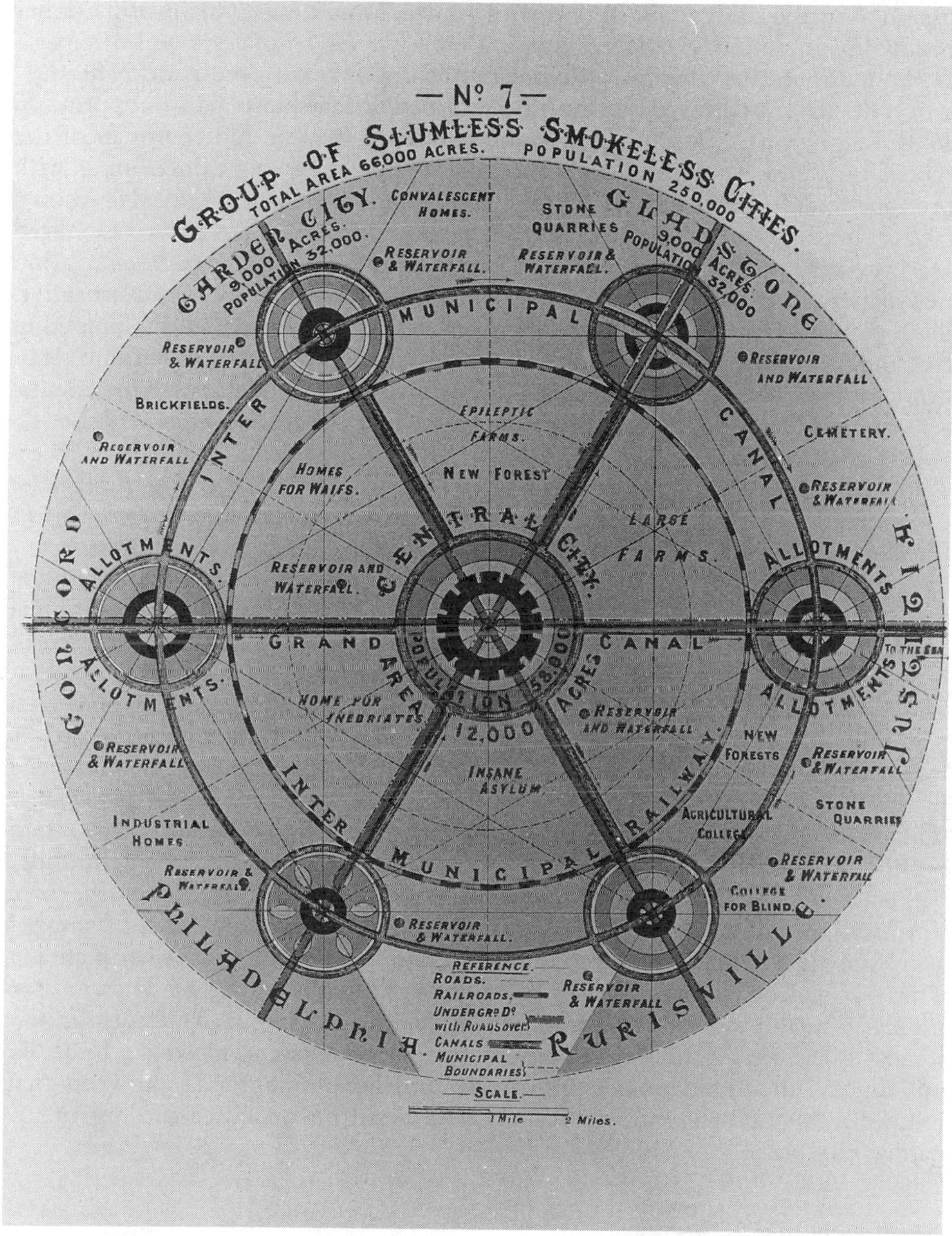

Figure 2.5 *Ebenezer Howard's book* To-morrow *(1898) articulated a vision of the urban future that addressed all the major social ills of late nineteenth-century society. The big, continuously built-up industrial city was to be replaced by a network of smaller garden cities, collectively known as the social city.*

potential in the movement, provided there was some toning down of its rather cranky image.

From about 1900 the initial dominance of the links with the Land Nationalisation Society interest was weakened and a new leadership began to appear. In 1901 Ralph Neville, a prominent London lawyer, became the chairman of the Association and he strengthened the links that were already developing with prominent industrialists like Cadbury, Lever and Thomas Idris, a well-known ginger beer manufacturer. Another extremely important recruit was Alfred Harmsworth, the proprietor of the *Daily Mail*, one of the first mass-circulation newspapers, catering very much for the new lower-middle-class suburbanites. Such patrons gave money and respectability to the garden city idea, something that the unworldly Howard would have found impossible. It soon became clear, however, that they saw the garden city primarily as a model of environmental reform and were less interested in Howard's wider reformism.

Letchworth Garden City

This shift was encouraged by the early experience of developing Letchworth (Miller, 1989). Although Howard and his immediate associates were able to buy the estate at agricultural value, it proved extraordinarily difficult to attract capital for such a Utopian venture. Compromises of Howard's principles of communal ownership and living were necessary, though this did not prevent Letchworth acquiring a reputation for social experiment. It was, though, the way that Letchworth was planned that gave it greatest claim to significance as the town planning movement took shape. Following a competition a plan was prepared by two young architects, Raymond Unwin and Barry Parker, both heavily influenced by the Arts and Crafts movement founded by William Morris (Day, 1981). Their scheme moved beyond the rather diagrammatic proposals in Howard's book to outline a carefully zoned plan with industrial areas segregated from housing, parkland and central shopping areas. As with other aspects of Letchworth, there were serious compromises in the implementation of this plan, but it expressed a holistic approach that became central to the wider notion of town planning which crystallized in the decade after 1900 (Sutcliffe, 1981a). There was a brilliance too in individual plan elements. Unwin and Parker were setting new standards at Letchworth in the design and planning of low-cost housing by 1905, building on the foundations of Bournville and Port Sunlight. In the years that followed, this element was increasingly detached from much of the rest of the garden city idea.

A town planning movement

The town planning umbrella

By 1905 the various initiatives of land and housing reform and the garden city were coalescing into a self-conscious town planning movement. The new term 'town planning' was apparently coined in 1905 by John S. Nettlefold, the Chairman of Birmingham's Housing Committee, in discussion with the same city's Medical Officer of Health, John R. Robertson (Adams, 1929; King, 1982).

***Figure 2.6** Letchworth Garden City, begun in 1903, embodied some of Howard's social ideals, but its major impact was as a model for environmental reform. Particularly important were the new standards for housing design and layout set by Letchworth's planners Raymond Unwin and Barry Parker.*

Nettlefold, a nephew of the radical 1870s' Birmingham Mayor, Joseph Chamberlain, and member of the city's Unitarian business élite, was a Unionist politician (in effect a radical Conservative, with policies very similar to the Liberals on urban and social matters) (Cherry, 1975a). His new label was quickly taken up, providing a politically neutral umbrella under which controversial reformist ideas could shelter, allowing them to regroup into something more widely acceptable and causing more overtly extreme ideas to be marginalized. All this was of course continuing a trend towards middle-class respectability which we have identified amongst the housing reformers and garden city advocates. And, in contrast to Howard's belief in the need to replace the big city, the new movement increasingly focused on ameliorating its housing and other problems.

German town extension planning

From about this time, a new idea for planned urban development began to appear, one which was to dominate actual planning policies in Britain until the 1940s. One of its earliest advocates was a Manchester reformer, Thomas Coghlan Horsfall, an active member of the National Housing Reform Council, who in 1904 published a book called *The Improvement of the Dwellings and Surroundings of the People: The Example of Germany* (Reynolds, 1952; Harrison, 1991). He showed how the Germans already had a well-developed system of town extension planning, *stadterweiterungen*. This provided a planning framework for urban expansion, reserving road lines and zoning land uses. It was also accompanied by cheap loans to builders of low-cost housing, municipal land purchase and cheap transport. The book had considerable impact, especially in Birmingham, and a municipal deputation, led by Nettlefold, duly went to Germany to see *stadterweiterungen* in action in 1905.

British town extension planning

While not denying the significance of garden city ideas, Nettlefold recognized that town extension was a solution which, with some changes in the law, could be readily applied to big cities like Birmingham. Some 2,200 miles of electric tramways were created throughout urban Britain over the years 1897–1906, amounting to nothing short of a revolution in cheap intra-urban transport (Finer, 1941). Because of this and the decline of agriculture an unprecedented quantity of extremely cheap land was becoming available for development in the urban fringe. As Nettlefold argued in his important book *Practical Housing* (1908), controls should immediately be imposed on this urban fringe to prevent land speculation and sweating and the associated cramped development. For preference the municipality should also own the land. There should also be positive measures to encourage the building of good-quality cheap housing, and here he looked to the co-partnership societies currently favoured by the National Housing Reform Council. Here, surely, was the solution to the problem of urban housing. If this could be implemented on a big enough scale, then it would allow densities in slum areas to be reduced by selective demolition, and remaining houses to be improved, avoiding any need for expensive municipal housing. 'It may be,' he commented, 'that Town Planning will show municipal house building to be unnecessary' (Nettlefold, 1908, p. 109).

A new model: the garden suburb

In English eyes there was, however, one weakness in the German model of town planning: its reliance on the high-density apartment block as the usual form of suburban expansion. It was here that the new residential ideals of Bournville, Port Sunlight and latterly Letchworth came into their own. The Garden City Association began increasingly to encourage the application of the residential ideals of Unwin and Parker to existing cities in the form of the garden suburb. This was something which was already starting to happen even as Letchworth was beginning. In effect they were accepting that it would be impractical to implement Howard's full social city strategy in the foreseeable future.

The earliest garden suburbs

During the Edwardian period there was a veritable flowering of garden suburb schemes (sometimes known as garden villages if there was an association with industry) (Culpin, 1913). A few, such as the LCC's schemes at Norbury, Old Oak in Acton, White Hart Lane in Tottenham and Totterdown Fields in Tooting, developed from 1900, were municipal schemes (Beattie, 1980). These were built at higher densities, however, and were much less impressive than the private initiatives. One of the first private schemes, slightly predating Letchworth, was New Earswick in York, a Bournville-type development initiated in 1902 by Joseph Rowntree, another Quaker cocoa manufacturer (Miller, 1992). As well as the Cadbury example, Joseph had been prompted by his son, Seebohm, who had just produced a seminal piece of social investigation based in the city, entitled *Poverty: A Study of Town Life* (Rowntree, 1901). Seebohm had already developed links with the garden city movement and it was he who brought in Unwin and Parker as architect–planners.

Hampstead Garden Suburb

Most famous was Hampstead Garden Suburb, in London, conceived in 1904, but begun in earnest from 1906, coinciding with the tube extension to nearby Golders Green (Miller and Gray, 1992). Initiated by wealthy philanthropists Dame Henrietta Barnett and her husband, the scheme quickly secured powerful and wealthy backing, in sharp contrast to Letchworth's still rather uncertain development. The scheme was planned by Raymond Unwin and Barry Parker, incorporating and developing their innovative work at Letchworth and New Earswick. It further benefited by more detailed architectural work from the well-known 'society' architect Edwin Lutyens. Its impact was tremendous, and the private Act passed to allow the variations to the local building by-laws necessary to give more freedom in site planning was an important legal precedent.

Many other schemes quickly followed (e.g. Harrison, 1981; Gaskell, 1981). Nettlefold himself presided over a co-partnership scheme at Harborne in Birmingham and nearby Bournville continued to expand on similar lines. Several were developed in Manchester and there were smaller initiatives in Coventry, Leicester, Liverpool and elsewhere. Within a very short period the garden suburb had secured a central position in the rapidly developing field of town planning. There was meanwhile a growing pressure on government to adopt measures that would allow town extension planning to work effectively by adding it to the formal powers of local authorities.

Other early dimensions

In addition to this town extension mainstream, there were other dimensions to the movement (Cherry, 1974b). A number of amenity preservation groups, particularly the Commons, Open Spaces and Footpaths Preservation Society, began to associate themselves with the new label. An important, if not always readily digestible, intellectual contribution came from the pioneer sociologist and biologist Patrick Geddes (Meller, 1990). Based in Edinburgh, Geddes was interested in the organic associations between people, culture, economy and the places they inhabited. Following the French social scientist Le Play, he expressed this relationship as Place–Work–Folk. From the 1880s he had been particularly interested in Octavia Hill's work in London and had become closely involved in a slum rehabilitation scheme in Edinburgh, based on a similar approach of a renewal based partly on the people themselves. Similar principles underlay his interest in civics, which developed a new approach to active citizenship, and his increasing advocacy of the civic and later regional surveys. These ideas came to increasing prominence outside Edinburgh from 1904, particularly through the early conferences of the new Sociological Society. For the moment, however, such ideas were secondary elements within the evolving notion of town planning.

THE BEGINNINGS OF STATUTORY TOWN PLANNING

The wider arguments for government action

Ideas and policy

We have so far concentrated on the development of ideas within the planning movement and its precursors. We have traced the logic of the arguments for town planning as its inventors understood it. We have noted how their social, economic, political and religious backgrounds shaped the nature of the solutions they advocated. This is only part of the explanation of the development of planning as an aspect of policy, however. We need also to ask why the relatively small group of people who originated these ideas were able to convince government of the need to pass legislation, moving town planning beyond the realm of ideas and voluntary action. To answer this question we need to understand something of the wider context of insecurity on the part of the British ruling class at this time (Masterman, 1909; Read, 1972).

International insecurity and British cities

One of the main sources of insecurity lay in the growing recognition that Britain was losing the position of world dominance it had enjoyed for much of the nineteenth century. The dramatic rise of Germany and the USA as world economic and military powers certainly frightened Britain's leaders. A particularly telling experience was the physical unfitness of many of the urban volunteers for military service during the Boer War, an imperial war with a strong undertone of Anglo–German rivalry. Recruiting stations in the bigger cities were obliged to reject between a quarter and half the volunteers on medical grounds; many more could be accepted only provisionally (Wohl, 1983). There were even more pessimistic claims that urbanization was bringing progressive degeneration of the British people.

Such concerns were reflected in the growing upper- and middle-class interest in eugenics, a branch of social genetics concerned with protecting and enhancing the hereditary quality of the population that began also to embrace social hygiene and environmental improvement for the masses. As well as prominent aristocrats, churchmen and professionals, its adherents included the Cadburys, the Chamberlains, Seebohm Rowntree, Patrick Geddes and other members of the nascent town planning movement (Garside, 1988; Hardy, 1991a; Aalen, 1992). The political significance of its concerns was manifest when an Inter-Departmental Committee on Physical Deterioration was set up by the Conservative government in 1903. The Committee's report the following year was critical of the environmental quality of urban life and made specific calls for ameliorative action, including greater control of the development of newer districts.

It was into this highly charged context that Horsfall's writings on German cities were received. Many observers were becoming predisposed to find reasons for Germany's spectacular rise to economic and military prominence after 1870 in its more advanced system of social provision. Perhaps, the argument went, if Britain adopted the German practices of planned town extension, low-cost housing (and technical education and all the rest), it would soon recover its former dominance. Certainly many of the more conservative advocates of town planning, such as Horsfall and Nettlefold, were quick to point this out. As Horsfall wrote in 1908,

> Unless we at once begin at least to protect the health of our people by making the towns in which most of them now live, more wholesome for body and mind, we may as well hand over our trade, our colonies, our whole influence in the world, to Germany without undergoing all the trouble of a struggle in which we condemn ourselves beforehand to certain failure.
>
> (cited in Ashworth, 1954, p. 169)

This militaristic and nationalistic tone was all a far cry from the gentle Utopianism of Morris or Howard. But it struck a raw nerve amongst Edwardian opinion-formers and decision-makers.

Industrial unrest and town planning ideas

At the same time there were also mounting fears about the stability of British society. A new labour movement was growing, using mass unionism as a force to change the status quo. There were a growing number of serious strikes from the 1880s, particularly after 1906 (Read, 1972). And it is clear that many of the moves towards social reform that characterized the first decade of the twentieth century were seen partly as ways of defusing this pressure. Such reforms, including the relief of unemployment, old age pensions and health insurance, began to address the main sources of economic insecurity in working-class life. The advocates of town planning did not shrink from pointing out the role that their new model for working-class life could play. George Cadbury Junior, in 1915, summed up sentiments that many middle-class reformers had shared for some years:

> The great stirrings of social unrest, which are such a striking manifestation in these days, are not controlled by considerations of finance only. The demand is not for a higher wage merely. In essence the demand is for a better way of life, for fuller opportunities, for the chance of self-expression in ways hitherto denied. Men ask for houses fit to live in – with gardens they can cultivate and air they can enjoy. They ask for a share in the good things of this life, in the things which elevate and inspire, and sooner or later that demand will be irresistible.
>
> (Cadbury, 1915, p. 136)

Social stability and town planning

What the town planners were essentially arguing is that society could be stabilized by widespread application of their principles. At the heart of the model they were purveying was the middle-class notion of home and family. Although there was some interest in women's co-operative forms of living within the movement, especially at Letchworth (Pearson, 1988), the town planning ideas that entered the mainstream of political discussion rested firmly on the notion of an increasingly nuclear family. Community and citizenship were encouraged but to supplement, not replace, the family. To insecure Edwardians, assailed by suffragettes and the gloomy reports of urban deprivation, the family itself seemed to be in danger of falling apart under the impact of pressures as varied as poverty, militant feminism and birth control.

The garden suburb offered a means of extending the upper-middle-class ideal of the family home down the social scale (Edwards, 1981). Nettlefold's garden suburb at Harborne was to be an 'Edgbaston . . . for the working men' (Harborne Tenants Ltd, 1907, p. 2). And Harmsworth saw clearly the appeal of the model for the widening lower-middle-class suburban readership of his newspapers. Central to all this was the garden. It served to delineate a specifically private family space separate from that of other families, in sharp contrast to the communal space of the courts and alleys of the older working-class areas. But it was also a source of physical, moral and economic reinforcement to family cohesion and therefore social stability. These sentiments, also from the Harborne prospectus, were entirely typical: 'in cleaner air, with open space near to their doors, with gardens where the family labour will produce vegetables, fruit and flowers, the people will develop a sense of home life and interest in nature which will form the best security against the temptations of drink and gambling' (*ibid.*, p. 2).

Here then was a vision of cosy domesticity and social order to counter all the sources of insecurity that troubled the British ruling class in the last years of peace.

A peaceful path to real reform?

Moreover, it seemed, at least as the town planners portrayed it, that this vision could be implemented without heavy cost to any important interests (Sutcliffe, 1981a). The rates could be kept down, compared to a municipal housing programme. There would be no need for central subsidies. And the working class would benefit by better housing, the industrialist by a better workforce and greater certainty about co-ordinating his own investment plans with those of others. Existing suburbanites could avoid the devaluing effects of new slums being built near them. Even landed interests were supposed to benefit. Town planning had grown out of an attack on the existing system of land ownership and development. But now it was being argued that even these interests would benefit by a rationalized land market, which would reduce the uncertainties of the speculative development process.

Statutory planning in practice

The Housing, Town Planning Etc. Act 1909

It was against this background that the first town planning legislation was passed by the Liberal government (Ashworth, 1954; Minett, 1974; Sutcliffe, 1988). The arguments were put with great effectiveness by the National Housing Reform Council, which renamed itself the National Housing and Town Planning Council in 1909, and the Garden City Association, which became the Garden City and Town Planning Association in 1908. Despite Nettlefold's probable role in its drafting, the resultant Act was, in practice, a great disappointment to the movement. Its main long-term significance was to make town planning into a local government function, a decisive switch from the private efforts of early years. It was also important because it marked the official endorsement of town extension planning. Town planning schemes could (there was no compulsion) be prepared for land liable to be used for building development. These would essentially allow the use and density zoning of the new suburban areas, though there were no powers to plan already built-up areas.

The 1909 Act and landed interests

Yet it would still have been possible to do a great deal with the Act if these had been the only limitations. One of the other main weaknesses lay in the unwillingness to intervene decisively in the land market (McDougall, 1979). Even the great reforming government of the Liberals from 1906, in spite of its other attacks on landed privilege such as the famous 'People's Budget' of 1909, was reluctant to take sufficiently radical action. Despite its leadership by John Burns, a former socialist member of the LCC, the Local Government Board created a legal framework that was heavily compromised by urban property interests. Specifically, there was no central impetus for the really extensive municipal land purchase that Nettlefold had urged. And the measures for taxing land value increases (betterment) produced by planning actions proved extremely weak. The rate was set at 50 per cent, but proved virtually impossible to collect. Conversely property was protected from planning provisions by extremely complex consultation procedures and, more seriously, the compensation system. This allowed owners the full value of any reductions in land value they could prove were the result of planning actions. In practice the statutory town planning agenda would be largely set by property.

The 1909 Act and working-class housing

Such compromises did not help the effectiveness of town planning in solving the housing problem. The housing sections of the Act gave some limited encouragement to co-partnership societies, by increasing their access to cheap loans. In practice, however, this did very little to tackle a housing shortage that worsened noticeably in 1909–14. Despite John Burns's former leading role in working-class politics, the Act had little credibility within the developing labour movement. From the outset the leaders of working-class politics were sceptical about the progressive industrialists and middle-class reformers who pushed the cause of a reformed urban land market as the solution to all the ills of urban society. The 1909 Act seemed to confirm that the much vaunted battle between progressive industrial capital and reactionary landed capital, a key element of the rhetoric of Liberal reformism at this time, had little substance, certainly when measured as material welfare outcomes for the urban masses. Instead the labour movement redoubled the pressure for municipal housing and central subsidies.

The implementation of the Act

The permissive nature of the Act, its procedural complexity and the great doubts about its possible benefits limited its use. Only Birmingham, where so much of the early conceptual groundwork for planning had been done, used the Act as its initiators had hoped, though they shared the general disappointment (Nettlefold, 1914; Cadbury, 1915). The city engineer began surveys of the whole city, particularly the developing urban fringe, and a series of town planning schemes were instigated each covering parts of this area. Thus two of the first three town planning schemes to be approved in the whole country, in 1913, were Birmingham schemes. The first, for Quinton, Harborne and Edgbaston, was intended to continue Nettlefold's strategy of opening up the urban fringe for garden suburb development, largely at twelve houses per acre. The other Birmingham scheme, for the east of the city, was a mixed industrial and residential area. Another influential early scheme was for Ruislip–Northwood, a largely residential area in north-west London (Aldridge, 1915). But there were precious few others. By 1914 only fifty-six schemes had been begun in England and Wales, covering just over 50,000 acres. Progress in Scotland was even worse.

Professional implications

Despite such modest action, the Act was already having two important professional implications. Firstly it was moving town planning more decisively into the purview of the professional expert (Cherry, 1974b; Hawtree, 1981). Figures like architects Raymond Unwin (Miller, 1992) and Barry Parker and surveyor Thomas Adams (Simpson, 1985) had already demonstrated something of the potential role of town planners as professionals, rather than reformers. As the notion of town planning became a statutory function, however uncertainly, it was inevitable that this role would grow. This in turn affected the second dimension, because the Act had moved town planning from the voluntary sector into the local government arena. Here a rather different set of professionals, the borough engineers and surveyors, took hold of the new activity, giving it a rather different twist. As Birmingham's city engineer, Henry Stilgoe, remarked revealingly in 1910,

> I think the people who will administer it [the 1909 Act] will be the borough engineers and surveyors of this country. It is their right. They are the officials, the statutory officials, appointed under the Public Health Act, and without their co-operation, and, in fact, without their intimate knowledge of their districts, this Act cannot be put into proper and efficient working order . . . what we want is men who will come forward and with a bold scheme, leaving the little matters, good as they are, the small garden suburbs, to develop of themselves. We must drive forward the great engineering works. The others will follow. Those people who talk so much about town planning do not know what they want; would not know what to do with it if given the opportunity.
>
> (Stilgoe, 1910, p. 44)

Meanwhile the architects particularly were also asserting their rights to dominate the new activity in a major Royal Institute of British Architects conference in the same year. Surveyors and lawyers also declared their interests.

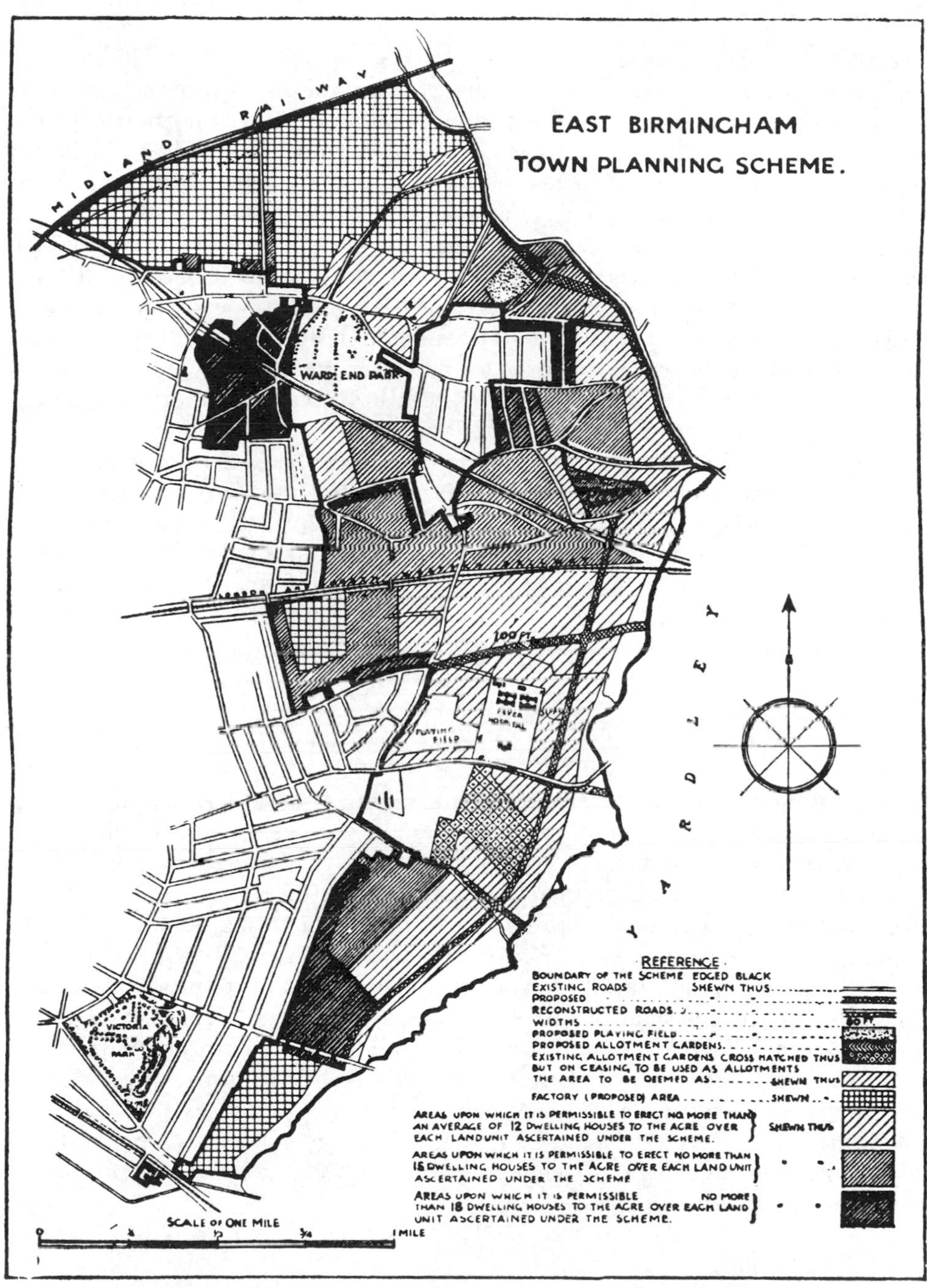

Figure 2.7 *Under the 1909 Act local authorities began to prepare statutory town planning schemes and Birmingham was one of the most active authorities. The prime concern here, as in other early schemes, was land use and density zoning and the reservation of road lines for suburban extension. Notice how the scheme proposed a gradual transition from already developed by-law housing areas (see Figure 2.2) to the new twelve house per acre standard, a solution which protected existing development land values.*

The Town Planning Institute

Gradually, out of this rivalry, the need for an inter-professional forum began to be recognized. The emergence of independent town planning education with the setting up of the Department of Civic Design at Liverpool University in 1909 was an important step (Wright, 1982). Its journal, the *Town Planning Review*, fostered a new sense of planning as a professional activity in its own right. The idea of a new professional body was pushed by Thomas Adams, who had been central to the evolving town planning movement at Letchworth, as one of the first town planning consultants and, latterly, first planning inspector at the Local Government Board. In 1913–14 the Town Planning Institute (TPI) was formed following the deliberations of an informal inter-professional group (Cherry, 1974b). Its structure represented a compromise between architects, engineers, surveyors and lawyers, but its main significance was that it was, above all, a body of professionals. Reformers and propagandists could be associated with it, but only professionals could be full members. As the TPI's historian, Gordon Cherry, has observed, 'From now on the course of town planning was to be very considerably affected by professionalism and the liberal, humanitarian and social reform element was partly squeezed into a professional frame. This "inner drive" both gained through technical expertise and lost through institutionalization' (Cherry, 1974b, p. 61).

Conceptual developments after 1909

Conceptual innovation and professionalization

This growing professionalization of town planning affected the development of ideas. There was less attention now to the grand strategies of urban reform and more to the development of a professional practice. Densities, zoning, site layouts and civic design now became the staples of planning theory and practice, transmitted through professional education to a small but growing number of practitioners. Patrick Geddes's emphasis on the need for surveys began to attract greater interest, though the shallowness and narrowness of statutory town planning schemes fell well short of what Geddes was advocating (Geddes, 1968, orig. 1915; Meller, 1990). More of the town planning literature of this period focused on professional ideas and practice, and there was less purely propagandist writing. There was a consolidation and articulation of design principles that had been established in some of the pioneer schemes. Foreign influences were revealed and, in many cases, further extended. And there were some significant British extensions of earlier initiatives.

The German aesthetic

One of the most notable trends was the interest in the traditional German approach to town design. This was distinct from the British envy of institutional arrangements described by Horsfall and drew particularly on the example of smaller historic towns. The main theorist of this approach was the Viennese architect Camillo Sitte, whose seminal 1889 book, *City Planning according to Artistic Principles*, though not translated in English until much later, became known through a rather bad French translation of 1902 (Sitte, 1965). Sitte emphasized the informality, absence of rigid layouts and careful use of streets and public spaces in older European towns. It was an approach that had a huge influence on Raymond Unwin, who had absorbed much from the medievalism of William Morris's Arts and Crafts movement in the 1890s (Miller, 1992). By 1909 the impact of medieval German towns such as Nuremburg and Rothenburg became clearly apparent at Hampstead Garden Suburb and, through emulation, other early garden suburbs. It was, however, his 1909 book, *Town Planning in Practice*, that articulated these ideas in a comprehensive way with a lasting impact on the aesthetic of British planning.

Other foreign influences
Meanwhile, however, important ideas pointing in a rather different aesthetic direction also entered the rapidly evolving mainstream of British town planning thought. Unwin's ideas of informality became accepted within a primarily residential and suburban context. But the planning of central areas, where the opportunity arose, seemed to require grander statements. The revival of classicist ideas under the influence of the French Ecole des Beaux Arts, saw a growing Edwardian interest in Baroque architecture that occasionally had wider planning significance, for example in the creation of the Cathays Park civic centre at Cardiff from 1901. There was also a growing interest in the American 'City Beautiful' planning approach, largely introduced via the new Civic Design Department at Liverpool (Wright, 1982). This had important similarities to the Beaux Arts tradition. It embodied a highly monumental, classically inspired formal approach, making extensive use of landscaping. Major plans such as that by Daniel Burnham for Chicago (1909) were reviewed in the professional press. In sharp contrast with the small-scale suburban emphasis of the 1909 Act, however, most such American plans were intended as major exercises in civic boosterism (Sutcliffe, 1981a). There were few opportunities to plan on such a scale in Britain before 1914. Despite a 'City Beautiful'-inspired replanning of Port Sunlight, suburban planning was dominated by the informal aesthetic pioneered by Unwin.

Nothing Gained by Overcrowding
Paradoxically though, it was Unwin who pointed the way to a formalization of this informality in specific residential density standards. After a variety of density figures in early suburbs, Unwin presented a brilliantly argued justification of the figure of twelve houses per acre (thirty per hectare) in his booklet *Nothing Gained by Overcrowding* (Unwin, 1912). He showed how the local by-laws, with their insistence on needlessly wide streets, encouraged the packing of narrow-fronted by-law housing on this expensive frontage, negating the opportunities offered by generally falling land costs. In contrast he showed how a more spacious approach, especially if by-laws were relaxed, would allow lower densities with private gardens and public open space. It was an argument directed at the ordinary tenant and builders, developers, surveyors and engineers. By this stage the town planning movement needed little convincing. It was written in the language of land and development economics, showing that garden suburbs did not need to be just

philanthropic gestures – they could be profitable and still cheap to rent. The influence of these arguments, as we will see in the next chapter, was immense.

OVERVIEW

By 1914 therefore town planning had arrived. It had been one of the great hopes for early twentieth-century urban reform, a cure for all the insecurities that assailed the ruling classes. But the hopes were exaggerated. When rhetoric was converted into legislative reality in 1909, Liberal, middle-class reformism proved itself unwilling to tackle the power of land where it really mattered, in the urban arena. Town planning therefore was turned into something that was less an attack on landed interests, more a process of conciliating urban land and development interests into accepting an overall approach. Unwin's 1912 booklet was centrally based on this premiss, in effect accepting that profit-seeking within the urban land market would continue to be the primary engine of urban development.

This was already far removed from the Utopian socialism of land nationalizers like Howard whose ideas had given birth to town planning. As the radical ideas of land reform and co-operation were dissembled and remodelled into a new practice of town planning, they lost their social reformism and became instead a model for less fundamental, if still important, environmental reforms. Coincidentally they were incorporated into the mainstream of bourgeois reformism, espoused by progressive industrialists and later, as the ideas were converted into policies, professional practice. This institutionalization was the beginning of a process that ultimately was to deny the traditions of independent radical thought from which town planning ideas had sprung. For the moment, however, the achievements of this institutionalization were so puny that the town planning movement retained much of its autonomy. The events of the next few years also reinvigorated some of the original spirit of radicalism that had spawned it.

3

WIDENING CONCEPTIONS AND POLICY SHIFTS 1914–39

INTRODUCTION

The twenty-five years between the outbreaks of the First and Second World Wars were extremely important for the evolution of all three dimensions of planning. For the first time planning was having real, if not always recognized, impacts on the way towns and cities were actually developing. The town extension policies which had been embodied in the 1909 Act were now being implemented on a vast scale, though by mechanisms that had precious little to do with any formal planning actions. Meanwhile the conceptual bases of planning were being widened to embrace new spatial problems and rehearse new policy solutions.

This second phase in the development of planning thought had various origins. In part it was simply a reaction against mass suburbanization. Thus Howard's social city was reconceptualized as a model for the management of urban growth involving planned decentralization and containment. Another source was the heightened social and political awareness of the long-standing problem of urban obsolescence, which encouraged a growing emphasis on planned redevelopment, increasingly under modernist influences. Entirely new concerns like the growing numbers of motor vehicles also stimulated planning thinking in a variety of ways. The most important new direction was the emerging concern with a national planning approach and regional balance reflecting the acute regional unevenness of the inter-war Depression years.

In contrast to this important conceptual activity, extending the range of planning ideas, there were only modest shifts in actual policies. After a short phase of political idealism associated with the later war years and those immediately post 1918, the inter-war period was dominated by a pragmatic but ultimately cautious Conservatism, with two short periods of minority Labour governments (Mowat, 1955). Attitudes to public intervention were dominated by the 'Treasury view' which held that government spending, however justifiable on social criteria, could only harm the economy and should therefore be avoided in a time of economic distress (Ward, 1988). Accordingly the new conceptual agendas of the planning movement were addressed only in partial or limited ways. There were some important local initiatives, but these too were necessarily limited in their wider impacts. Overall therefore the period saw the conceptual development and rehearsal of a new policy agenda for planning, but a postponement of any committed government action to address the items on it in a comprehensive way. At the outset though it seemed as if things might be different, as grand visions of a post-war world began to develop.

Town Extension Planning and Mass Suburbanization

Housing policies and town planning

Political context

From 1916 post-war reconstruction became an important government priority (Marwick, 1967). War produced only very slight destruction of the physical fabric of Britain's towns and cities, in contrast to the experiences of parts of Belgium and France (and indeed Britain itself during the Second World War). Yet reconstruction concerns grew from the political need to maintain popular support for a war that was horrifically expensive in human lives and required a far higher level of mass loyalty and commitment within the industrial workforce than was ever necessary in peacetime. In effect, as was to happen during the Second World War, war heightened the bargaining power of working-class men and women. The 1917 Russian Revolution reminded the wartime coalition government what could happen if mass needs were not satisfactorily addressed.

Britain, however, was an altogether more robust political edifice than Russia. Moreover, the war also strengthened the bargaining power of the industrialists, whose commitment was equally necessary to running the war economy. What this meant was that the emergent reconstruction agenda had to improve the lot of the working classes, though not in ways that would prove too costly for these powerful business interests. It became increasingly clear that greater state intervention, particularly to improve social welfare, would be an important means of managing these potentially conflicting pressures.

The pressures for government housing policies

There were important signs of these new political realities during the war itself. Unlike during the Second World War, when town planning quickly assumed a major position in reconstruction debates, it was the question of housing that now became prominent, although of course it retained its strong connections with town planning. Thus the initial moves on rent controls, though not immediately linked to planning, were to be of profound long-term importance (Bowley, 1945; Melling, 1980). There was already, as we have noted, a serious pre-war housing problem. The war worsened this by bringing a virtual cessation of new housing construction and, more seriously, an increased demand for accommodation in the main industrial centres as production moved into higher gear to meet war requirements. In a virtually free housing market, this combination of fixed supply and increased demand quickly brought higher rents. In turn this threatened to have a knock-on effect by disrupting war production as workers pressed for higher wages. Higher rents themselves also directly produced serious disruption in some areas, particularly Clydeside, where there were rent strikes.

In 1915, therefore, the government passed the Rent and Mortgage Interest (War Restrictions) Act, which held rents at the pre-war level. It was a measure that benefited both sides of industry, essentially at the expense of the private landlords who lacked any strong political voice in central government. In the long term the move was important because it accustomed people to paying rents that were increasingly out of touch with market rents in a period of serious inflation. In turn this exacerbated the wartime disruption of the private housebuilders by seriously weakening their profit-based incentive to build. These specific short-term responses to the pressures of war and the continuing political strength of working-class interests, coming on top of the already serious pre-war housing position, pushed government into becoming a major provider of housing.

The Tudor Walters Report 1918

As the government moved down the path of granting Exchequer subsidies to local authorities to provide council housing on a greatly increased scale, the question of design standards assumed prominence (Swenarton, 1981). The emergent reconstruction ideal favoured standards somewhat higher than pre-war norms. It was here that the direct link with town planning became more obvious. In 1917 the Tudor Walters Committee was established to consider the methods and standards of building the new working-class houses. Its chair, Sir John Tudor Walters, was a Liberal MP who was also a director of the Hampstead Garden Suburb Trust. Several of its other members were associated with housing reform and planning. But its most important member was Raymond Unwin, the co-planner of Letchworth, Hampstead and New Earswick and by this time a senior housing and planning adviser with the Local Government Board (Miller, 1992).

The Report (Tudor Walters Committee, 1918), published a few weeks before the Armistice in autumn 1918, embodied much that was undiluted Unwin. What was proposed was a rationalized version of the garden-city-type cottage housing built at a density of twelve per acre (thirty per hectare), the standard Unwin had defined in his 1912 pamphlet *Nothing Gained by Overcrowding*. Following the Arts and Crafts origins of the garden city movement, the building technology was to be traditional, but using increasingly standardized materials. All the Unwin devices of attention to site and aspect were present. Space standards within dwellings were to be very high, better in fact than those for equivalent pre-war housing in the garden suburbs. The overall pattern for urban change was to be firmly based on pre-war town extension – the garden suburb rather than the garden city – and housing schemes were to be developed in conjunction with statutory town planning schemes.

Housing, Town Planning Etc. Act 1919

The Tudor Walters recommendations were embodied in the 1919 housing manual (LGB, 1919) and underpinned the Housing, Town Planning Etc. Act of the same year (Swenarton, 1981). This measure is usually known as the Addison Act after Dr Christopher Addison, the radical Liberal who was a wartime Minister of Reconstruction and then became the first Minister of Health, responsible for implementing large parts of the reconstruction agenda, including housing and planning. The Act became the main vehicle of Prime Minister Lloyd George's 1918 election pledge to make Britain 'a land fit for heroes to live in'. Its most important innovation was to provide central government housing subsidies. These were extremely generous and limited local authority loss to the product of a one (old)-penny rate, regardless of the final housing costs. The Act also included proposals to make the preparation of town planning schemes compulsory for towns of over 20,000 population, showing a clear intent to develop housing within a statutory planning framework (Cherry, 1974a).

The 1923 and 1924 Housing Acts

In fact the high cost of the housing subsidies and very high standard municipal garden suburb estates that were built under the Addison 'homes for heroes' programme soon proved too much for the Conservative majority in Lloyd George's coalition government (Bowley, 1945; Daunton, 1984; Smith and Whysall, 1990). Addison himself was dismissed and the programme was dropped in 1921. More modest fixed housing subsidies were reintroduced in the Housing Etc. Act 1923 introduced by Conservative Minister of Health, Neville Chamberlain, who had been closely associated with early town planning initiatives in Birmingham. The following year John Wheatley, the Clydeside socialist Min-

Figure 3.1 *During the 1920s the principles of planned town extension were implemented mainly through the state-assisted council housing programme. This picture shows the Allens Cross estate in Birmingham, built from 1929 under the Wheatley Act. Detailed design and layouts of such schemes were usually more austere and standardized than the pre-1914 garden suburbs and there was less emphasis on community development. Such estates were generally popular with the better-off workers who lived on them, however.*

ister of Health in Ramsay MacDonald's first minority Labour government, introduced more generous subsidies in a further Housing Act. Both these Acts represented retreats from the most generous of the Tudor Walters standards, though twelve houses to the acre, suburban cottage-type housing had by now become the benchmark standard for such schemes (LCC, 1937; Manzoni, 1939).

Later housing legislation

Until 1929 the two Acts operated in tandem, with the Wheatley subsidy largely used by local authorities, the Chamberlain subsidy by private builders. In 1929 the Chamberlain subsidy was abolished by MacDonald's second Labour government. Arthur Greenwood, Labour's Minister of Health, strengthened the Wheatley subsidy, which had been reduced from the 1924 level, and added a new concern with municipal slum clearance and redevelopment in the Housing Act 1930 (Yelling, 1992). This became a dominant element in housing policy from 1933, when the Wheatley subsidy was abolished by the Conservative-dominated National Government. The approach of targeting housing subsidies on the less well off was continued when the Housing Act 1935 gave subsidies for the abatement of overcrowding.

We will return to examine important aspects of housing policy in the 1930s in the section on the origins of planned redevelopment. For the present we can note that the bulk of housing built under these 1930 and 1935 Acts was also garden-suburb-type housing, again following the basic twelve per acre density standard. Yet the pressure to build at lower rents for poorer tenants but with lower subsidies meant that the resultant housing was a rather degraded version of Tudor Walters' standards. Although individual authorities continued to produce good results, there was less attention to site and aspect detailing than Unwin had wanted. The space standards of individual houses were also much reduced. Overall such schemes were significantly worse than those built under the Wheatley and especially the Addison legislation (NCDP, 1976).

Private housing and town planning policies

The private housebuilding boom

The main political rationale for the shift in housing policies in the 1930s was the tremendous buoyancy of private housebuilding (Richardson and Aldcroft, 1968; Oliver, Davis and Bentley, 1981). In 1930 for the first time over 100,000 private houses were built in one year without subsidy in Britain. Despite the Depression, the figure continued upwards, remaining above 250,000 per annum during the five years 1934–8 inclusive and peaking at over 275,000 in 1935 and 1936. Even in 1939, when war brought severe curtailment of housebuilding, just over 200,000 private dwellings were completed. The overwhelming majority of these were suburban houses, usually in semi-detached or short terraced form, built at densities of twelve to the acre or less, with great emphasis on private gardens.

The main sources of this boom reflected a combination of demand and supply factors. On the demand side, the lower cost of living in the 1930s' Depression years increased the disposable income of those who remained in employment.

Demographically the tendency for smaller families also had a similar effect. Meanwhile there was a powerful shift in social preferences towards owner-occupation, a vision which became strongly associated with the garden housing of the new suburbs, especially in the advertisements of the development industry (Gold and Gold, 1990, 1994). On the supply side there were historically very low interest rates, the rapid expansion of the building societies, cheap labour and relatively cheap building materials. Around all the major cities undeveloped land was readily available and very cheap, reflecting the generally depressed state of agriculture. In the early 1920s land prices of about £300 an acre, sometimes much less, were common in provincial urban-fringe locations, a little higher in London. And there was a general willingness of local authorities and public utilities to service developments with water, gas, electricity, sewerage and roads. Above all, public transport had a decisive impact in opening up the urban fringe for residential development.

The role of public transport

In London the public transport undertakings, such as the Southern Railway, the Underground Group and later the London Passenger Transport Board (LPTB) under the leadership of Frank Pick played a key role in facilitating the spread of suburbia (Jackson, 1991). Underground extensions such as that from Golders Green to Edgware in 1926 had quite consciously opened up a whole sector of the urban fringe for development, which in turn generated more passengers. The suburban message was then promoted in advertisements: 'Stake your claim in Edgware. Omar Khayyam's recipe for turning the wilderness into paradise hardly fits the English climate, but provision has been made at Edgware of an alternative recipe which at least will turn pleasant undulating fields into happy homes' (Graves and Hodge, 1971, p. 168).

The formula was used over and over again. And similar tendencies were apparent in other cities, especially those where electrified commuter rail systems were being created. Everywhere though the role of public transport was a decisive element in the private suburban development process.

Developments in statutory planning

But what, given such powerful formative factors in promoting private suburban growth, was the role of statutory planning? Before considering this we must review the main developments in planning policies since 1919. The record was not impressive. The compulsory element of town planning provisions in the Housing, Town Planning Etc. Act 1919 had never proved practicable, largely because, as feared by some planners, there were simply too few of them to operate the new system. Moreover, the housing section of the Ministry of Health had, for quite understandable reasons, given subsidies to countless council housing schemes without insisting that these be embedded within statutory planning schemes, as was originally intended. Deadlines for planning schemes had regularly been extended and progress under the Act proved very slow (Cherry, 1974b). By 1928, 98 of the 262 urban authorities with populations of over 20,000, for whom planning was theoretically compulsory, had still submitted no proposals, and very few schemes were actually near completion.

Figure 3.2 *By the 1930s the private housing boom saw the garden suburban imagery of the early planning movement fully adopted by private developers, building societies and commuter transport operators. Notice here how the suburban residential ideals of space, light, nature and fresh air, in contrast to the gloomy by-law terraces, are directly linked with the promotion of owner-occupation*

In recognition of the growing complexity of housing policies, housing and town planning had been separated as objects of legislation. The Housing Etc. Act 1923 was the last (until 1986) to incorporate planning proposals (extending the remit of planning to include areas of special historic and architectural interest). The Housing Act 1924 marked a legislative parting of the ways, however, and in 1925 a Town Planning Act was passed to consolidate existing legislation. Important planning provisions were also included in the Local Government Act 1929, allowing county councils to assume responsibility for statutory planning activities. In practice though, the system remained very similar to that created in 1909. The second Labour government came to power with a commitment to strengthen statutory town planning and extend its role.

Town and Country Planning Bill 1931

In 1931 Arthur Greenwood introduced a Town and Country Planning Bill, the first to consider rural issues, but also intended to confer greatly enhanced powers to plan urban areas (Ward, 1974; Sheail, 1981). Thus entirely new powers were included to allow planning to deal with already fully developed areas, which we will examine in the section on the origins of planned redevelopment. Of more general importance were the proposals to reduce the gap between compensation and betterment, which had remained unevenly balanced at 100 per cent compensation to 50 per cent betterment since 1909. More seriously the short period allowed for betterment collection meant that it was rarely recovered. Now the new bill included betterment proposals of 75 per cent and there was some prospect of a more workable system. Moreover, front-bench spokesmen on all sides welcomed the bill and gave it general approval.

Town and Country Planning Act 1932

At this point, though, the Labour government fell in the economic, financial and ultimately political crisis of 1931 (Skidelsky, 1967). The subsequent general election gave a huge endorsement to the Conservative-dominated National Government, which secured 60.5 per cent of the poll and 554 of the 615 seats on a mandate of cutting government commitments. The new Conservative Minister of Health, Sir Edward Hilton-Young, himself a moderate Conservative, reintroduced the bill into a very different House of Commons to that which had considered it the previous year. The comments of a new Conservative member were typical:

> In spite of what the Parliamentary Secretary has said 'this is the 1925 Act over again, why grumble at it; it was a Tory measure', that is what is the matter with it. Our party was as bad as most others in passing Socialist measures. There are many new Members in this House, and a good deal that has been done in the past we may take the opportunity of undoing.
>
> (Ward, 1974, p. 686)

And that, broadly speaking, is what happened. Many extra complications were introduced into planning procedures and the new proposals in the bill were hedged about with restrictions. Planning ceased, even formally, to be a compulsory obligation and a new concept of 'static areas' was introduced that heavily compromised the extension of planning to all areas built and unbuilt that Greenwood had intended.

Statutory planning under the 1932 Act
Paradoxically though, despite its unpromising beginnings, the Act was more successful than any previous legislation. Thus the area covered by planning schemes at various stages of completion grew from over 9 million acres in 1933 (when the 1932 Act became operative) to nearly 26.5 million in 1939 (Barlow Commission, 1940). The numbers of approved schemes grew much more modestly, however, from 94 to 143. This reflected a growing willingness to engage with local planning activity, particularly by county councils, but a parallel scepticism about the need to go through all the cumbersome procedures of actually finalizing schemes. Interim development control in conjunction with a draft scheme became a favoured mechanism by which local authorities sought to influence the process of development.

Planning schemes and the private housebuilder
Much of the increase in the planned area simply reflected the spread of planning in the countryside (Sheail, 1981), but the main thrust of the statutory process remained in town extension planning. The town planning scheme typically allowed local planning authorities to introduce outline land use and density zoning and reserve road lines into the developing suburban areas. Often the schemes simply reflected and co-ordinated the expressed desires of individual landowners and developers, while incorporating the borough or county engineer's road schemes. A good deal of persuasive discussion went on to try and convince recalcitrant property interests, usually on density questions. Planners though had little more than moral or technical authority in such negotiations. Essentially they were handicapped in imposing any planning outcomes on any landowner or developer who disagreed with their proposals by the compensation provisions of the planning legislation, which could saddle their authority with heavy claims. When authorities wished to vary the development process significantly, the more predictable course of action was to buy the necessary land themselves, rather than work through the planning system.

Accordingly, despite occasional bleating from property interests, there is little evidence that the pre-1939 planning system was a serious obstacle to their operations. Development control, even for an approved scheme, was usually a formality since a scheme itself effectively permitted development to the extent of its proposals. Even in the occasional instances when permission was refused, enforcement powers were very limited. Overall the system contrasted sharply with the much tighter planning system which was created after 1945, when there have been frequent complaints about excessive planning constraints on the supply of development land. In fact there was a strong inter-war tendency to 'overzone' land for development. Thus Unwin estimated that the residential land zoned in draft planning schemes in 1937, when Britain had an actual population of just 46 million, would have accommodated a population of 291 million! (Barlow Commission, 1940, p. 113). Moreover the more rational development of whole areas that planning promised offered real benefits to individual developers. Thus it was common for private housing schemes to be advertised as conforming to the requirements of a local planning scheme.

Restriction of Ribbon Development Act 1935
There was one aspect of suburban development which caused particular public and political concern in the 1930s' building boom. This was the tendency for arterial roads to be subject to ill-considered ribbon development of cheaper housing extending out well beyond the more continuously built-up areas (Cullingworth, 1975, pp. 167–77). The practice saved the developers the cost of building proper residential roads, but it damaged scenic quality and undermined the efficiency and safety of main roads by mixing local and through traffic. It was disliked even by many of the right-wing opponents of town and country planning.

Theoretically the 1932 Act offered all the powers necessary to stop ribbon development. The practice, though, was rather different and a new Act specifically focused on this proven abuse was passed. It gave highway authorities (counties and county boroughs) immediate control over main roads, regardless of whether a planning scheme (normally prepared by a county district or county borough) was in force. Moreover it had the tremendous advantage of allowing immediate enforcement, which was not possible under interim planning development control. Basically it allowed the whole question of planning in the vicinity of major roads to be tackled more effectively. Yet, despite what the widening planning lobby would have liked, it did not challenge the predominant emphasis on peripheral suburban development as the main strategy for urban growth.

Planning solution or problem?: impacts of suburbanization

Overall dimensions
In all some 4.3 million dwellings were built in Britain between the wars, a greater addition to the housing stock than in any previous comparable period (Richardson and Aldcroft, 1968). Of these almost 31 per cent (nearly 1.33 million) were provided by local authorities. Eleven per cent (a little over 0.47 million) were built by private enterprise with state assistance (including housing associations). Over 58 per cent (just over 2.5 million) were provided by unaided private enterprise. As we have shown, the great majority of these dwellings were built in the new suburban areas at lower densities than ever before. This suburbanization occurred everywhere, though it was most pronounced around the big conurbations, especially in the more prosperous south and midlands. In particular the distribution of private building closely reflected the wider economic changes in these regions.

Spatial impacts
There was accordingly a huge conversion of rural land to urban, particularly residential, uses though no precise record was kept of the extent of this (Johnson, 1974). It has been estimated that the annual average transfer of agricultural to urban uses in England and Wales rose from some 9,100 hectares per annum in

1922–6 to 21,100 hectares in 1926–31. From 1931 to 1939 it was running at an astonishing 25,100 hectares per annum. This represented a total of over 340,000 hectares converted between 1922 and 1939, amounting to an increase of roughly 40 per cent in the total urban area. Not surprisingly there was a growing concern at the speed and extent of the change.

Impacts for agriculture

By the later 1930s there were growing fears about the implications of the loss of agricultural land, given the growing threat of war. Certainly suburbanization, especially around London, had often been at the expense of highly productive land (Robson, 1939). Thus London County Council (LCC)'s giant estate at Becontree-Dagenham had taken some of the best market gardening land in Essex and there were similar losses in west London. The whole issue was highlighted during the 1930s by the influential Land Utilisation Survey of Britain, headed by the geographer, L. D. Stamp (1962). There were, however, some weaknesses in the agricultural protectionists' arguments, since they took no account of the potential food production of suburban gardens, which proved very important during the war years.

Transport and congestion costs

One consequence of suburbanization was simply that the big cities got even bigger. This, combined with a rise in the numbers of motor vehicles from just over 2 million in 1928 to just over 3 million in 1939, including an increase from just under 0.9 million to just over 2 million motor cars, contributed to mounting congestion problems, especially in central areas. Moreover, longer journeys to work for the suburbanite absorbed appreciable amounts of time and money. The average London family was spending £15 per annum on fares by 1938, representing 8 per cent of working-class family income (Barlow Commission, 1940). Clearly many suburban families were spending rather more than this. And the advantages to the transport undertakings of promoting suburban development were also diminishing. Thus it was increasingly noticeable that new traffic in the outer suburbs was being partly offset by the reduced revenue from the inner areas. There was also the growing and costly necessity to retain passenger stock simply to meet peak loadings for long suburban journeys to work, identified as a serious problem in Birmingham as early as 1932 (Finer, 1941, pp. 357–8). And by the late 1930s Frank Pick of the LPTB was rethinking his strategy of promoting suburban growth around London (Hall, 1988, p. 66).

The social costs

Suburban living undoubtedly proved immensely attractive and popular with the majority of people who were able to bear the extra costs of living there. As well as extra spending on transport, mortgage repayments or rents were usually higher than in the inner city (Young, 1934; Durant, 1939; Williams, 1939; BVT, 1941; Jevons and Madge, 1946). Moreover suburban areas could not offer the bargain shopping and cheaper food that was available in the street markets of the older districts. And the low-density suburban homes cost more to heat than the higher-density terraces. These costs bore particularly heavily on the working-class suburbanites, those in the cheaper private housing and especially the council slum clearance estates of the 1930s. For them lower-density garden suburban living did not always produce the health improvements promised by the pre-1914 planners. A famous study in Stockton showed how health actually deteriorated because very poor families from the slums were obliged to spend less on food in order to make ends meet (M'Gonigle and Kirby, 1936).

Suburban life

Even those with sufficient money to enjoy suburban living found that social and community facilities were often inadequate, especially in the early years. There was also a growing critique of the social coldness and soullessness of suburban living. Petty social distinctions between the smarter and cheaper private estates were rife. Council estates were usually beyond the pale and were resisted in many areas of private suburbia. In a few cases, notably at Downham near Bromley and Cutteslowe in Oxford, unofficial walls were erected by residents or developers to close roads between the two kinds of area (Collison, 1963; Weightman and Humphries, 1984). Even within socially homogeneous areas, a more private and less communal life seemed to be emerging. As a housing survey of Birmingham in the late 1930s noted, 'The neighbours are not so handy, or not so obviously neighbours at all, since they are up trim paths and behind trim curtains' (BVT, 1941, p. 96).

To many, particularly in private suburbia, this was less of a problem, as a new mass family and home-based culture of domestic life developed (Gittins, 1982; Weightman and Humphries, 1984; Jackson, 1991). It was based on the notion of a comfortable home away from the city, occupied by a smaller, nuclear family and a widening range of consumer durables. It relied much more on home-based leisure pursuits like gardening or the radio. Yet, even for adherents of this new suburban culture, the dynamic and seemingly unstoppable suburbanization process was unsettling, particularly when their homes ceased to be on the edge of the city. The rural landscape and sense of being in the country, a quality that developers frequently emphasized, rapidly became more remote (though not as remote as it had become for inner-city dwellers of course). These issues formed yet another strand in the emergent social critique of suburbia.

Moreover it was a way of life that was increasingly subject to intellectual and left-wing criticism for the way it encouraged a sense of escape from the social realities of both the cities and the wider world (Oliver, Davis and Bentley, 1981). Thus the famous socialist writer George Orwell, returning from fighting the Fascists in the Spanish Civil War and conscious of the imminence of a wider struggle, remarked, not unkindly, that the 'huge peaceful wilderness of outer London' seemed, along with most of the rest of the country, to be 'sleeping the deep, deep sleep of England' (Orwell, 1962, orig. 1938, p. 221). Such ideological perspectives were important because, in conjunction with the other manifest problems of suburbanization, they had an important influence on many of the younger planners and architects of the period. And they certainly contributed to the growing interest in the higher-density and supposedly more communal forms of housing being advocated by modernist architects and planners whose work will be discussed more fully later in this chapter.

Planned extension or unplanned sprawl?

The mounting critique of suburbanization during the 1930s frequently represented it as unplanned sprawl. Yet in its essentials it did exactly what the dominant strand of pre-1914 planning opinion had wanted. The main agencies of this change, the municipal housing authorities and the speculative builders, were certainly not those that Cadbury, Nettlefold and the other pioneers had expected or wanted. But did that matter? Inter-war suburbanization had extended the towns, providing huge quantities of good-quality and relatively cheap garden housing for a much wider section of the population than hitherto. Moreover the design inspiration for the basic form of the housing certainly came from the garden city tradition (Sharp, 1932). The line of influence was directly traceable for the municipal garden suburbs via the Tudor Walters Report of 1918. It was still apparent, if more indirectly, for private suburbia. Moreover statutory planning played a significant, if not major, role in allowing mass suburbanization and providing some rudimentary framework within which it could happen. All this is not to deny that there was a debit side that was increasingly apparent by the late 1930s. But it would be wrong to imply that this arose because the suburbs were not planned in any fundamental sense. We have only to compare inter-war London with the French capital for example to see what unplanned suburbia was really like (Evenson, 1979). In Paris large numbers of self built and frequently ramshackle suburban dwellings were put on plots with no co-ordination whatsoever and lacking even the most basic infrastructure. Dirt roads and lack of mains drainage were typical of many areas. Although environments like this did exist in inter-war Britain, in some rural and coastal areas (Hardy and Ward, 1984), the inter-war suburbs of British cities were, by comparison, models of ordered development. We must then reject the charge that the suburbs were not planned. What was happening was that, for the first but not the last time, a planning solution was in its mass implementation turning into a planning problem. It was this recognition that drove the search for alternative models for accommodating urban growth.

New Models for Managing Urban Growth

Planned decentralization

A conceptual refit

As we noted in the previous chapter there was from the outset a strong decentralist tradition in British planning represented by the garden city movement. This had, however, become largely deflected into 'garden suburb revisionism', part of town extension, before 1914. Ebenezer Howard's grand vision of a social city network and a withering away of the large concentrated cities was largely forgotten. Towards the end of the First World War, however, there was a revival in the purer strand of garden city thinking, though with some important differences from Howard's original ideals (Hardy, 1991a). In 1918 a small but important book, *New Towns after the War*, was published as a contribution to the reconstruction debate. The authors were a group of garden city enthusiasts who called themselves the New Townsmen. They included Howard himself, but the book was essentially the work of two of his younger supporters, C. B. Purdom and F. J. Osborn (New Townsmen, 1918).

Conceptually it was important because it marked the first attempt to transform the garden city idea into the New Town. The key innovation was a much greater reliance on the state in initiating and undertaking New Town development, compared to the voluntarist, co-operative framework originally proposed by Howard. In addition, it was also becoming increasingly clear from this and other contemporary work, particularly by Purdom, that there was also to be an important though subtle shift away from the social city regional model. Instead the wider strategic concept was of new or satellite towns (as they were increasingly called in the inter-war years) used as a conscious strategy for metropolitan decentralization (Purdom, 1925). The objective was no longer to replace the big city but to accommodate metropolitan growth in an alternative form to continuous suburbanization.

The idea received clear endorsement in the Ministry of Health's Unhealthy Areas Committee Reports of 1920 and 1921 (Chamberlain Committee, 1920, 1921). The Committee was chaired by Neville Chamberlain (Cherry, 1980) and had a number of prominent members of the planning and housing reform movements. But its proposals, though conceptually important, had no immediate impact on policies. Foreign examples, particularly the work of Ernst May (a former associate of Unwin's) at Breslau and Frankfurt in pre-Nazi Germany, were to be influential in advancing the idea amongst British planners later in the 1920s (Hall, 1988; Ward, 1992a). It was eventually to become an idea of tremendous importance for the long-term development of British planning policy.

A second garden city

Meanwhile, contradicting some of the views he had endorsed as a New Townsman, Howard initiated a second experiment in more traditional garden city development (Filler, 1986; de Soissons, 1988). Fundamentally Howard did not believe the movement could afford to wait for the state to shift its ground. While his fellow authors wanted to concentrate on a propagandist campaign, Howard's response was that they were wasting their time: 'If you wait for the authorities to build new towns you will be older than Methuselah before they start. The only way to get anything done is to do it yourself' (Beevers, 1988, p. 160).

Given the failure of the New Townsmen to persuade government, even in its most idealistic moment under Addison, to bend in their direction, he perhaps had a point. But without consulting his sympathizers he went off and bought a site. Though horrified, they could not afford to discredit the movement by not supporting him. Thus Welwyn Garden City was born in 1920. Even more than Letchworth it provided a model of good planning on principles that were now very familiar. There were many examples of learning from the first experiment and generally there were fewer compromises in its development. Its planner, Louis de Soissons, was able to maintain a consistently high design standard, especially in the centre which had a rather formal, neo-Georgian feel. Overall it became itself an important exemplar of the planning idea during the inter-war and wartime periods.

Municipal satellite towns

Despite Howard's pessimism, some municipalities also adopted elements of the emerging model of metropolitan decentralization in some of their large estates. Most prominent were Wythenshawe in Manchester and Speke in Liverpool. The former was the more ambitious, a scheme for some 25,000 dwellings, 20 per cent of them private sector (Deakin, 1989). It also had the more authentic links with the garden city tradition in that it was planned, from 1927, by Barry Parker, Raymond Unwin's former partner at Letchworth. Speke was smaller, planned during the 1930s for ultimately about 7,000 dwellings, all of them municipal (HDL, 1937). Its neo-Georgian styling, designed during the 1930s by Liverpool city architect, Lancelot H. Keay, was reminiscent of Welwyn, though had a rather higher proportion of flats and much poorer open-space standards.

Both schemes were very innovative in that they aimed for a high degree of self-containment, offering full community facilities and local employment. This last was a particularly ambitious aim, not achieved (or attempted) even in Ernst May's much-admired Frankfurt satellites. Occasionally a degree of self-containment had arisen fairly spontaneously within suburban areas. At Slough, for example, a new community grew around a pioneering industrial estate (see also below) or at Becontree-Dagenham, where the Ford motor company moved in 1931, providing much-needed local employment adjoining the giant LCC estate. Wythenshawe and Speke were planned from the outset with this in mind. But both showed some of the major difficulties in achieving planned decentralization through municipal action. A particular problem was securing sufficient land to allow the satellite to grow with even a very modest degree of physical separation from the 'parent city'. Both authorities had to obtain major boundary extensions, which seriously delayed development and was fiercely contested in the case of Wythenshawe. In both cases they were far from complete by 1939 and had not yet fulfilled their distinctive self-containment objectives.

Other decentralist proposals

The decentralist idea was given further endorsement in a 1931 report of the Greater London Regional Planning Committee (GLRPC) (Miller, 1992). This body had been established in 1927 to develop planning proposals for the metropolitan area, and its reports, issued in the late 1920s and early 1930s, broke new ground in several important respects, as we will see. The GLRPC's technical adviser and author of these reports was Raymond Unwin, who had by now left the Ministry of Health. In the 1931 report he proposed a combination of fairly self-contained, planned suburban units, satellite towns and industrial garden cities as a means of accommodating London's growth. Along with other parts of the emergent planning agenda, these proposals were victim of crisis and retrenchment

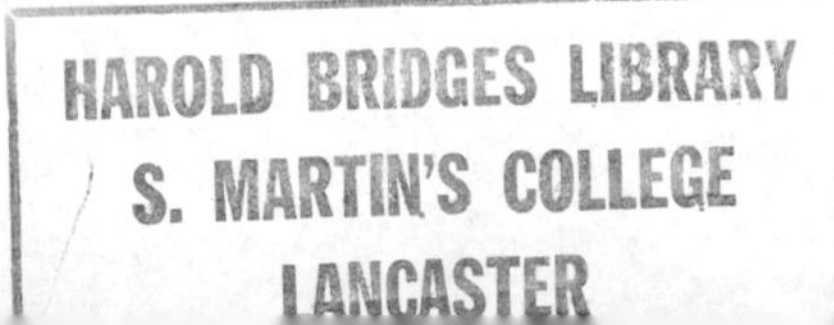

in the early 1930s. They did, however, help shape the growing debate about planning large urban areas in the 1930s.

The beginning of the decade also saw Unwin becoming a member of a Ministry of Health Committee set up under Lord Marley to examine garden cities and satellite towns. The group basically endorsed planned decentralization as an approach to regional and national planning and wanted to see a greatly strengthened planning framework to allow implementation. This was a conclusion that might have been received with some sympathy by Arthur Greenwood, the Labour minister who had initiated the inquiry. But it found no support in the changed political order after 1931 and was only grudgingly published four years later (Marley Committee, 1935).

Thus Britain's National Government remained profoundly opposed to extending its influence over urban change, an essential prerequisite of a decentralist programme. Meanwhile an altogether more sympathetic approach to planning and particularly garden city ideas was apparent in the USA. It represented the culmination of a remarkably creative period in American planning, producing two concepts that became extremely important to Britain's planning history. Throughout this period, Britain benefited by very strong professional and cultural links across the Atlantic. Figures such as Adams (Simpson, 1985) and Unwin (Miller, 1992) practised in both Britain and the USA and played an important role in spreading new ideas.

The neighbourhood unit

The garden city idea as originally articulated by Howard had included the notion of 'wards' within the town that would each have their own communal facilities. In the late 1920s Clarence Perry, a sociologist working on the New York Regional Plan, had taken the idea rather further and given it the formal title of the neighbourhood unit (Perry, 1939). It was to be a conscious grouping of schools, community facilities and shops within easy walking distance of housing in a way that would create a sense of social identity and attachment to the locality. In fact its higher aims of social engineering are of rather dubious provenance, but as British planners became more aware of it in the 1930s and especially the 1940s it became extremely popular. One of its first uses in Britain was by Barry Parker in his 1931 plan for Wythenshawe, discussed above.

The Radburn principle

Closely related to the neighbourhood unit was the distinctive residential layout developed by Henry Wright and Clarence Stein during 1927–9 (Stein, 1958). Radburn was actually a planned garden city project in New Jersey. Only one neighbourhood was ever built but it was sufficient to demonstrate the principle. It involved the complete segregation of pedestrian and vehicle circulation systems, making extensive use of culs-de-sac, 'inner parks' and pedestrian underpasses. Overall Radburn was an adaptation of the low-density garden city type of residential environment to bring it into the era of the motor vehicle.

Radburn and neighbourhood principles were used together in the green-belt town programme introduced by Roosevelt's Resettlement Administration set up under the 'New Deal'. The towns were small residential satellites with full

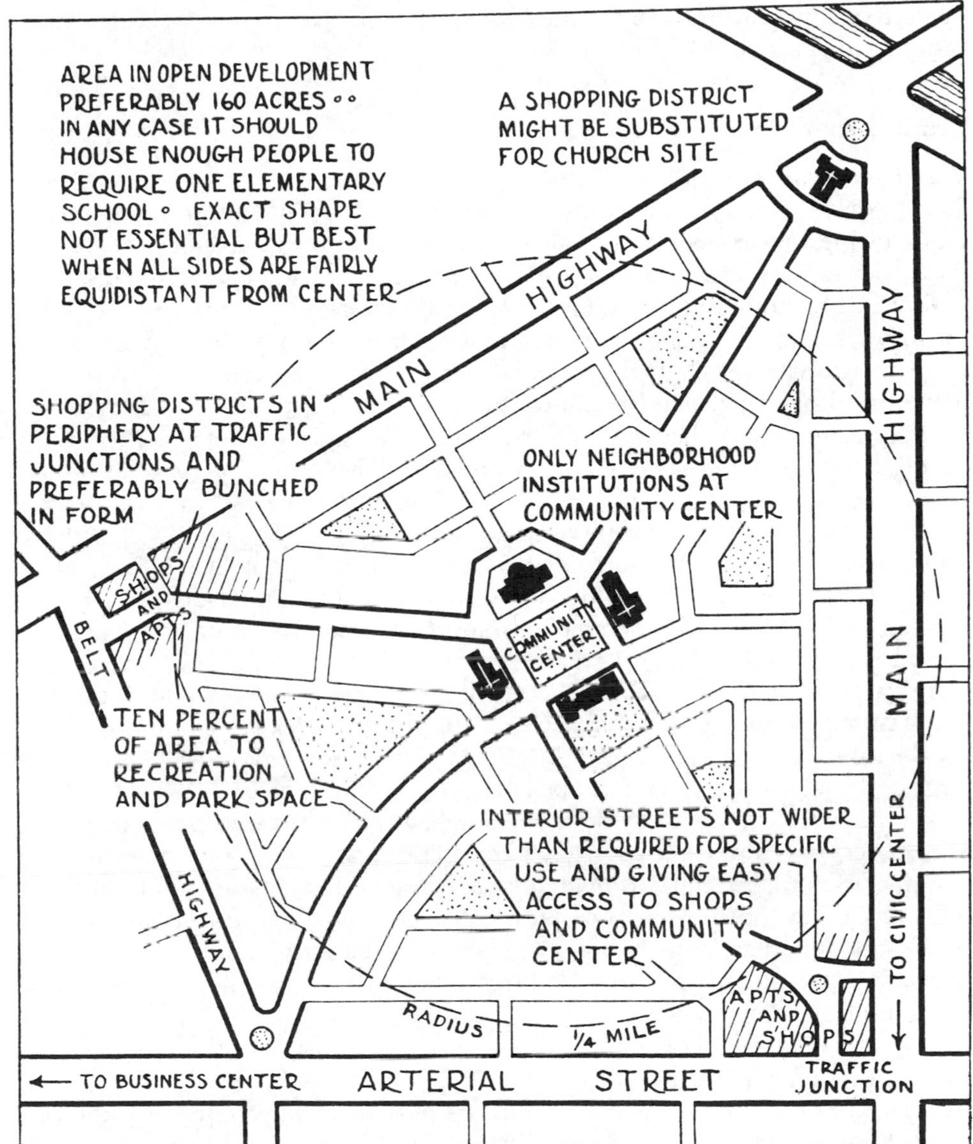

Figure 3.3 *In the late 1920s Clarence Perry, an American sociologist, developed the important concept of the neighbourhood as a physical model for grouping community facilities in residential areas thereby, it was hoped, promoting social coherence. With slight modifications it was widely (and rather uncritically) adopted in Britain in the 1940s and after.*

community facilities (though no industrial areas) in a generous green-belt setting that separated them from larger population centres. Although only three – Greenbelt (Maryland), Greenhills (Ohio) and Greendale (Wisconsin) – were ever built, they attracted considerable interest in Britain because of their novel planning features. The Radburn principle was, however, slow to be adopted, partly because motor vehicle ownership was far lower than in the USA. It only began to be used on any scale in Britain in the 1950s. By contrast the green-belt ideal made more headway in 1930s' Britain.

Planned containment

The pressures for restraint

The corollary of metropolitan decentralization was containment, stopping the big city growing so that its growth could be channelled into garden cities or satellites. As we saw in the previous chapter, this was an idea that had begun to be discussed before 1914, though not quite in the form it later came to be understood. The idea now began to involve the spatial limitation of urban growth, giving a firm edge to the city to allow the preservation of rural landscapes for scenic and recreational enjoyment and retention of agricultural land. It was a notion that advanced significantly between the two World Wars. As part of the decentralist model it was pushed by the Garden City and Town Planning Association and there were small agricultural belts around Letchworth and Welwyn. Moreover during the 1920s there was a mounting preservationist lobby reacting against all kinds of development that impinged on the countryside (Williams-Ellis, 1975, orig. 1928). The Council for the Preservation of Rural England (CPRE) was formed in 1926 and soon became an important umbrella group, co-ordinating the efforts of many different bodies.

It was, though, very difficult to do anything to implement green belts through the statutory planning system because of the risk of heavy compensation claims for the loss of development rights. But the very cheapness of land that encouraged large-scale suburbanization also opened up possibilities for large-scale purchases by municipalities or voluntary bodies to sterilize areas from development. Also, it sometimes proved possible to secure agreement to schemes preventing building, in some areas where land was held in large private estates. Politically there was a widening lobby for the green-belt idea, especially (and paradoxically) from many of the new suburban areas. Many of the new suburbanites certainly felt they were being cheated out of their initial direct relationship with the countryside by continuing building activity.

Early preservation initiatives

Many early policy actions were partial rather than being consciously directed towards the achievement of planned containment (Sheail, 1981; Elson, 1986). By the 1920s, for example, Birmingham had acquired a large part of the Lickey Hills on its south-western boundaries and there were similar initiatives in many other large provincial cities, especially Sheffield, Glasgow and Leeds. By 1934 Brighton had bought some 4,000 hectares of the South Downs to prevent building development. Voluntary actions were also important. In Birmingham the Bournville Village Trust and the National Trust dedicated large rural areas south of the city to perpetual undeveloped status. From 1927 the newly formed Oxford Preservation Trust raised money to retain important green areas around that city.

Although all these schemes shared a common desire to stop building development, there was a good deal of variation in practice about the prime objectives of

Figure 3.4 *The other important American planning innovation of this period was the Radburn layout, perfected by Henry Wright and Clarence Stein. Heavily influenced by Perry's ideas, Radburn principles involved the separation of pedestrian and motor traffic, effectively updating the garden city vision for the motor age.*

such reservations on the urban fringe. In Brighton, for example, the main motivation was securing the purity of water supplies for the town, and similar concerns were apparent in the Pennines around Greater Manchester and Sheffield. In Oxford scenic quality and a particular concern with protecting the setting of the historic city were paramount. Attitudes to public enjoyment of these preserved areas varied a great deal. Birmingham had always seen public access as a key consideration in the Lickeys, so much so that it operated a tram route to serve them. Access was much more limited in the Pennines, however, even to publicly owned land, though Sheffield made some areas available (Rothman, 1982). By further contrast, Brighton Corporation was prepared to allow public enjoyment (notably motor racing) of a type that sharply contradicted the preservationist aims of the adjoining authorities. These kinds of confusions and conflicts were to be inherited as the incremental acquisition of open land on the urban fringe gave way to the more formal concept of a containing green belt.

The concept of a containment strategy

The Unhealthy Areas Committee had broadly endorsed the idea of a continuous agricultural belt in 1920–1, though with no immediate impact on policies. Gradually the idea was being changed from a continuous narrow belt of parkland within the city to something that could actually redirect urban development as part of a broader strategy for growth management involving planned decentralization. It took some time before this was articulated fully, though there were important interim stages. Barry Parker's first proposals for Wythenshawe included a narrow green belt to separate satellite and parent city. He also adopted the American parkway concept for the main road developments which were set in a generous open space reservation, attractively landscaped to create high scenic quality.

These developments were important in demonstrating the ways green belts could be treated as objects of design. Yet again it was Raymond Unwin who outlined its full strategic role, in conjunction with planned decentralization, as a means of managing metropolitan growth and change (Miller, 1992). Thus his first report for the Greater London Regional Planning Committee in 1929 incorporated schematic proposals for a 'green girdle' around the built-up area with green wedges along river valleys. Beyond were to be the new satellite towns, referred to earlier. There was, however, more short-term action on the matter of the 'green girdle' than on the remainder of Unwin's work.

The origins of green-belt policies

Although the operations of GLRPC were curtailed in 1931, the quest for a London green belt was taken up by the LCC following Labour's victory in the 1934 elections (Gibbon and Bell, 1939). The following year Herbert Morrison, the new Council leader, unveiled proposals to implement a green belt around the capital (Sharp, 1980; Ward, 1991). The LCC would provide funds (initially up to £2 million) to allow the purchase or acquisition of development rights on undeveloped land. Adjoining authorities, including Middlesex, Surrey and the County Borough of Croydon also reinforced the LCC's programme with their

Figure 3.5 *The Labour victory in the London County Council elections in 1934 brought decisive moves towards implementing Unwin's proposed green girdle.* Punch *adopted a Shakespearean analogy, with Labour's London leader, Herbert Morrison, portrayed as Oberon, enlisting Puck's aid to put a green belt around the capital. Reproduced by permission of Punch.*

own vigorous programmes providing ample testimony of the popularity of the scheme in urban-fringe areas. It did not quite produce the continuous area which Unwin had wanted, but he greatly approved of what was being done. Moreover the manner of its implementation was easier than that which his proposals had implied, since it involved fewer authorities. There was a legal loophole in the 1935 scheme, however, requiring the Green Belt (London and Home Counties) Act to be passed in 1938.

Eventually some 27,600 acres were acquired under the scheme, rather less than the 70,000 acres originally intended. But changes far more dramatic than Morrison or Unwin ever dreamed possible occurred in the following decade, superseding this original scheme, so it would be wrong to imply it failed. More serious though was that the exact objectives of the scheme in practice remained rather confused, echoing in some respects the confusions apparent in earlier and more incremental preservationist initiatives. The LCC, as befitted what was by now an inner urban authority, was particularly concerned with recreation and access for its residents. Yet its concerns were not always reflected in its practice, particularly since adjoining authorities were not usually anxious to see widespread public access. But it would be wrong to imply that there was any widespread concern about these issues of social equity. The green belt was a widely approved initiative which by 1939 was being tentatively adopted as a model for other cities. There had, however, been greater tangible activity in the field of planned redevelopment, to which we now turn.

Origins of Planned Redevelopment

The need for renewal

The slum problem

Although the early planners had used the slum extensively in their arguments in favour of planning before 1914, little had actually been done to deal with the problems of obsolescence. Despite the demolition of some 548,000 dwellings between the wars, there were still over 3 million pre-1855 dwellings surviving by 1938 (Richardson and Aldcroft, 1968). They had been built by what was a relatively poor industrializing economy. There had been no effective form of building control for most such dwellings. Building standards were often such that their expected life would have been a few decades at the most, even with a standard of maintenance that most had not received. Moreover many were designed in a physical form that was already seen as very unsatisfactory by 1914.

Local examples

In 1913 Birmingham, for example, had over 43,000 back-to-back houses (BVT, 1941). It also had over 42,000 dwellings without separate water supplies, sinks or drains. Over 58,000 dwellings had no separate water closet. These facilities had to be shared in the common courts on to which this housing fronted. And Birmingham, it should be stressed, was far from being the worst. The smaller city of Leeds had 78,000 back-to-backs in 1914, representing 71 per cent of its housing stock (Finnegan, 1984). Unlike other cities it continued to allow them to be built well into the twentieth century, though about half had already been built before building controls were introduced. London suffered from a variety of slum types including multi-occupied and decrepit larger property with shared facilities (LCC, 1937). It also had areas of small, poorly constructed working-class terraced housing, especially in the East End. Urban Scotland and Tyneside were typified by large numbers of exceptionally small dwellings, often with poor, shared facilities and normally overcrowded. Thus in Newcastle in 1921 almost 37 per cent of families lived in dwellings of one or two rooms. In nearby South Shields the figure was over 43 per cent (Mess, 1928).

The wider problems

But although housing was the most obvious manifestation of obsolescence, much of the remaining environment of these inner slum areas was also in need of renewal. Old factories and cramped workshops were often interspersed with slum housing, contributing to industrial inefficiency and pollution. Street corner shops and beer houses were often in a similar physical state to the housing. Parks and allotments were usually lacking and schools were often inadequate, especially in open-space provision. Road and street layouts were typically rather chaotic, reflecting an unplanned development process that predated and was increasingly inappropriate for the motor vehicle.

Was clearance the only option?

All these problems meant that wholesale clearance was the only viable policy for dealing with such areas (Yelling, 1992). Some local consideration was given to reconditioning and improvement policies, on the lines of pre-1914 'slum patching' and 'closet conversion' schemes. The latter, for example, were important on Tyneside in converting the last urban areas that relied extensively on ash closets after 1924. The problem was that such initiatives usually depended on landlords being prepared to contribute some of the costs. This was rarely practicable, since most slum landlords were people of small means, though a few housing associations did improve older property. In addition, the actual business of restoration was more difficult than it is today in the absence of the sophisticated chemicals to eliminate damp and rot. And, quite simply, most of the slum areas were too far gone to be salvaged.

The central areas

Although the slums posed much the most serious redevelopment problem, there were also mounting demands to renew and expand the central areas of towns and cities. Central commercial, retailing and administrative areas had been created in the later nineteenth centuries, but they were much smaller than today and often still hemmed in by slum areas of the type described above. The pressures for change in the city centres in this period have been rather dwarfed by the even more dramatic changes of the post-1945 period. But important changes were beginning to occur even in these years.

Thus the growth of a much more consumer-oriented society between the wars strengthened the demands for greater central shopping provision. New multiple stores like Marks and Spencer, Littlewoods, Boots and Woolworths began to make their mark in central locations (Jefferys, 1954). Bigger and grander department stores were also being built, especially in the provincial cities. The continued development of commercial and financial office-based activities created a similar pressure for new central space, as the volume of business for the insurance companies, banks and building societies increased (Weightman and Humphries, 1984). There was also a great expansion of the public services, especially local government. This generated further demands for central office space and other facilities. Changes in transport, towards the motor vehicle, were also creating mounting congestion problems in the Victorian streets of the central areas and generating demands for new central facilities like bus stations.

Ideas for the redeveloped city

Traditional influences

During the inter-war period there were two main sets of ideas available in approaching the design of redeveloped higher-density urban spaces: traditionalism, usually derived from classical architectural principles or more recent variants, and modernism. In general the British approach to urban design was rather more conservative than in other European countries. Accordingly more traditional ideas, usually interpreted broadly as a restrained version of the Beaux Arts approach, with its strong sense of formality and symmetry, or even more in neo-Georgian styling, tended to dominate many of the redevelopment proposals in both central and inner areas.

Central area redevelopments

There were actually very few large central area schemes in this period, but those that there were underline the strength of traditional design concepts in planned redevelopment. Thus the Headrow scheme in Leeds developed from the early 1920s (Ravetz, 1974a), or the slightly later Ferensway scheme in Hull were relatively unusual as large street-widening and redevelopment projects, though they clearly show how such planning problems were approached. Both used very formal facades fronting on to widened central streets. The many smaller central schemes usually betrayed broadly similar traditional influences, sometimes mixed with the classically inspired monumentality of the American 'City Beautiful' movement. Such mixes of traditional ideas were apparent in many of the projects for new 'civic centres' that were pursued in several towns and cities, for example in Birmingham, to provide distinct administrative areas of civic buildings surrounded by gardens within the central area.

Modernism

(i) Le Corbusier: An altogether more challenging set of ideas was developing outside Britain: in France, pre-Nazi Germany, Austria and elsewhere (Larsson,

1984). This was the emergent functionalist and consciously international approach of modernism. During the 1920s and early 1930s this took important steps forward as an urban planning project with the publication of a series of hypothetical plans by a Swiss-born self-trained architect. His name was Charles-Edouard Jeanneret, though by this time he was known as Le Corbusier (1946, orig. 1927). His Ville Contemporaire (1922), Plan Voisin (1925) and Ville Radieuse (1933) together represented a vision of a rational, functional approach to urban design (Le Corbusier, 1922, 1925). Important developments in his ideas are apparent, notably from a capitalist, class-segregated 'Contemporary City' to a syndicalist and egalitarian 'Radiant City'. Most of the essential elements of the city as a physical entity survived this transition, however.

The basic assumption of his work was that the big city itself could be perfected. The 'Contemporary City' had a population of 3 million, a hundred times larger than Howard's garden city. Utopianism in planning no longer had to look to an escape from the big city. It was, however, to be a drastically remodelled city, based on very strict zoning principles and abolishing many traditional facets of the urban scene such as the corridor street, lined with buildings. The central areas, dominated by huge office towers, were to be major interchanges for road, rail and air. Huge superhighways were to be built right into the central city. Beyond this were first luxury apartment blocks set in parkland for the élite and then the lower-density satellite towns for workers and factories. The Plan Voisin broadly interpreted this ideal within the specific context of central Paris, making few concessions to the existing city. His final articulation of the ideal, the 'Radiant City', had an entirely uniform housing area in the central area, composed of giant 'Unite' slab apartment blocks and devoid of any social distinctions, with the office and administrative areas to the north and industrial areas to the south.

(ii) German examples: Le Corbusier did not develop his ideas in isolation. Other architects were working along similar lines and an international group called the Congres Internationaux de l'Architecture Moderne (CIAM) was founded in 1928. Le Corbusier's main role at this time was the development of a theoretical programme for the city. Meanwhile architects in other European countries were creating more immediate models. In Weimar Germany socialist architects such as Walter Gropius, Bruno Taut, Martin Wagner and Hans Scharoun were developing the design of residential areas in social housing schemes such as Britz and Siemenstadt in the Berlin suburbs designed between 1925 and 1931. Particularly important was their use of modern apartment blocks in ways that developed attractive higher-density alternatives to the twelve cottage-type houses per acre of Britain (Hass-Klau, 1990, pp. 72–7). Their community facilities, the care taken to separate pedestrians and traffic, and their provision of gardens and open space were all particularly impressive.

(iii) Other influences: The social housing schemes of pre-Nazi Vienna in the 1920s and early 1930s were also important (Denby, 1938). Large apartment complexes were developed by the socialist-controlled municipality, most famously

Figure 3.6 *Modernist architectural imagery began to influence British planning thinking in the 1930s and had a marked influence on ideas for the redeveloped city. These 1930s' tower blocks at the Cité de la Muette at Drancy in the north-east suburbs of Paris were featured in several important British planning and architectural publications of the period, though there were no actual British examples of this form of housing until after 1945.*

at the Karl Marx Hof, with excellent community facilities. And there were several notable French schemes, especially in the late 1930s. Most important were Drancy-la-Muette in Paris and Villeurbanne near Lyons which included tower blocks of social housing of sixteen and eighteen storeys respectively. And the practice of architects in other countries, particularly Holland, Czechoslovakia and Sweden, also contributed significantly to the further development of modernism as a project for urban planning, especially in the design of housing areas. In the Soviet Union the concept of the linear city emerged (Miliutin, 1974). This revived ideas of the Spanish engineer Soria y Mata, who in the 1880s had proposed a linear urban form, and combined them with Ebenezer Howard's garden city. The result reflected the exciting mood of early post-revolutionary social and design experiment in Russia that attracted great interest from the modernists further west.

Modernism and British planning

It was only during the 1930s that the ideas of modernism began to influence British planning (Ravetz, 1974b). There was a mounting ideological opposition to traditionalism, reinforced by the wider critique of suburbanization. This received powerful reinforcement as the rise of the Nazis in Germany and the increasing cultural conservatism of Stalinism in Russia brought many of the seminal figures of the modern movement into Britain during the 1930s. The MARS (Modern Architectural Research) Group was established to provide a British forum on the model of CIAM (Johnson-Marshall, 1966). MARS was heavily influenced partly by Le Corbusier's schemes and Soviet ideas of the linear city as later became apparent in the revolutionary MARS plan for London, unveiled in 1942. But it was the shifts in housing policy towards slum clearance and the increasing use of flats in redevelopment that did most to introduce the concepts of modernism into British planning practice.

The 'Great Crusade' 1930–9

Early attempts at planned redevelopment

Considering the seriousness of the problems, it is remarkable that so little was done about the slum areas for so long. In fact, as we have seen, the problems of absolute shortage took priority in 1920s' housing policy. Neville Chamberlain's Unhealthy Areas Committee had addressed the slum question in 1920–1, but was largely ignored. Only a very few authorities took any significant action on slum clearance in the 1920s, particularly the LCC and Liverpool (Sutcliffe, 1974b; Yelling, 1992). The former produced an important scheme at Ossulston Street (near St Pancras Station) in 1926 that echoed some of the features of the early Viennese estates. Its final form was, however, a good deal more conventional than originally intended, a testimony to the prevalent design conservatism of the period. The favoured solution in both London and Liverpool was the neo-Georgian, traditionally built block with balcony access. Estate layouts generally showed little of the design innovation that was apparent in German and other schemes of the same period.

The Five-Year Programmes

The Housing Act 1930 and the abandonment of the Wheatley general-needs subsidy in 1933 marked, as we noted earlier, a decisive switch in housing policy. Local authorities had to submit Five-Year Programmes for clearance and renewal in 1934 (Richardson and Aldcroft, 1968). Reflecting the still-dominant policy of restraint in public finance, many of these targets were very low. Despite all its slum problems, Birmingham proposed a target of only 4,500 demolitions, though Manchester proposed 15,000 and Leeds (which was roughly half the size of Birmingham) a heroic 30,000 (Ravetz, 1974a; Finnegan, 1984). Nationally the target was a little over 333,000. And, as public expenditure restrictions were eased from the mid-1930s, the drive to clear the slums acquired greater momentum, becoming the 'Great Crusade' (to borrow the title of a widely circulated 1936 promotional film). A much more ambitious revised target of over 600,000 was introduced, though it proved over-optimistic.

The political economy of redevelopment

Against the political background of economic and financial orthodoxy during the 1930s, the mounting impetus to clear the slums, involving a great deal of government intervention, appears as a paradox. And it certainly puzzled George Orwell (1970, pp. 214–16) when he arrived in Liverpool in 1936, admired its new Viennese-style workers' flats, and then discovered that they were the work of a Conservative council. There were certainly energetic Labour councils who pushed the slum clearance programme, especially Leeds, Sheffield and London, but the loose alliance of forces that underlay the 'Great Crusade' was much more complex than either municipal politics or the strength of the reformism of the planning and housing movements.

There was a growing business lobby for slum clearance. Sometimes this was expressed locally in desires to renew the city's social capital, or extend the central

Figure 3.7 *The most typical flats built in the redevelopment areas of the big cities in the 1930s were medium-rise blocks, sometimes (as here) reflecting the influence of the pre-1934 schemes of socialist Vienna, though without ever matching all the latter's progressive community features. This photograph, taken c.1937, shows the recently completed Gerard Gardens scheme in Liverpool.*

commercial areas, reflecting concerns with business prosperity that overrode the, by now, rather marginalized private landlords of the slums. Moreover, the recognition of the huge new opportunities for business represented by the production and use of the new dwellings was very important. We can detect this in the great national interest taken in slum clearance by the building, building materials and energy industries. Model flats projects were sponsored by the concrete, steel, gas and other industries and important films were made by them to promote the slum clearance message. Quite simply a move from a small, ill-equipped slum dwelling to a fully equipped modern flat or house implied a massive increase on the underconsumption represented by the slums. This was not just cement or bricks or glass or paint but also stoves, bathroom fittings, lighting systems, etc. The scope for growing business for such industries represented by planned redevelopment was immense.

The redevelopment process in the 1930s

Of course, as we noted earlier, the bulk of the rehoused populations moved to suburban estates. The densities of the slum areas (and of course it was the very worst slum areas that were being tackled at this time) were usually so high that it was never possible to rehouse an equivalent population on the site. Clearance also left large areas of land that could be reused in a variety of ways. Non-housing use was common in smaller centres. Wakefield, for example, had embarked on a scheme to extend its central shopping area and create a new bus station on slum-cleared sites by the late 1930s. In the bigger cities most inner city sites were necessarily reused mainly for housing, however. Inevitably this soon raised the question of new roads and social facilities.

By the end of the 1930s towns and cities were beginning to declare redevelopment areas under the Housing Acts. This enabled them to consider the replanning of wider areas, albeit mainly for housing, rather than simply working on the basis of individual clearance areas. Birmingham, for example, declared a 267-acre redevelopment area in Duddeston and Nechells in 1937 and was actively considering other large areas (Manzoni, 1939). The LCC similarly declared a large area in Stepney and Liverpool had five smaller redevelopment areas by 1937 (LCC, 1937; HDL, 1937). Yet the limitations of general planning and housing powers for land acquisition made the implementation of wholesale replanning very difficult. Some cities, like Birmingham in 1939, also began the process of statutory planning for their non-suburban areas at this time, but the weaknesses of the 1932 Act were even more exposed in dealing with the non-housing aspects of redevelopment areas. Even the acquisition and demolition of houses proved very slow, and by 1939 only a small part of the overall problem had been tackled. But the redevelopment areas were important in initiating a policy concern that was largely completed under post-war planning and housing legislation.

Modernism and redevelopment

Much the same conclusion about having a preview of the policies and approaches of the post-1945 period hold true for the impact of modernism (Ravetz, 1974b). The bulk of flats built in redeveloped areas in the 1930s were very traditional in

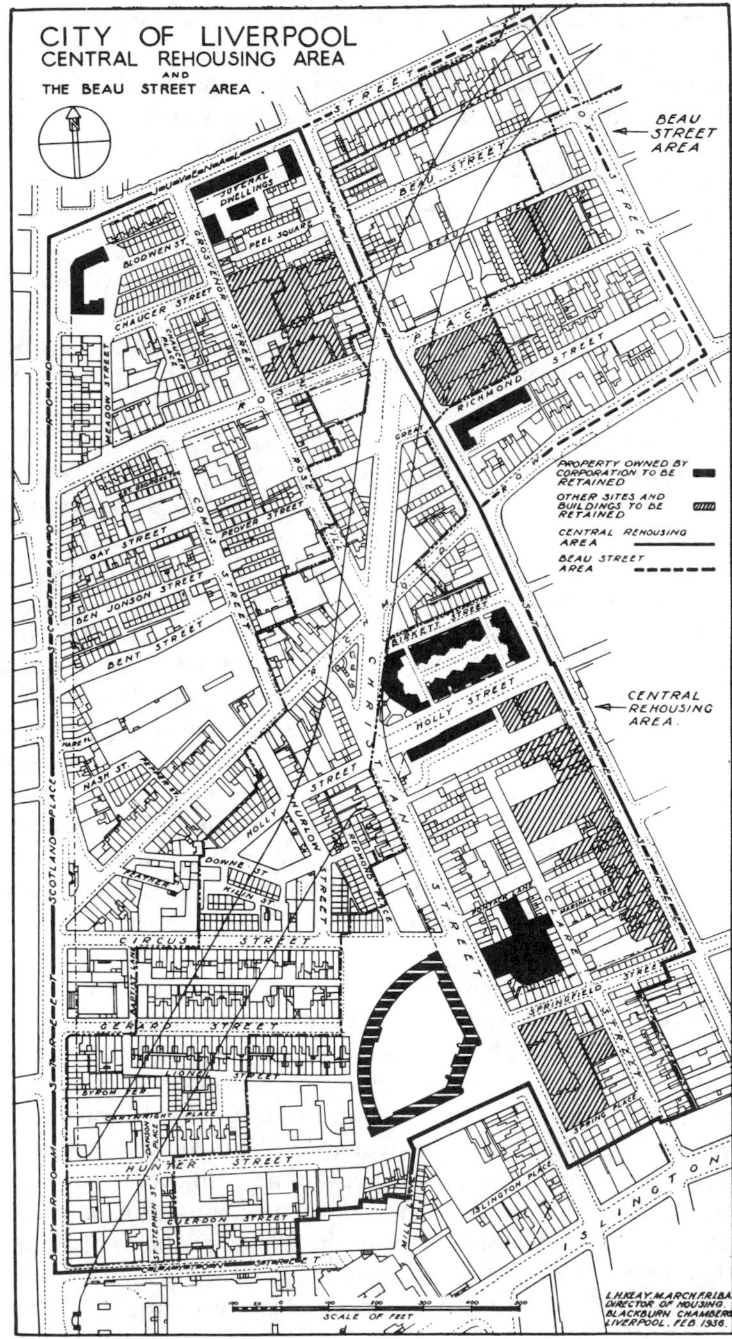

Figure 3.8 *By the late 1930s, slum clearance was being integrated in wider projects of comprehensive redevelopment involving new housing community facilities and road systems. Gerard Gardens (Figure 3.7), visible as the fan-shaped building in the south of the area, was conceived as part of a larger scheme of redevelopment.*

style. The LCC and Liverpool, the two great inter-war municipal flatbuilders of England and Wales, continued to favour neo-Georgian designs, though there were some exceptions. Gerard Gardens (1937) in Liverpool was an early experiment in a modernist design showing strong continental influences (HDL, 1937). Birmingham, the most reluctant flatbuilder, finally produced a compromised modernist design for its 1939 scheme at Emily Street (St Martin's Flats) (Sutcliffe, 1974a). Manchester had already pointed the way with Kennett House (1935) and claimed another first at its Greenwood Estate (1938) with the invention of the mixed development of houses and flats that became officially endorsed in the 1940s.

More dramatically modern in conception was Kensal House (1937), designed by CIAM member Maxwell Fry and very innovative in communal facilities. It was a demonstration scheme produced by the Gas Light and Coke Company in north Kensington, housing tenants from slum clearance areas. Most profoundly impressive of all was Quarry Hill (1938–41) in Leeds, a huge 938-flat scheme built on the same French building system as Drancy-La-Muette and inspired by the Karl Marx Hof in Vienna (Ravetz, 1974a). Built on a sloping central site, and eight storeys at the highest point, it incorporated many important innovations including lifts, a sophisticated waste-disposal system and communal laundries. Here was the most profound and complete pre-1939 expression of what modernism could do in the solution of Britain's urban problems. And it was in schemes like this that planners, architects and housing reformers increasingly saw the future. Developments at the regional level of planning were similarly beginning to rehearse future policy options.

Origins of Balanced Regional Development

Regional planning

Regional surveys

As we saw in the previous chapter the Geddesian concept of the region, wider than the individual town or city, had featured in the initial body of ideas that had made up town planning. Despite this, the statutory framework of planning created in 1909 had ignored this dimension and made districts and boroughs into the sole planning authorities. The town planning scheme was very localized and did not oblige planning authorities to think about wider questions. There were some signs of a revival of regionalism in the later years of the First World War, however. The most important was the South Wales Regional Survey established by central government as a response to wartime labour unrest on this important coalfield (MH, 1921). It had been felt that part of the problem was the poor environment and the absence of any social or economic counterbalance to mining in the communities where the miners lived. The committee quite sensibly moved beyond the survey role and its report, issued in 1921, was in effect Britain's first regional plan, notable because of proposals for socially and economically diverse satellite towns as an alternative to pithead villages.

Regional planning initiatives

The 1920s saw a great deal of regional advisory planning activity (Cherry, 1974b; Massey, 1989). Central government officials recognized the limitations of a planning system that was based on very narrow spatial units and encouraged *ad hoc* groups of local authorities to come together to think collectively about the planning of their areas. By 1931 eleven years of this encouragement had produced 104 such regional planning bodies in England and Wales, largely in the main urban areas. Such regional planning only had an advisory status and relied on its proposals being adopted within the statutory planning schemes of its constituent members. Moreover the actual areas covered by such initiatives varied tremendously. Some, like the Midlands Joint Town Planning Advisory Council, established in 1923, covered a vast area of some 1,565 square miles with a population of some 2.5 million. At the other extreme were urban regions, covering the immediate sphere of influence of a particular centre. The model here was the Doncaster Regional Plan initiated in 1920 to make proposals for planned development associated with this new coalfield area. There were many intermediate types of area, however, sometimes with boundaries that overlapped or were entirely contained within other regional planning areas.

The process of regional planning

But despite the absence of any overall framework, there were many advantages in this approach to regional planning. For one thing it tended to produce much more attractive documents than the town planning schemes, which had to be written in a highly legalistic style. The general standard of the documents was set by the most prominent regional planner of this period, Patrick Abercrombie (Dix, 1981). His elegant prose made a sharp contrast with the rather dreary planning schemes. Conceptually the plans were a derivation and a development of the Geddesian tradition of Place–Work–Folk. This attention to the major themes of change and continuity brought the plans considerable publicity in the provincial press. And the plans provided a focus for a rather wider public debate than the town planning schemes, which tended to attract interest only from local property interests.

The implementation of regional plans

In many cases the regional advisory proposals did encourage co-ordination in local scheme preparation. The problem was that the only powers behind these regional plans were those of persuasion. Even the modest powers of the statutory planning process were available to implement the regional plans only if the local authorities who exercised these statutory powers were agreeable. A classic case here was in Manchester in the late 1920s where two regional planning committees developed alternative proposals for the area which became the Wythenshawe satellite town. The North Cheshire Regional Planning Committee reflected the fierce opposition to the proposal in Cheshire (and their successful resistance of a proposed boundary extension in 1927). By contrast the Manchester and District Regional Planning Committee embodied the emergent satellite town idea, but was only able to begin to implement it when Manchester, which already owned most of the land, finally secured a boundary extension in 1930 (Deakin, 1989).

But despite such problems, this was something on which action could be taken comparatively quickly. The most radical proposals to come out of the regional plans were virtually impossible to implement in the short term. Rather, they often served to implant planning proposals that acquired longer-term significance. The most striking example here was in Raymond Unwin's radical proposals to the Greater London Regional Planning Committee, noted above, which were only partially implemented. The green-belt scheme grew from his recommendations, but his most radical proposals for decentralization served mainly to rehearse new planning solutions that were taken further in the 1940s.

Regional planning and public spending

Finance was often a key consideration in the implementation of regional plans (Ward, 1986). Many regional proposals involved public expenditure, often on the part of agencies other than the statutory planning authorities. With a generally cautious central attitude to public spending in the 1920s, it was inevitable that few such proposals would be implemented. The 1929 general election saw unemployment and a major public works programme feature as significant issues, however (Skidelsky, 1967). The pace was set by the Liberals and their important 1928 publication *We Can Conquer Unemployment*, based directly on the work of the famous economist John Maynard Keynes. What he and the Liberals were specifically arguing was that it was possible to use public works to put money in the pockets of the unemployed which they would spend, boosting aggregate demand in the economy.

In fact the Liberals came third in the 1929 election, but they held the balance of power and the Labour government depended on their support. Moreover some prominent Labour politicians (most notably Sir Oswald Mosley) were very interested in a public works-led programme to deal with unemployment, while Labour itself had no real alternative strategy, so these Keynesian Liberal ideas continued to reverberate in the policy process. The link with the regional plans is that they represented, in the eyes of many Liberals at least, an unparalleled shopping list of public works proposals. It was then against this background that an official committee was set up in 1931, chaired by Lord Chelmsford, to examine the possibilities of using the regional plans in this way, especially in the depressed areas (Chelmsford Committee, 1931).

The regional problem

By this time the major dimensions of the regional problem were clearly evident (Law, 1981). The older coalfield-based, export-oriented industries of coal, cotton, steel and shipbuilding had been experiencing serious decline since the collapse of the post-war boom in the early 1920s. By contrast a whole set of newer consumer-based industries and services like motor vehicles, electrical engineering, food processing, building materials and retailing were showing rapid growth, largely in the south and midlands. This unevenness was manifest in patterns of inter-regional migration and, above all, unemployment. In 1931, for example, unemployment (recorded in the Census) varied from 7.8 per cent in the south east to 19.2 per cent in the northern region, with the north west, Scotland and Wales all over 16 per cent. And it was getting worse.

Regional plans and the Chelmsford Committee

This background of mounting economic crisis meant that greater than usual political courage was necessary to implement a radical and fiscally unorthodox Keynesian programme. The problem was that it would have needed to rely on heavy government borrowing, which ran counter to the 'Treasury View' of public finance and would have frightened the money markets (Ward, 1986). Moreover Ministry of Health officials particularly were very worried about forming too direct a link between a highly controversial public works programme of planning and the more regulatory frameworks of statutory planning. In the event the Chelmsford Committee took a cautious line, which was heavily reinforced by the political changes of 1931. We have already noted how the Town and Country Planning Bill 1932 was badly mauled by the Conservatives. Any incorporation of positive public works proposals would almost certainly have been fatal.

Regional plans and the depressed areas

The result of this separation was that mainstream planning had little to offer the depressed areas of the north and west of Britain. Under the orthodox view of public finance that continued to dominate over most of the 1930s, areas of private disinvestment represented the most serious risk for public investment. The Ministry of Health carefully controlled the approvals of capital spending and maintained a secret loan sanction blacklist (known as List Q) that discriminated heavily against such areas (Ward, 1984). Partly for this reason capital spending tended to be rather lower in the depressed areas than elsewhere. Thus new road schemes that were proposed in regional plans in depressed areas such as that for South Tyneside (STJTPC, 1928) remained unbuilt, though they were of great long-term importance.

Given this apparent unwillingness to implement public spending proposals in such areas, many planners questioned the validity of pursuing the normal kind of planning. If no new private-sector investment in industry was likely to occur, with no hope of major public works, what was the point of planning town extensions (or indeed garden cities or satellite towns)? Some of the later regional plans, for example Abercrombie and Kelly's Cumbrian Regional Planning Scheme of 1932, virtually admitted this. Occasionally the sentiment was expressed in more forthright terms:

> [The Planning Officer] . . . may be planning in a 'distressed area'. And as he puts the delicate bands of colour on his map that will allow Coketown to expand from its present 20,000 to 200,000 inhabitants, the thought suddenly strikes him that the place has been damned and dead for years, with no hope of resurrection. Actually what he should be planning for it are cemeteries to hold the wretched people who have been left, workless and poverty stricken, because of the lack of industrial planning, which he is not allowed to undertake.
>
> (cited in Ward, 1986, p. 16)

By this time though there were the first signs of a policy response to the problems of such areas (Ward, 1988).

Regional policy

Responses to the depressed area problem

We have already noted the Ministry of Health's financially orthodox response to the problem of the depressed areas. This was reinforced because the Ministry also took responsibility for locally funded poor relief, which tended to be very high in such areas, sometimes accounting for half the local expenditure budget. The underlying approach was defined by the Permanent Secretary of the Ministry, advising his Minister, Neville Chamberlain, in 1925: 'we and the other Departments should not agree to additions to the capital liabilities in these areas. One wants in effect to discourage people from sticking to them and to drive them into shifting for themselves by going elsewhere' (cited in Ward, 1988, p. 175).

In 1927–8 this was taken further in the Industrial Transference Scheme developed by the Ministry of Labour (Pitfield, 1978). This subsidized the movement of unemployed workers (and their families in some cases) out of high unemployment areas. This policy was seen as one of lubricating a 'natural' response to spatial unevenness and indeed far more people moved from such areas than were covered by the scheme.

Within this unpromising policy framework set by central government, many areas tried to attract new investment (Ward, 1990). Some reluctant encouragement for this came in the Local Authorities (Publicity) Act of 1931. Development Boards of various kinds were set up in Lancashire, Scotland, south Wales, Cumberland and north-east England from 1930 and there were many other local initiatives. The Ministry of Health was worried about the risks of municipal competition for new investment subsidized out of local taxation. Meanwhile, however, the Board of Trade hoped that such local initiatives might provide an alibi for their own inaction on the depressed area question during the worst Depression years. Ultimately, though, it was very difficult to mount effective local promotion and development policies while capital spending in such areas was inhibited by central constraints.

Special Areas Act 1934

As the Depression of the early 1930s began to ease nationally it was clear that recovery was barely touching the export coalfield areas of outer Britain (Fogarty, 1945; Ward, 1988). These areas had of course also been very badly depressed during the 1920s, and the likelihood of spontaneous improvement now seemed remote. In particular, outmigration from these areas had been greatly reduced during the worst Depression years, simply because there were fewer jobs elsewhere. And there was now increasing publicity for the plight of such areas reflecting a mounting public and political sympathy. Ministry of Labour figures showed the shipbuilding town of Jarrow on Tyneside, for example, with average unemployment of 67.8 per cent in 1934. In the nearby town of Gateshead it was

Figure 3.9 The appalling unemployment and dereliction of the depressed areas began to touch the consciences of middle-class public opinion by the mid-1930s. King Edward VIII highlighted the problem during his short reign. Visiting the derelict steel works at Dowlais, which had employed 9,000 workers, he said that 'something must be done to find them work'.

44.2 per cent. In Cumberland, Maryport recorded a figure of 57 per cent. The south Wales coal and steel town of Merthyr Tydfil had 61.9 per cent unemployment. And there were many other examples, usually of smaller mining villages, above that figure, though many sizeable towns had unemployment above 30 per cent.

It was political expediency which pushed the National Government into passing the Special Areas (Development and Improvement) Act 1934 (Booth, 1978; Pitfield, 1978). The National Government, firm in its philosophical commitment to economic orthodoxy, had absolutely no belief that such actions could be anything other than a social gesture. The legislation designated four areas – south Wales, Durham and Tyneside, west Cumberland and a wide belt of central Scotland – as so-called Special Areas, overseen by two Commissioners (one for England and Wales, another for Scotland). Extra financial assistance was to be available to these areas, though on rather limited terms. Moreover nothing was to be done to ease the spending straitjacket of local authorities in the depressed areas. Some useful projects to clear derelict land, improve housing and health facilities and provide infrastructure were initiated. But the scale of expenditure was miniscule in comparison with the size of the problem.

Special Areas Acts 1936 and 1937

During 1935 the pressure for more action on the Special Areas intensified. The imminence of the 1935 general election was important because many of the depressed areas (including Jarrow) had returned Conservative MPs in 1931. Not surprisingly they were rather worried about keeping their seats. Accordingly an industrial dimension was added to the policy and implemented in the 1936 and 1937 legislation. The first provided a framework for charitable money to be used to finance new small business projects via a body called the Special Areas Reconstruction Association. The second took this further and provided direct central government aid, and extended the scope of such assistance beyond the Special Areas proper (to include for example areas like Lancashire). All this was done with as little real commitment as the original legislation. The amount of spending, though increasing, remained paltry in comparison with the scale of the problems. Qualitatively though the principle of giving special assistance to industry in such areas was an important policy innovation of profound long-term importance.

Industrial planning

This shifting of the Special Areas initiative into the beginnings of an industrial location policy in 1936 and 1937 began to bring more obvious linkages with mainstream planning policies. In 1936 the important policy of developing new planned industrial estates in the Special Areas began, providing low-cost modern premises for new industries. Although there were references back to Trafford Park and the industrial estates developed at Letchworth and later Welwyn, the conceptual inspiration was the Slough Trading Estate (Cassell, 1991). This had been developed as a fully serviced industrial estate by a private company from the early 1920s. It had attracted an impressive range of the new lighter growth industries of the inter-war years and was regarded as one of the symbols of economic success in the period.

Now a slight variant of the model was used at new Special Area estates at Team Valley (Gateshead), Treforest (near Pontypridd) and Hillington (near Glasgow). Like Slough they were large schemes; Team Valley, the biggest, was over 700 acres (Loebl, 1988). Their size and location, near main routes, was with maximum publicity value in mind, as at Slough. Development was carefully planned, with full road and rail transport infrastructure, though without the total package of electricity and process steam provided by the Slough estate company. Communal facilities, including canteens, were provided by the government-owned estate companies which developed them. Most importantly a range of standard factories were available 'off the peg', though sites for self-build were also provided.

Other smaller industrial sites were also developed in other parts of the Special Areas, usually closer to the worst local unemployment blackspots, but these were less sophisticated than the big estates which themselves became important models of planned industrial development, widely emulated in the post-war period. Several local authorities were also actively planning industrial estate developments at this time. Although these were generally less impressive than the Special Area estates, those at Liverpool and Manchester were important because of their integration in the satellite towns of Speke and Wythenshawe, noted earlier. Ideas for developing similar relationships within a more comprehensive approach to regional development planning in the Special Areas, particularly in Durham, were also being explored by the later 1930s. Like much else in this period, however, this only bore fruit after 1939.

Special Areas and the wider planning debate

One of the key reasons why so much qualitative innovation in policy was crammed into such a heavily constrained programme was due to the personality of the first Special Area Commissioner for England and Wales, Sir Malcolm Stewart. A successful businessman in the brick and cement industries, one of the major growth sectors of the inter-war years, he did not behave exactly as his orthodox political masters hoped. As early as January 1935 the Chancellor, Neville Chamberlain, noted: 'I am afraid we have made a mistake with our commissioner and I anticipate more trouble' (Booth, 1978, p. 149). Stewart constantly pushed his powers to the limit and used his three annual reports to press all sections of political and influential opinion for enlarged powers. Most important of all was his final report where he argued that the problem of the Special Areas could not hope to be solved without an embargo on new industrial development in Greater London (Barlow Commission, 1940). In doing this he authoritatively established a basis for a comprehensive reconsideration of planning policy.

Towards a Comprehensive Approach

The pressures for change

The end of laissez-faire

The late 1930s saw an important shift in business attitudes towards the role of government intervention beginning to take place (Marwick, 1964). The fact that a prominent industrialist like Stewart could embrace increasing state involvement with private industry quite so enthusiastically hints at this shift. Similarly the manifest commitment of wide sections of industry for the 'Great Crusade' of planned redevelopment provides further evidence. Nor were these isolated instances. We could list many individual businessmen, including captains of both growing and depressed industries, who wanted to see a stronger government role in many aspects of economic life. Thus Israel Sieff of Marks and Spencer, Lionel Hichens of Cammell Laird and Sir Basil Blackett of the Bank of England were all prominent advocates of planning. Similar shifts were taking place within the wider establishment. Thus we can find Civil Servants (though not yet the most senior), progressive Conservative politicians (for example the young Harold Macmillan) and leading professionals looking to a more active government role.

These shifts were rather more important than those within the Labour movement, because they represented changes in the thinking of the dominant political interests of the period. But the Labour movement's virtually complete abandonment of any project for the wholesale socialist replacement of capitalism was not without significance. Spearheaded by Hugh Dalton, Ernest Bevin and others, the

Labour Party and trades unions became increasingly interested in the successful management of capitalism as a precondition for a welfare state (Saville, 1988). This involved a much greater reliance on economic planning, including all levels of spatial planning.

Intellectual advocates of change

A wider planning movement was beginning to take shape in the 1930s that was providing an intellectual basis for this shift. Most important was John Maynard Keynes who, with the publication of *The General Theory of Employment, Interest and Money* in 1936, demonstrated convincing arguments for an overall approach to macro-economic management based on demand orchestration (Keynes, 1973). Building on his earlier advocacy of public works, his arguments finally challenged the orthodox approach to public expenditure on which the whole philosophy of minimal state intervention during the 1930s' economic crisis was based. He gave scant attention to spatial matters (though he recognized a need for some regional economic policies) (Lonie and Begg, 1979), but his work provided the political space within which more specific planning proposals could develop.

These were advocated by both old and new planning groups. Amongst the older groups, the Garden Cities and Town Planning Association conducted a campaign for a comprehensive planning approach, under the vigorous leadership of Frederic Osborn. Newer groups, especially Political and Economic Planning (PEP), founded in 1931, provided a valuable and convincing series of reports and commentaries on many policy areas, including authoritative studies on housing (PEP, 1934) and industrial location (PEP, 1939). And many local surveys highlighted the arguments for a more considered approach to planning matters.

Sources of change

The sources of this shift towards more comprehensive planning were not simply a product of intellectual changes and research, however (Parsons, 1986). They embraced the much wider concerns about the social and economic price of suburbanization in congestion costs, loss of amenity and agricultural land noted above. They embraced too the mounting enthusiasms for planned redevelopment. To these were added a series of worries about unemployment and the depressed areas. Genuine humanitarian concerns, especially for the depressed areas, were certainly present. The famous Jarrow Crusade of 1936 brought forth a tremendous surge of sympathy for such areas (Wilkinson, 1939). But this was underpinned by real fears about the destabilizing effects of unemployment on the social and political fabric. The prospect of 'Love on the Dole', of people raising families in depressed areas without ever having worked, was particularly shocking to many in the comfortable classes. More threatening perhaps was the spectre of an alienated and embittered underclass, permanently unemployed and prey to political extremism. The rise of the Nazis in Germany and the apparent advances in the mid-1930s of the British Union of Fascists, led by the former Labour advocate of Keynesian public works, Sir Oswald Mosley, seemed to underline this threat. The Communists were also very active in the Unemployed Workers movement (Hannington, 1937, 1977).

And by the late 1930s there were other concerns that appealed directly to business self-interest (Ward, 1988). The continued growth of the consumer boom industries depended on ending the underconsumption represented by unemployment and poverty. Moreover labour costs were rising significantly in the prosperous regions in the late 1930s' boom, especially when rearmament began. The mounting threat of war also highlighted the great strategic importance of the badly run-down traditional industries and industrial communities of the depressed areas. Overall business interests were increasingly looking to government to reintegrate the depressed areas into the national economic and social mainstream. Meanwhile the increasingly imminent threat of aerial warfare underlined the folly of continued metropolitan expansion. Bigger cities made easier bombing targets and the failure to encourage new militarily important industries like motor and aircraft manufacturing to grow outside the vulnerable south and midlands began to look very foolish (Hornby, 1958). Overall there was a growing demand for a more national approach.

A new beginning?

Overall policy shifts

Slowly and tentatively the government was shifting its ground, particularly where the arguments for a more interventionist approach were undeniable, especially to business interests. Apart from the quickening interest in slum clearance, noted above, the most striking instance was in roads planning. Throughout the 1930s unfavourable comparisons were being made between highway development in this country and the imaginative approaches apparent in Nazi Germany, Fascist Italy and the USA (Charlesworth, 1984). A key problem in Britain was the localized control even of the main national routes. This meant that despite all the Ministry of Transport's policy guidance, advice and grants, the creation of a truly modern national road system depended on the financial capacity and willingness of the local highway authorities (essentially the counties). Many of these, especially in the rural and depressed areas, were unable to sustain major programmes. Therefore the nationalization of control of these routes was a precondition of any major improvement programme. The Trunk Roads Act 1936 put over 4,500 miles of major roads under direct Ministry of Transport control. This allowed a truly national programme to be devised, paralleling the major exercise in roads planning for London produced in 1937 by Sir Charles Bressey and the architect Sir Edwin Lutyens. But of course these initiatives came too late to have any real impact before the outbreak of war.

The Barlow Commission

All the mounting concerns for a more comprehensive and national approach to planning came together in 1937 when Neville Chamberlain, by now Prime Minister, set up the Royal Commission on the Distribution of the Industrial Population. It was headed by a former Minister of Labour, Sir Montague Barlow, though from the planning viewpoint much the most important member was the most experienced and respected regional planner of the inter-war period, Patrick Abercrombie. Other members were drawn from business, the professions, Civil Service and the trades unions. Their terms of reference were wide ranging but fully reflected the moves towards comprehensivity we have identified:

> to inquire into the causes which have influenced the present geographical distribution of the industrial population of Great Britain and the probable direction of any change in that distribution in the future; to consider what social, economic or strategical disadvantages arise from the concentration of industries or of the industrial population in large towns or in particular areas of the country; and to report what remedial measures if any should be taken in the national interest.
>
> (Barlow Commission, 1940, p. 1)

This was in many ways a direct response to Sir Malcolm Stewart's parting shot as Special Areas Commissioner. And it was a breakthrough in the sense that it showed that government was at least looking at the urban and regional problems of Britain as a whole. Chamberlain was widely praised for his statesman-like action (though as PEP pointed out, it would have been truly statesman like if he had done it ten years earlier). But it did not necessarily signal a commitment to significant change. Parsons (1986) particularly has stressed how sceptical Chamberlain and many Conservatives remained about any real policy shifts. Although the pressures for more powerful planning policies were impressive and growing, the forces that were limiting change also remained strong. The evidence presented to the Barlow Commission during 1938 highlighted these limits.

The limits to change

There was still a good deal of caution about drastic change (Ward, 1988). Despite the commitment of particular businessmen, the Federation of British Industries was reluctant to see anything beyond a government information and advisory role on industrial location. Similarly, although there was acceptance of a social case for more planning, by government departments, the important economic case, implying locational controls, was not accepted, especially by the Board of Trade (and, though it did not give evidence, the Treasury). Even some prosperous provincial cities, like Birmingham, which had pioneered local planning, were deeply sceptical of national planning controls especially on industrial location. Its civic leaders were acutely aware that their avoidance of the inter-war recession had rested on the freedom of the new industries. To this list we could add urban property. Remarkably no evidence was presented to the Commission by any body representative of such interests. This was despite the fact that the Commission was actively considering a significant shift in the balance of public and private power in the urban land market. But we can be certain that they opposed much of what this shift implied, as we will see in the next chapter, when the tentative shifts of the late 1930s received the powerful reinforcement of war.

4

A New Orthodoxy of Planning 1939–52

Introduction

The outbreak of the Second World War in September 1939 appeared to dash hopes of any immediate action on the planning front. The quickening governmental interest in planning matters that had been apparent in the last years of peace, particularly in the work of the Barlow Commission, seemed to have been put into suspended animation. At home the new and urgent wartime priorities of evacuation, air raid protection, rationing, and a whole panoply of government emergency controls took precedence. Yet it soon became clear that physical and land use planning was not as irrelevant to the war effort as was immediately supposed. In fact the war turned out to be the catalyst that finally overcame political objections to the creation of a comprehensive physical and land use planning system.

The 1940s thus became the most critical decade in the development of British planning policy. They saw the dramatic emergence of an active and mass consensus for planning during the war years, widening the basis of its support. This was paralleled by the development of a new and comprehensive policy agenda and, after some delay, its translation into a new policy framework. Conceptually too there were important innovations apparent in the plans prepared during this period. By 1952, therefore, an entirely new planning system and practice of planning had been created that was to survive, with some important modifications, to the 1970s. Yet the mass and active consensus of wartime had narrowed and shifted its character to a more enduring form by the early 1950s. This chapter shows how and why these important changes took place and begins to consider some of their more important consequences.

War and the Emergence of a Mass Consensus for Planning

The war economy

State controls

During the Second World War, Britain became a centrally directed state economy to an even greater extent than had occurred in the 1914–18 war (Calder, 1971).

By 1943 state spending, under a Conservative-dominated wartime coalition government, accounted for almost three-quarters of the gross national product, higher than at any time before or since. Behind this lay an extraordinary range of state controls that would have been totally unacceptable in peacetime, but were accepted with equanimity and a surprising degree of popular enthusiasm in the equally extraordinary circumstances of war.

As never before the state operated its own factories and directed the work of private businesses. The normal operations of the stock and money markets were curtailed. Private property, including land, could be and was appropriated at short notice for government purposes. The use of labour was heavily controlled on an almost military basis to facilitate war production. Rationing and the queue became the official allocation mechanisms for many ordinary commodities, reducing (despite the black marketeers) the active role of money and the price mechanism. Even in their own homes, householders could at short notice find themselves having to accommodate complete strangers. And to those in the armed forces the hand of the state was even more dominant.

Full employment

The effects of this wartime collectivism were much more beneficial for ordinary working people than they might sound today. After the mass unemployment of the 1930s, the most positive effect of the state-directed war machine was that it brought full employment. This was of profound long-term importance because it showed beyond doubt that government could ensure that everyone had a job, if the political will was there. This was something governments between the wars had conspicuously failed to do, and the contrast did not go unnoticed. Full employment also strengthened the position of working people both in their workplace and in society generally. The war economy and indeed the whole war effort depended on the active support of the ordinary workers to a far greater extent than was common in peacetime (Marwick, 1970). Government and employers necessarily became aware that they had to address the needs and demands of working people.

Industrial location policies

Another related issue that was to be of long-term importance in planning history was the way wartime production was steered to the traditional depressed areas (Booth, 1982; Parsons, 1986). Such moves, which had begun under rearmament but were now intensified, reflected the need to make best use of labour surpluses and avoid areas at greatest risk of bombing. The control mechanism was a Board of Trade licence, introduced by the Location of Industry (Restriction) Order of 1941, under the wartime Defence Regulations. Initially the licence was to apply to businesses using premises of over 3,000 square feet, but this limitation was dropped the following year. The importance of these controls was, of course, that they again demonstrated what was possible when the political will was present. Moreover they showed businessmen that government industrial location policies could work in their interests by helping them to make maximum use of cheaper labour.

Figure 4.1 Bombing destroyed lives, homes and communities. The poorest districts, near docks and factories, were usually the worst affected, leaving bewilderment and confusion amongst survivors. In such circumstances firm plans for rebuilding and a commitment to create something better were seen as imperative to maintain civilian morale. The aftermath of the Blitz also fostered a society and a state that were more caring than before 1939, with lasting effects on the political agenda of post-war Britain.

Land and the built environment

Bombing

War also brought major changes to the urban and rural fabric of Britain (Calder, 1971). Much the most dramatic was the destruction produced by bombing. In the autumn, winter and spring of 1940–1 the *Luftwaffe* launched a strategic bombing offensive, 'the Blitz', against London and major provincial cities. Although this was the most concentrated period of attack, intermittent bombing occurred throughout the rest of the war. The 'Baedeker raids' (named after the then famous tourist guides) concentrated on historic towns. Flying bomb and rocket attacks continued through until the final months of hostilities.

In total 475,000 dwellings were destroyed or made permanently uninhabitable in England and Wales alone. Destruction was particularly severe in some towns and cities (Hasegawa, 1992). The most badly affected were London, especially east London, Liverpool, Bristol, Southampton, Plymouth, Hull, Belfast, Coventry, Clydebank, Sheffield, Exeter and Canterbury. Virtually every city of any size (and many smaller towns) experienced at least some damage. Nor was there any certainty of immunity in those few centres that emerged unscathed.

The urban land market

The overall impact of war on the urban land market was nothing short of catastrophic. Although a few London property agents profited from the demands of bombed-out businesses (Marriott, 1969), bombing obviously did not encourage property values to recover (Ambrose, 1986). A few astute individuals were shrewd enough to acquire valuable redevelopment sites at knock-down prices, but generally urban property interests looked for government to come to their rescue. The immediate problems were addressed in war damage legislation from 1941. There was, though, no immediate rush to redevelop. Property and development interests looked for major injections of confidence and certainty into the urban land market before being prepared to initiate any redevelopment on bombed sites (Ward, 1975; Backwell and Dickens, 1978). The only agency capable of providing this confidence during wartime was the state.

Building licences

The government's major preoccupation in the immediate aftermath of bombing was patching up repairable buildings and ensuring that only new development essential for the war effort (such as factories or emergency housing) was undertaken. Again a special licensing system, administered by the Ministry of Works and Buildings, was introduced in 1941, surviving until 1954 (Kohan, 1952). The building licence enabled government to make a decision about whether any development project was justified on the grounds of its economic or social value to society. This was a quite separate question from whether it accorded with any town planning criteria. It was not in fact administered by planning authorities but by the district valuers who worked directly for central government. Yet the existence of this far-reaching building control system parallel to planning was to prove extremely important as the new more permanent post-war planning system was created.

Rural land

The countryside generally benefited by the wartime state support of agriculture in order to boost home food production. The risks of government expropriation of rural land for wartime purposes or eviction of unco-operative farmers diminished some of these benefits, but generally not for the majority. In sharp contrast to peacetime, when agriculture had been every bit as depressed as coal or cotton, lavish assistance was available for improvement and drainage. It was in fact yet another instance of state involvement in the economy protecting the position of important interest groups, in this case the farmers and rural landowners.

The demand for planning

Business interests and planning

As the preceding discussion has shown, the war and many of the short-term government measures that were introduced to facilitate the war effort helped prepare the way for a more permanent planning system. Important interest groups such as the industrialists and farmers were brought into much closer collaboration with government than was usual in peacetime. Moreover they found the relationship profitable, even when it involved unprecedented restrictions. All this predisposed them to add their general, if not always unqualified, support to the emerging wartime government commitments to a post-war system of comprehensive physical and land use planning that began to emerge after the Blitz.

Even before 1939 the rural landowners, represented by the Central (later the Country) Landowners Association, had shown a considerable measure of support for planning. They were now joined by the industrialists. Thus we find the Federation of British Industries commenting in the editorial of its journal in July 1941:

> Bombs have made builders of us all. We have hoped, and continue to hope that, in terms of bricks and streets, the tragedy of today may become tomorrow's opportunity. . . .
>
> Industry, while seeking as much freedom for its choice of sites as is compatible with national (and local) interests, will agree that there ought to be some sense of orderliness and proportion in the planning of industrial towns . . .
>
> (cited in Ward, 1975, p. 13)

Urban property and development interests had been the traditional opponents of a more interventionist planning system. Thus the opposition of the National Federation of Property Owners to state acquisition of development rights under active consideration in May 1941 was quite predictable. As we have seen, however, such interests had been greatly weakened by war. Their admission that bombing had created a 'new and unprecedented situation' (*Estates Gazette*, CXXXVII, 1941, p. 473) that required new national planning powers was effectively an admission that they needed planning in these areas (Ward, 1975; Ambrose, 1986).

Wider social pressures

The extent of popular and media interest in planning issues during the war years was also quite extraordinary. To some extent this flowed directly from the changes that war brought to society by breaking the narrowness and fragmented nature of peacetime social existence (Calder, 1971). It brought people from all classes of society into close contact with each other as part of a common endeavour, often while coping with shared hardship and loss. An early example was the evacuation scheme (intended to decrease the concentration of population in cities) which took mothers and children from poor inner-city working-class districts and billeted them with wealthy residents of the urban fringe and countryside. The armed forces and other forms of national service (for example in the mines or on the land) also 'mixed up' society in ways that were quite unknown in peacetime.

These and similar experiences gave ordinary people throughout society insights into the tremendous divisions and inequalities which then existed in Britain. And amongst the middle and upper classes, who learned most from such revelations, they encouraged a growing sense of the unfairness of many of these traditional distinctions and a commitment that things would have to be better when the war was over. Bombing did a great deal to strengthen this developing mood, since it mainly affected the poorer working-class districts that adjoined the factories and docks. Those who had gained least out of British democracy were now losing their homes, communities and lives in its struggles. A genuine commitment on the part of wide sections of middle- and upper-class opinion grew out of this to create something better, as a kind of repayment for wartime suffering.

The class dimension and morale

There was, however, more to the emergent social commitment to planning than the middle and upper classes simply doing 'the decent thing' for the poorer members of society. A critical variable was the greatly strengthened power of the working classes because of full employment. Indeed in wartime their bargaining power was heightened further, because the whole war effort depended on the active commitment and co-operation of working people to ensure the smooth and efficient operation of the fighting services and the war economy (Marwick, 1970). This assumed still greater importance during the period of the Blitz, because of the widespread belief amongst military strategists that a strategic bombing offensive against cities might well produce a catastrophic decline in civilian morale. Thus the wartime diarist Harold Nicolson, who had a junior ministerial post at the Ministry of Information, wrote after the first raids: 'Everyone is worried about the feeling in the East End, where there is much bitterness. It is even said the King and Queen were booed the other day when they visited the destroyed areas' (1967, p. 114).

The worst fears were not borne out, and civilian morale held remarkably well, as it did later under the much heavier bombardment meted out to German cities. But government received a sharp reminder that it needed to act decisively to sustain morale at this time when Britain's cities were, in effect, the 'front line' of the battle. Planning and physical reconstruction were, as planning bodies such as the Town Planning Institute were quick to point out, obvious ways of doing this. Yet the working classes were not themselves actively demanding town planning itself to any great extent. In fact they had the much more tangible wants of decent housing and neighbourhoods and secure employment. Planning was portrayed as a means to those ends but importantly it also offered a way of reconciling them with the requirements of the other interest groups, notably industry and land, who had now lent at least some backing to the planning idea.

Propaganda

It was, then, against this background that Churchill's wartime government actively began to sponsor the notion of a comprehensive planning system. In the months and years that followed, a huge amount of effort was put into the promotion of the planning idea. Maps, models, exhibitions, radio broadcasts and films were all used to put over the message (Gold and Ward, 1994). Newspapers and magazines gave large-scale coverage to planning matters on an unprecedented scale. Despite wartime shortages, a large number of books appeared to feed the public and political appetite for knowledge about planning. Thomas Sharp's brilliantly argued Penguin paperback, *Town Planning* (1940), was a best-seller (Stansfield, 1981). Official agencies like the (hugely important) Army Bureau of Current Affairs and private bodies like the Town and Country Planning Association (the former Garden City and Town Planning Association) organized lunchtime talks and discussion groups (Hardy, 1991a). Almost invariably it was a message which was sold by professionals, middle-class reformers or junior officers, usually of centre-left political sympathies. These were becoming the guardians and the evangelists of the new consensus for planning. Meanwhile within government a group of people with broadly similar social backgrounds were actually formulating the details of the new system. That they were able to do this owed much to the dramatic high-level political shifts of 1940–2.

Formulating the Post-war Planning Agenda 1940–2

The Barlow Report

Context

It is conceivable that even without the war some strengthening of planning powers might well have occurred. The most tangible sign of the mounting pressures to act was the Royal Commission on the Distribution of the Industrial Population (the Barlow Commission) appointed two years before war broke out. As we noted in the previous chapter, however, Neville Chamberlain's government was reluctant to move decisively. The act of setting up a Royal Commission with a timescale of several years was itself a symptom of this indecision (Parsons, 1986). The outbreak of war initially allowed this indecision to continue, although the Civil Defence Act 1939 had actually reflected part of the spirit of the Barlow message, by charging town and country planning with the duty of avoiding dangerous concentrations of population. More seriously, however, the publication of the Commission's report, completed by September 1939, was deferred. Following parliamentary pressure, it was finally published in January 1940 (Barlow Commission, 1940).

Findings

Although the Barlow Report has good claim to be considered the single most important policy document in British planning history, this was not immediately apparent when it was issued. Analytically it was impressive. There was a detailed study of the patterns and processes of change in the distribution of population and employment in recent times and the problems it posed. It considered the increased regional unevenness as between the depressed and buoyant regions of Britain, the huge scale of inter-war suburbanization of both people and jobs, the increasing tendency to redevelop inner slum areas and the problems of the countryside. It also authoritatively spelled out why these spatial phenomena, separately or together, were national problems and why existing planning powers were inadequate.

The analysis was thorough so far as it went but, with hindsight, it is clear that it gave too little attention to service employment as a determinant of the geographical distribution of population. The emphasis was almost completely on the impacts of mining and manufacturing industries, with the geography of services seen as deriving largely from the patterns of these primary and secondary industries. In addition, though perhaps more understandably, the Commission failed to recognize the growing significance of large multi-locational firms as a factor in regional development.

Recommendations

More immediately, however, it was the slightly indecisive and cautious nature of the recommendations that struck those in the planning movement. Majority and minority reports were issued, reflecting a failure to agree not so much about the nature of the problem as on the solution and, more especially, the means to achieve it. Both called for a central planning authority and accepted the planning and other arguments that the present patterns of development were largely unsatisfactory in their cumulative effects. Regional balance in the distribution of employment, planned metropolitan decentralization and the continuance of planned redevelopment, all to be major elements of post-war strategic planning policy, were endorsed as planning objectives.

There was disagreement about the extent to which new industrial development should be restricted. The majority wanted it to be restricted only around London, whereas the minority, including Patrick Abercrombie, wanted restrictions to apply to a wider area. They also wanted inducements to be available to assist industrial development in the depressed areas. More generally they wanted a more powerful central ministry than that proposed by the majority. It was these minority proposals which began to seem more appropriate in the changed circumstances of war. But it took the Blitz to convince the government of this.

Lord Reith and the Ministry of Works

Reith's appointment

We have already noted the new pressures that the Blitz imposed on government. One of its first responses was to create a new Ministry of Works and Buildings to oversee the immediate building activity that would be required in October 1940. It was headed by an austere and autocratic Scot, the newly ennobled Lord Reith who, as Sir John Reith, had set up the BBC during the 1920s and 1930s (Addison, 1975). He was not a professional politician and had no real party allegiance. Like several other outstanding and dynamic managers he had been brought in to help run the war machine. Partly because of his background he lacked many basic political skills such as knowing when to compromise and how to acquire allies. His independent and intransigent personality was to be a crucial factor in the fluid political situation created by bombing. More than any other single figure it was he who translated the emergent demands for interventionist reconstruction measures and the weakness of those interests traditionally opposed to planning into the beginnings of real policies.

Reith's planning responsibilities

In the discussions about his remit, Reith secured a vaguely defined responsibility for the long-term reconstruction of town and country (Cullingworth, 1975). This inevitably involved planning, though at this stage statutory planning remained the responsibility of the Ministry of Health. Matters were further confused when another more senior minister was appointed in January 1941 with overall responsibility for post-war planning and reconstruction. Reith, however, was untroubled by the niceties of these ministerial precedents and overlapping responsibilities (or the misgivings of his own Civil Servants about linking works and planning too closely) and began to push the planning agenda at a quite extraordinary speed.

'Plan boldly'

One not untypical incident gives a particular flavour of Reith's impact at this time. Early in 1941 he met a municipal deputation from the blitzed city of Coventry, who were beginning to wonder how to tackle the immense task of rebuilding, given the absence of any powers other than the rather weak 1932 Act. Without any assurance yet that the rest of the government would agree to his rapidly evolving proposals for a comprehensive planning system, Reith seized the moment: 'I told them that if I were in their position I would plan boldly and comprehensively, and that I would not at that stage worry about finance or local boundaries. They had not expected such advice. . . . But it was what they wanted and it put new heart into them' (Reith, 1949, p. 424).

And he gave the same advice to other blitzed cities, seeing them as test cases which would be used to determine what kind of planning powers would be needed. It was a decisive moment in the history of British planning policies. Suddenly all the traditional limitations on planning action were, on ministerial advice, to be ignored (Mason and Tiratsoo, 1990).

Another more politically visible symptom of how much things had changed was his establishment of the Expert Committee on Compensation and Betterment, headed by the Honourable Mr Justice Uthwatt. Its brief was to advise what powers the government should acquire in this traditionally most sensitive dimension of town planning: the relationship between the state and the land market.

Government commitment

Meanwhile the Cabinet (of which Reith was not a member) decided to back Reith's proposals for a more comprehensive approach to planning. Given the mounting popular and political pressure for such moves, an 'action-man' figure like Reith was politically useful. In February 1941 he was able to make an important statement, expressing the basis of government thinking:

(1) that the principle of planning will be accepted as national policy and that some central planning authority will be required;
(2) that this authority will proceed on a positive policy for such matters as agriculture, industrial development and transport;
(3) that some services will require treatment on a national basis, some regionally and some locally.

(cited in Cullingworth, 1975, p. 61)

This marked a clear policy break with the past. Government was here committing itself to the basic principles of a planning policy radically different from what had gone before, even if it was still far from clear what the exact form of the policy would be.

'Moving too fast'?

The immediate political pressures on the government to be seen to be taking action on the planning front diminished as the Blitz ended and the nature of the war shifted to involve Russia, the USA and Japan during the remainder of 1941 (Ward, 1975). Yet Reith was not easily stoppable and was busily engaged in developing proposals for a central planning agency and the necessary legal framework for the new comprehensive planning policies. Not surprisingly, he was finding himself in continual conflict with other ministers who were jealous of their own post-war responsibilities which Reith now appeared to be usurping.

Nor was the opposition purely internal. The land market also began to show the first signs of recovery as the likelihood of an invasion receded. Thus the interim proposals of the Uthwatt Committee for state acquisition of development rights on undeveloped land provoked a sharp response from the National Federation of Property Owners: 'it would mean, in effect, the nationalisation of a part of an owner's interest in land, and thus strike at the principle of private enterprise in property' (*Estates Gazette*, CXXXVII, 1941, p. 582).

For the moment though, Reith was in a strong enough position to ignore these kind of pressures (Ambrose, 1986). In August 1941 he established the Scott Committee on Land Utilisation in Rural Areas to advise him on other aspects of planning policies. Throughout 1941 he was initiating important plan-making activity, especially for London. And after intense ministerial discussions he secured the overall authority for physical planning that he had been seeking. The Ministry of Health's statutory planning powers were transferred and he became Minister of Works and Planning in February 1942. The central planning authority had appeared. But Reith himself was sacked within a few days of the decision, a victim of mounting Conservative fears about the radical nature of the planning agenda he was formulating and his personal lack of political friends (Reith, 1949).

The Uthwatt and Scott Reports

After Reith

Although one man had played a key role at a critical moment, and was now being sacrificed for political expediency, there was by early 1942 a huge momentum for planning and reconstruction issues. Property interests were certainly re-emerging as a significant lobby, seemingly somewhat horrified at what they had conceded over the preceding months. The wider social backing for planning was still strengthening, however, particularly as more specific proposals began to emerge and entered popular discussion through propagandist efforts. Industry also remained generally sympathetic, broadly welcoming the appearance of practical schemes and proposals. In the longer term the most important of these were the Scott Report and the Uthwatt Final Report, which appeared in summer 1942.

The Scott Report

The Report of the Committee on Land Utilisation in Rural Areas (Scott Committee, 1942) produced what the official historian has called 'a remarkable hotch-potch' of specific proposals and recommendations related to rural land and life (Cullingworth, 1975, p. 41). At the heart of the report was the assumption that the countryside was a precious asset that should be protected, particularly its major industry, agriculture. It was a view that exactly embodied what its vice-chairman, the geographer L. Dudley Stamp, had been arguing in the 1930s when he had been horrified at the loss of farm land to urban uses (Stamp, 1962). And of course it made a great deal more sense in wartime, when productive farming was critical for national survival and the countryside itself had become a powerful symbol of the homeland that was being fought for. (We can detect such sentiments for example even in Vera Lynn's immensely popular wartime song 'The White Cliffs of Dover', which contains lines that are replete with rural imagery such as, 'The shepherd will tend his sheep/the valley will bloom again'.)

Virtually all the Scott Report's majority recommendations (and there were over a hundred of them) derive from the presumption of preferential treatment for farming and the countryside. Amongst the most significant were proposals for national parks, nature reserves, green belts, strict controls on any industrial or other developments in the countryside (with particular concerns to ensure compactness and avoidance of good-quality agricultural land) and a whole series of measures to revive the quality of village life. Only S. R. Dennison, a distinguished economist, who issued a lengthy minority report, saw fit to challenge the majority assumptions as a basis for long-term planning. Little notice was taken of these views, however, which now appear a good deal more far-sighted than perhaps they did in 1942. By contrast the majority recommendations received much public attention and attracted widespread approval.

The Uthwatt Report

In contrast to the diversity of Scott's concerns, the Final Report of the Expert Committee on Compensation and Betterment (Uthwatt Committee, 1942), though written in the dry professional language of the lawyer, focused exclusively on one topic of the highest political significance (Ambrose, 1986). Its concern was to ensure that the state system of planning had the powers in relation to private landed interests to be able to implement the kinds of policies which Scott (and Barlow) had recommended. At the heart of the report were its proposals for compensation and betterment. Compensation was, of course, the obligation of the planning authority to compensate private landowners who suffered through planning action. The discussion of betterment, in essence increases in land value, referred to the obligation by private landowners to pay at least some of that increase to the state, on the grounds that planning decisions or actions would be responsible for creating at least part of it.

The Final Report confirmed the approach signalled in its Interim Report (Uthwatt Committee, 1941) that had so alarmed the property lobby. It was clear the Committee favoured a highly interventionist solution that avoided many of the problems of the pre-war planning system by relying heavily on direct state ownership of land or development rights. Thus it proposed that development rights in all undeveloped land should be immediately acquired by the state with compensation based on March 1939 development values. Furthermore the land itself would be purchased outright by the state (at undeveloped value) when development was set to occur. If private development was envisaged the land would then be leased out at its development value, so that betterment (and freehold of the land) was retained by the state.

The proposals for built-up areas, still largely outside any statutory planning controls, avoided the outright nationalization that was to be applied elsewhere. The same principles of state intervention were apparent, however. As a first stage planning controls were to be imposed in such areas, so that the right to change land use or redevelop were effectively nationalized. Paralleling the recommendations for newly developing areas, redevelopment was also to be brought directly under state (that is local authority) control. Thus the local authorities would be granted stronger compulsory purchase powers for bombed or obsolescent areas that needed wholesale reconstruction. Again private developers might become involved in such redevelopments, but land in redevelopment areas would only be disposed of on a leasehold basis that ensured betterment stayed with the local authority. Finally a betterment levy of perhaps 75 per cent would be collected on already developed sites on a regular basis.

Popular and political reactions

The reactions to Uthwatt tell us a good deal about the continuing strength of the new consensus for planning that had been forged since 1940. In property circles there were cries of rage, for example from the Housebuilders Association, who complained about 'sweeping and Dictator-like proposals' (*Estates Gazette*, CXL, 1942, p. 536). More widely there was widespread press support even from many Conservative newspapers (*ibid.*, p. 273). Labour's *Daily Herald*, though generally supportive, even criticized the report for being too modest and wanted full land nationalization. Partly as a result of this high level of popular interest, the fate of Uthwatt, with Scott and Barlow, was coming to be seen as a test case of the government's commitment to building a better world after the war.

Partial Implementation of the Planning Agenda 1943–5

Wartime planning legislation

Government commitment

By the later years of the war there were many doubts about the extent of the commitment of Churchill's government to action on the high reconstruction ideals that had been encouraged in the early war years. Increasingly, popular opinion was looking forward to the implementation of Uthwatt, Scott, Barlow and other proposals, particularly the hugely important Beveridge Report on Social Insurance and Allied Services (which set out proposals for the welfare state and was also published in 1942). There was a growing anxiety, fostered by the left, to ensure that the country was not 'cheated', as had happened between the two World Wars, when Lloyd George's promise of a 'land fit for heroes to live in' had given way to unemployment, poverty and another war (Calder, 1971; Addison, 1975; Hennessy, 1993).

There is little doubt that Churchill himself also wanted to avoid this sense of disillusionment, though his preferred solution was to try and retreat from what this clutch of reports seemed to represent. As final victory began to look inevitable following the late 1942 allied successes in the Coral Sea, at Stalingrad and El Alamein, the morale-boosting arguments for post-war visions disappeared. Traditional Conservative concerns for ensuring the health of the normal processes of wealth creation began to take precedence. In January and February 1943 Churchill gave explicit warnings to his Cabinet about the 'dangerous optimism' that was developing about post-war reconstruction: 'While not disheartening our people by dwelling on the dark side of things, Ministers should, in my view, be careful not to raise false hopes, as was done last time by speeches about "homes for heroes" etc.' (Churchill, 1951, pp. 861–2). Set against this background it is easy to understand why most of the major legislative planks of the new planning orthodoxy were still not in place by 1945. But some significant partial steps were taken.

Ministry of Town and Country Planning

One of the most important steps was the creation of a second (and more enduring) central planning authority. Reith's successor as Minister of Works and Planning had not been interested in the planning remit and after a great deal of inter-departmental wrangling it was decided to hive off planning to a new small Ministry of Town and Country Planning, created in early 1943. In many respects a ministry with a specific remit to deal with issues that had a highly technical dimension was a sensible solution. Reith had brought in a very talented group of planners, including William Holford (Cherry and Penny, 1986), Gordon Stephenson (Stephenson, 1992), Thomas Sharp and John Dower, who now joined the new Ministry. The problem was that its Minister, W. S. ('Shakes') Morrison, had no Cabinet position and had relatively little 'clout' in dealing with other established ministries. Specifically, there were no powers over finance

WALLFLOWERS
July 16, 1943: Three reconstruction reports strangely neglected by the Government.

Figure 4.2 High public expectations about post-war planning had been engendered during 1940–2. By 1943 there was a widening view, expressed here by the famous radical cartoonist Vicky, that vested interests were blocking any decisive action on the trilogy of wartime reports. Meanwhile the new Minister of Town and Country Planning, W. S. Morrison, was able to do little to fulfil the promise of his new ministry. © Vicky/Solo

(essential for action on Uthwatt), housing, industrial location, transport or agriculture. It was, as the prominent planning propagandist Frederic Osborn put it privately, 'grudgingly equipped with charming but weak men unable to fight for the necessary powers' (Hughes, 1971, p. 51). Had there been a firm overall government commitment to act, this need not necessarily have been a problem, as was demonstrated after 1945. W. S. Morrison did not have this, however, for the reasons we have already outlined.

The Interim Development Act 1943

The first legislative fruit of the new Ministry, The Town and Country Planning (Interim Development) Act, seemed to confirm these fears (Ward, 1975). Superficially it was a great step forward because it implemented a key requirement of the new planning agenda by putting all the country under planning control. In June 1942 only 3 per cent of the area of Britain had been covered by operative planning schemes, with a further 48 per cent subject to resolutions to prepare schemes and interim control. At a stroke therefore this Act put the remaining area on to a similar basis. What weakened it was that it merely applied a streamlined and very slightly strengthened version of the 1932 Act, which everyone was now agreed was a very unsatisfactory basis for planning. One prominent (Labour) member of the Cabinet referred to the interim development proposals as a 'farce' in ministerial discussions and its parliamentary reception was scarcely more complimentary (Cullingworth, 1975, p. 82).

The 'Blitz and Blight' Act 1944

By 1944 the Ministry, and the new Ministry of Reconstruction, which handled post-war planning issues in the Cabinet, were still struggling with the Uthwatt proposals (Woolton, 1959). These were of course the key to a comprehensive planning system that could implement the specific policies that Barlow, Scott and by now the blitzed cities were proposing. Decisive action was prevented essentially by a combination of departmental and property interests, the latter strongly represented within the Conservative Party who still dominated both the coalition government and Parliament. A 'strategic retreat', abandoning any pretence of seeking to implement Uthwatt or Uthwatt-style proposals, was being contemplated by March 1944 (Cullingworth, 1975, p. 112).

Despite these problems it did prove possible to legislate on the aspect of Uthwatt that dealt with redevelopment areas. This was the planning question that was most pressing as specific planning proposals began to be formulated for the blitzed cities (where Reith's advice to 'plan boldly' had been taken seriously) (Hasegawa, 1992). And it was the area where all interested parties accepted that new powers were needed. In practice though it proved extraordinarily difficult to secure parliamentary approval. Extensive amendments and redrafting were necessary while it was being considered. Despite this, the result satisfied no one. Conservative and landed interests saw it as going too far; for many others it still left too many important matters unsettled.

The Town and Country Planning Act 1944 was the result, soon dubbed the 'Blitz and Blight Act' since its purpose was to give compulsory purchase powers to permit the wholesale replanning of war-damaged and obsolescent areas. It marked the formal beginnings of positive planning as opposed to the purely regulatory statutory planning of the 1932 and 1943 Acts. More specifically it introduced the redevelopment area, an important statutory planning device that, in a strengthened form after 1947, became known as the comprehensive development area. For the present, however, this quite tentative move into more interventionist planning marked the limits of political possibility as far as the wartime coalition was concerned.

The Control of Land Use White Paper 1944

On the same day as the 1944 bill was published, the government issued a White Paper that set out the rest of its thinking on the issues raised by Uthwatt, so far

CONQUEST ABROAD — SURRENDER AT HOME

Figure 4.3 By 1944 there was still greater cynicism about the reassertion of private property interests compromising the post-war planning agenda. Once again Vicky satirized Morrison's impotence, reflecting widespread disbelief that the White Paper would lead to any real legislative action. © Vicky/Solo

as it had been able to reach agreement on them (MTCP and SO, 1944). Basically the paper diluted the most controversial of Uthwatt's proposals, particularly for the outright state purchase of land about to be developed, and made other more technical modifications. The key point, however, was that there was never any intention that legislation would immediately follow (Cullingworth, 1975). It was rather put forward as a focus for public discussion (and, in effect to take the pressure off government for a while). In the circumstances it was inevitably regarded with some suspicion.

Plans and reports 1943–5

The wartime plans

Much more significant was the plan-making activity which had largely begun during the Reith era. The plans themselves were beginning to appear by the later war years, though many more appeared after 1945 (Hasegawa, 1992). Particularly important among the wartime publications were the *County of London Plan* (1943), prepared by Patrick Abercrombie and the LCC chief architect, J. H. Forshaw, and Abercrombie's *The Greater London Plan 1944*, to be examined in more detail below (Forshaw and Abercrombie, 1943; Abercrombie, 1945). Other important plans that Abercrombie (Dix, 1981) was involved in, were those for Plymouth (with J. Paton Watson) (1943), and for Hull (with Sir Edwin Lutyens) (1945) (Watson and Abercrombie, 1943; Lutyens and Abercrombie, 1945). Much the most controversial was City Architect Donald Gibson's 1945 plan for central Coventry, which had appeared in outline as early as 1941 (Gregory,

1973). Although their character varied they all captured something of the idealism that Reith particularly had encouraged. In some cases, notably the idealistic Coventry plan, Ministry of Town and Country Planning officials felt that 'Lord Reith's "plan boldly" . . . has been worked to death', and actively tried to undermine many of its proposals (Mason and Tiratsoo, 1990, p. 106). But more generally these plans were important in the framing of the 1944 Act and many of them incorporated important, even dramatic, conceptual innovations.

Conceptual innovations

The key concern of most of them was planned redevelopment of central and inner areas. A marked break with pre-war precedents was apparent. Thus the models of redevelopment these plans offered invariably involved far more drastic simplifications of land use zones than had been contemplated before Uthwatt. Backyard and workshop industries interspersed with other uses were now to be replaced by new industrial estates. Shopping and other community facilities were also to be grouped. The principles of the modern movement in architecture became much more pronounced, especially in the new residential zones. Site layouts and housing designs began increasingly to reflect the innovations of pre-Nazi Germany, 1930s' France and Scandinavia. 'Mixed development' of houses and flats with public open space and some private gardens, was particularly popular (Bullock, 1987). More traditional civic design approaches, based on classicism, were also apparent in many of the plans (e.g. Plymouth and Hull), particularly those for central areas. Indeed Coventry's uncompromisingly modernist central area plan was an exception.

Much of the physical restructuring of residential areas and community facilities was underpinned by the community principle of the neighbourhood unit, developed in the USA during the 1920s as discussed in the last chapter. It became a major and remarkably enduring planning principle in Britain from the 1940s. By contrast, the other major American contribution to residential planning, the Radburn layout, allowing complete segregation of pedestrian and motor traffic, was conspicuous by its absence. Instead systems of partial segregation were increasingly favoured, removing through traffic and encouraging the creation of precincts to which only local traffic was admitted. There was an important conceptual advance on this front in wartime Britain. In 1942 Scotland Yard's chief traffic policeman, H. Alker Tripp, developed an extended case for replanning on the precinct principle. Tripp's ideas also depended on a clearer road hierarchy for metropolitan areas, allowing through traffic to move more freely because it was separated from local traffic. These principles of hierarchy and precinct duly appeared in the 1943 County of London Plan and elements of these ideas influenced other plans of the time.

The Dudley Report 1944

Some of the important innovations within the wartime plans were also incorporated in the Dudley Report (1944) – *The Design of Dwellings* – and the *Housing Manual* (MH & MW, 1944) of the same year. They promised new higher

***Figure 4.4** Mixed development of flats and housing in redevelopment areas became a new wartime model for site development, shown here in Greater London Plan proposals for West Ham. The flats were higher than any municipal blocks then existing. Both they and the (almost) flat-roofed long terraces built at right angles to the road show a strong continental modernist design influence.*

housing standards than pre-war (though as was by now familar the later manual diluted the earlier proposals). More importantly from the planning viewpoint, they gave official endorsement of the principles of mixed development and the neighbourhood unit that had been expressed in the County of London Plan of the previous year.

The Greater London Plan 1944

Most famous and influential of all the wartime plans was the Greater London Plan of 1944 (Abercrombie, 1945), prepared at the initial request of Lord Reith. Unlike the other plans published in wartime, it focused on a metropolitan region, so that it considered the full range of urban planning problems and solutions. In one document it elegantly and forcefully articulated three of what were to become the four main strategic concepts of post-war planning orthodoxy, namely decentralization, containment and redevelopment. It also rested explicitly on the fourth – regional balance. Not all of this was entirely new of course and the planning ideas had been well rehearsed before 1939. But the authoritative linking together of these proposals made it into one of the most significant plans ever produced anywhere.

In it London was represented as four rings: the inner, tightly developed, ring; the lower-density suburban ring; the green belt; and outer country ring. Following the Barlow recommendations, the plan proposed a ban on additional industrial development within the plan area, though extensive factory relocation was envisaged. Planned redevelopment would occur within the inner ring, as had already been spelled out in the County of London Plan. The green belt was to be greatly enlarged from the pre-war efforts and enforced by planning controls rather than public ownership. The most radical proposal, however, was the planned decentralization of over a million people from the inner ring, mainly to the outer ring. In addition to the expansion of existing towns, eight completely new towns

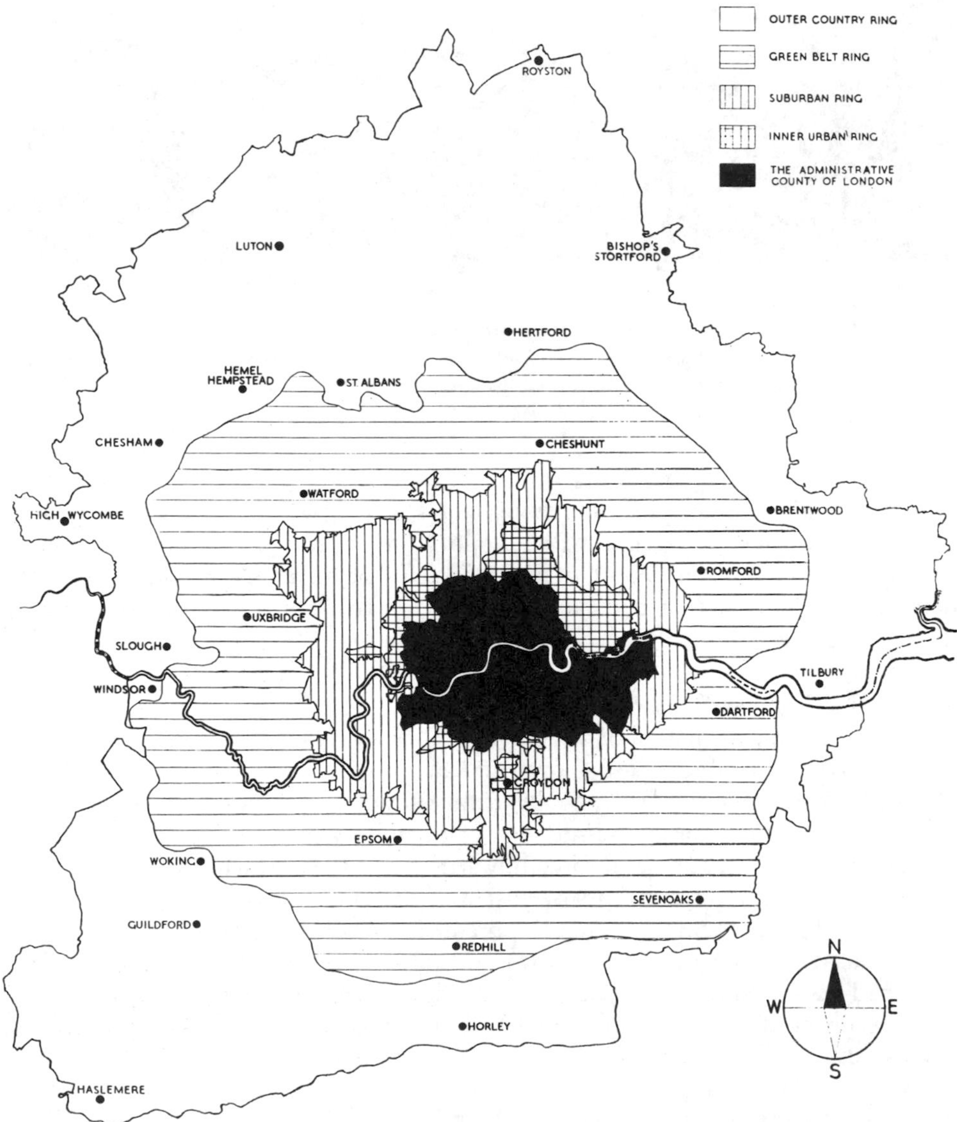

Figure 4.5 *One of the most important aspects of the Greater London Plan was the proposal for a large and continuous containing green belt, to halt continuous suburban spread. Further development was to be concentrated largely at planned expansion points in the outer country ring.*

were proposed to house a population of over 400,000, with full local employment and community facilities. Taken together these historic proposals were a foretaste of British post-war planning. All that was missing was the legal power to implement it, although the wartime building and factory licensing controls temporarily prevented development that would prejudice its proposals.

Regional policy

Employment Policy White Paper 1944

Even before the war ended, however, action was being taken on one of the most fundamental aspects of post-war planning: industrial location. The agenda for action on this derived from Barlow. Wartime experiences had, though, added a new and important dimension in the form of the all-party commitment to policies to ensure full employment, essentially through Keynesian macro-economic policies of demand management (Parsons, 1986). This was doubly important for spatial planning. Firstly it was clear that extra measures would be needed to assist the traditional depressed areas (Fogarty, 1945). Less obviously though, the agreement of more buoyant areas to restrictions on further industrial development would be easier to secure if they were made less vulnerable to trade-cycle-related unemployment in their existing industries. Such considerations had influenced Birmingham's opposition to locational controls in the 1930s (Ward, 1988). Accordingly the inclusion of a section on industrial location in the White Paper was entirely logical. What was more surprising, in view of the government's

Figure 4.6 *The most impressive proposals for planned expansion were eight entirely new satellite towns. The plan developed detailed proposals for one of the satellites (never built) at Ongar in Essex. This view shows the proposed town centre, anticipating the post-war move towards pedestrianized shopping precincts. Notice how shopping is firmly stereotyped as a female role, with Carmen Miranda starring at the Ritz cinema to relieve the boredom.*

failure to produce satisfactory legislation on any other part of the 1940–2 planning agenda, is that enduring legislation was passed on this topic before the 1945 election.

Distribution of Industry Act 1945

There were essentially two reasons for this apparent inconsistency. One was that the economic interest group most affected by this measure was industry rather than property. As we have already noted, the manufacturers were not unsympathetic to planned industrial location which had worked well for them during the war years. And by 1945 a large measure of trust existed between industry and government, based on close wartime co-operation (Booth, 1982). The other reason was that the President of the Board of Trade, Hugh Dalton, was a Labour politician who represented a pre-war depressed area (Dalton, 1957). He was sufficiently anxious to secure effective legislation before the election (which Labour was by no means certain of winning) to threaten resignation over the issue.

The Act was passed immediately before the dissolution of Parliament. Basically it extended the powers of the pre-war Special Areas legislation. It allowed financial assistance to firms in the scheduled development areas, provided for the development of factories and key worker housing and the improvement of derelict land. Spatially too the new Development Areas were extensions of the old Special Areas, soon joined by Merseyside (McCrone, 1969). Other features marked a clearer break with the past. Most important was the provision for negative controls on industrial developments over 3,000 square feet. The building licence was initially used as the control mechanism, though a more enduring instrument, the industrial development certificate (IDC), was introduced in 1947. Finally, not the least important feature was that the Act confirmed the grip of the Board of Trade on this aspect of spatial policy, even though it was obviously central to the other aspects of physical and land use planning. But the significance of this dichotomy was initially overshadowed by the tremendous burst of activity in the Ministry of Town and Country Planning after the 1945 election.

Creating the New Orthodoxy 1945–7

The Labour government and planning

Attlee's government

The rejection of Churchill, the wartime leader, and the scale of Labour's landslide victory in the 1945 general election was remarkable (Sissons and French, 1964; Morgan, 1984; Hennessy, 1993). The first election for ten very important years, it returned a House of Commons in which the power of business and property interests was dramatically reduced. Conservative representation was halved to 213 seats, compared to Labour's 393. At the first meeting of the new Parliament Labour MPs shocked the hitherto established order by singing 'The Red Flag' in the House of Commons. Yet this was to be no socialist revolution. The new Prime Minister, Clement Attlee, had been Churchill's trusted deputy in the wartime coalition government. His middle-class, public school background and unassuming manner engendered widespread trust, especially amongst the suburban middle classes whose votes had assured Labour's handsome victory.

The new government was committed to an impressive programme, though not so much outright socialism as limited state ownership, a welfare state, and planning in the widest sense. These were objectives for which public support was overwhelming and their implementation marked a decisive break in Britain's political trajectory, setting the tone for all that followed, at least until the 1980s. There were specific commitments to resolve the issues outstanding from Uthwatt. The advocates of town planning therefore had justifiable confidence that the battle for physical and land use planning, largely postponed during the later war years, would now be won.

Lewis Silkin

Although Labour had committed itself to merging the ministries responsible for town and country planning and housing and local government, the desire to act swiftly meant the move was postponed. Lewis Silkin was appointed Minister for Town and Country Planning, which remained a non-Cabinet post. Silkin was a distinctly uncharismatic figure and rather a boring speaker. Yet he possessed a profound and detailed understanding of the technicalities of physical and land use planning, which was rather rare amongst politicians (Hardy, 1991b; Hennessy, 1993). This understanding had grown from his background as a solicitor and local London politician. He had chaired the LCC Housing and Town Planning Committees and had presided over an important Labour Party wartime policy committee on both these topics. Given high level government commitment to producing a new planning system, Silkin was probably the ideal political instrument to achieve this objective.

Sources of continuity

But, despite the obvious political break in 1945, it is important to understand that there were strong elements of continuity. The wartime consensus for planning still existed. Although the property market had strengthened, landed and development interests remained relatively weak. Immediate post-war shortages of all kinds of materials ensured the continuance of the wartime building licence system, which held private development in check. Within the Ministry of Town and Country Planning (and other ministries) the officials who were involved in the as yet unresolved policy issues remained the same. In such a highly technical subject like town and country planning their role was extremely important. More specifically the official and professional work on technical reports and plans initiated in the war period continued. The wartime propaganda effort for planning also continued apace with a spate of important new official films appearing in 1945–8. The emphasis of these was, however, shifting from selling an idea of the future to selling actual proposals (Gold and Ward, 1994).

The New Towns

The background

Although there had been nothing specific in Labour's election manifesto, the Labour Party was particularly attracted to the New Town idea, which would serve as an uncompromised model of the better post-war world that planning could create (Osborn and Whittick, 1977). Frederic Osborn, the energetic secretary of the Town and Country Planning Association, the main New Town pressure group, was a party member, and the Deputy Prime Minister, Herbert Morrison, was very enthusiastic. Paradoxically though, Silkin was thought to be rather sceptical about planned decentralization and had given little emphasis to the idea during Labour's wartime discussions. The growing number of plans recommending planned decentralist solutions was important, however, reinforcing party enthusiasm. The Greater London Plan was joined by the Clyde Valley Plan of 1946 (Abercrombie and Matthew, 1949) and regional proposals for south Wales were making a similar point. Accordingly Silkin took his first action on New Towns.

The Reith Committee

In 1945 he recruited Lord Reith to head a committee, including Osborn, to decide on a suitable planning and development framework for New Towns (Cullingworth, 1979). Its three reports (Reith Committee, 1946a, 1946b, 1946c) recommended a model for such settlements of 30,000–50,000 population. Each town was to be developed by a development corporation (that could be central government appointed or sponsored by local authorities) and which would have wide-ranging powers. Occasional exceptions of New Towns developed by 'authorized associations' were contemplated, but essentially this was a state-dominated model; the offers of private builders to develop these towns were rejected. All land in the designated areas would be vested in the development corporations, with compulsory purchase powers if needed.

In planning terms the towns would be self-contained in the sense that a balance of jobs to workers was being encouraged, and that full social amenities would be provided. The principles of neighbourhood planning that had appeared in the wartime plans were endorsed and social class balance was to be sought even within individual neighbourhoods. This notion that the neighbourhood idea should now incorporate a social class mix objective was pushing it conceptually beyond what its American originator, Clarence Perry, had intended. Such idealism was a reflection perhaps of the strong perceptions of social cohesion which had grown up in wartime Britain.

The New Towns Act 1946

In sharp contrast to wartime experiences, Silkin accepted virtually everything that the Reith Committee said and immediately legislated on it. The Act provided an enabling structure for the creation of New Towns, so that it did not directly focus on detailed planning matters, which were reinforced in policy guidance. With the Reith Committee's blessing Silkin had in fact already begun the designation of the first New Town at Stevenage (Balchin, 1980). He was acting in advance of the Act, employing a previously unused part of the Town and Country Planning Act 1932. This unorthodox move was running into legal difficulties, so speed was essential. Moreover the bill attracted virtually no opposition in Parliament, though Lord Hinchingbrooke objected to its heavily statist character and predicted it would, like all state experiments, wreak 'havoc, bitterness and grave social damage' (Cullingworth, 1979, p. 25).

Early New Town designations

Events at Stevenage during 1946–7 appeared to lend some credence to his views (Orlans, 1952). New Towns policy had opened with a public relations disaster. The local residents of the small pre existing town of Stevenage objected strongly to the government acting in, as they saw it, a high-handed and dictatorial way. The famous novelist E. M. Forster became involved and presented the issue as a symptom of a deeper malaise in a radio broadcast (Forster, 1946). Rather more graphically Silkin found Stevenage's railway station name board had been replaced with the word SILKINGRAD when he arrived for a public meeting in the town. Eventually, however, his right to proceed with the New Town was established by judgement of the House of Lords.

The 1946 Act provided a surer basis for action and the Ministry soon learned the need to prepare the way with local discussions. Around London a spate of designations followed Stevenage, at Crawley, Hemel Hempstead, Harlow (Gibberd *et al.*, 1980) (all 1947), Hatfield and Welwyn Garden City (where the Garden City Company was nationalized) (Filler, 1986; de Soissons, 1988) in 1948, and Basildon and Bracknell (Ogilvy, 1975) in 1949. All these were primarily intended to implement Abercrombie's planned metropolitan decentralization strategy proposed in the Greater London Plan. Yet only Stevenage and Harlow were on sites he suggested. (He had proposed the expansion of Hatfield and Welwyn, though not specifically as New Towns.) Outside the Greater London area other New Towns were designated at Newton Aycliffe (1947) (Bowden, 1978) and Peterlee (1948) (Patton, 1978) in County Durham, at East Kilbride (1947) (Mitchell, 1967; Smith, 1979) and Glenrothes (1948) in Scotland, Cwmbran (1949) (Riden, 1988) in south Wales and Corby, Northamptonshire (1950). Most of these designations had important regional development objectives, though planned decentralization was the main purpose of East Kilbride, a site which had been recommended in the Clyde Valley Plan of 1946.

Early New Town plans

These early New Towns (often referred to as Mark I) followed the Reith Committee recommendations in most respects (Schaffer, 1970; Aldridge, 1979). Their target populations (that is pre-existing population plus newcomers plus subsequent natural increase) at designation varied from 10,000 at Newton Aycliffe to 80,000 at Basildon, though all were subsequently increased at least once over the following two decades. The average target size of the London New Towns at designation was almost 50,000, in line with Reith, though only Crawley actually had this as its original target. Bracknell at 25,000 was the smallest. The provincial

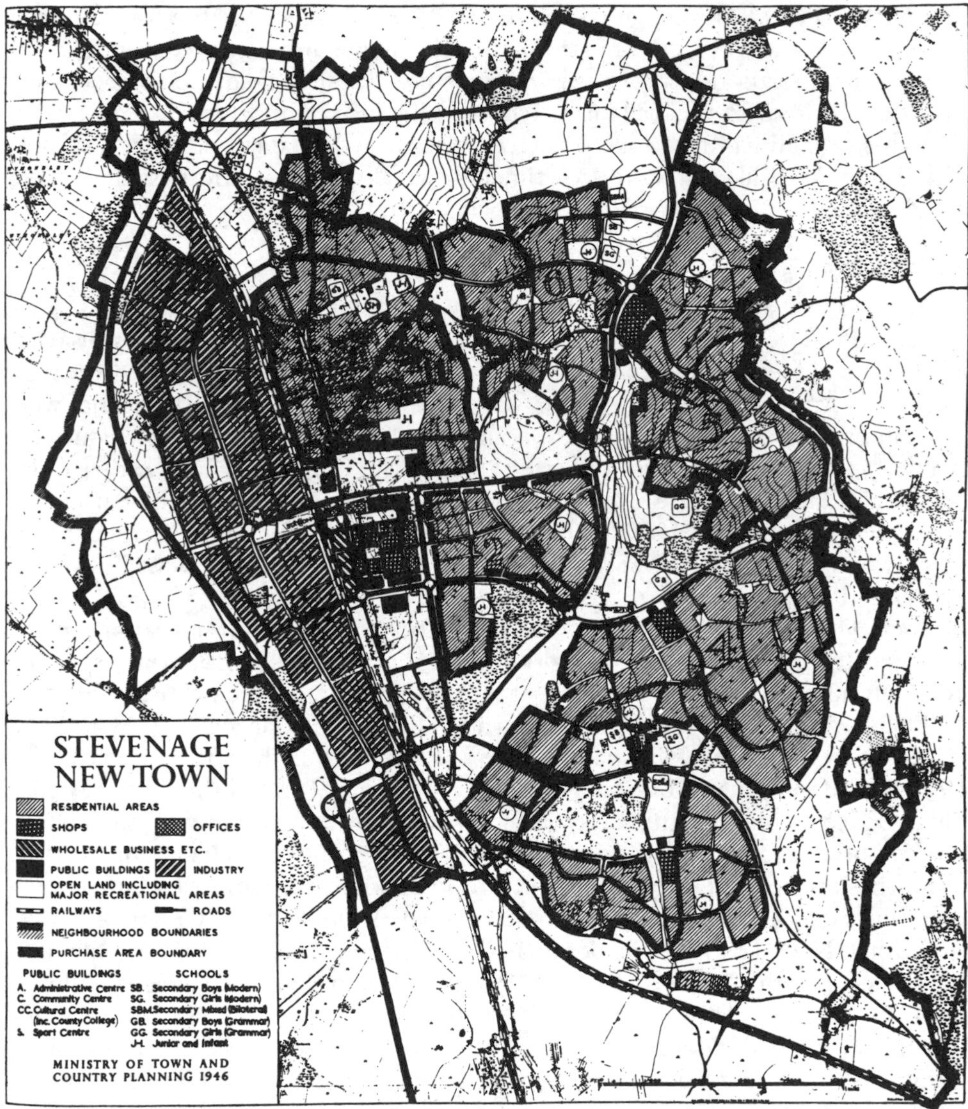

Figure 4.7 *Stevenage was the first New Town and its plan, prepared initially by planners in the Ministry of Town and Country Planning, broke new ground. Notice the strong (and quite typical) emphasis on neighbourhood principles and the large single industrial area between railway and main arterial road. Not visible on this general plan was the pioneering pedestrianized central shopping area.*

New Towns had a target average of 32,000, with East Kilbride topping the list at 45,000.

The plans were typically prepared by architect–planner consultants, though sometimes incorporating earlier local or Ministry of Town and Country Planning work. All were classic statements of the new planning orthodoxy, designed to provide a physical setting for the new model citizens of post-war Britain. Neighbourhood principles, with grouped communal facilities, were usual, though planners followed wartime plans and placed these centrally in neighbourhoods rather than at main road junctions as Perry had proposed. As Reith had recommended, the plans tried to avoid the spatial segregation of social classes that typified existing cities. This concern was widely reflected in the speeches of politicians and decision-makers at this time. These 1947 comments, by the Chairman of the new Stevenage Development Corporation, were quite typical:

> We want to revive that social structure which existed in the old English village, where the rich lived next door to the not so rich and everyone knew everybody . . . the man who wants a bigger house will be able to have it, but he will not be able to have it apart from the smaller houses.
>
> (Orlans, 1952, p. 82)

Much stricter land use zoning by activity was also typical, with strict separation of housing and industrial areas. Physical design standards and open-space provision were exemplary. Above all was the presumption of employment self-containment, reflected in careful planning for industrial development, which internationally was the most distinctive feature of the British New Towns.

New Towns and the garden city tradition

The conception of the New Town that was appearing was in most respects a state-led version of pre-war garden city thinking, which had itself moved some way from Howard's original prescription (Ward, 1992a). Newer detailed planning concepts had obviously been incorporated. But the most important contrast was that the notion of the co-operatively run social city superseding the big city, as Howard had expressed it, was gone. In its place official policy endorsement was now given to the metropolitan decentralist or regional growth focus model that left the existing settlement pattern relatively unchanged. It was this shift, especially the metropolitan decentralist strategy, which underpinned the new significance given to the concept of self-containment, to allow the new towns to detach themselves from the magnetism of the giant city.

Early New Town progress

Overall the New Town progamme was intended at this time to house over half a million people (Cullingworth, 1979). In fact by the end of 1950 only 592 houses had been built in all of them combined; in Stevenage, the first to be started, the figure was a pitiful twenty-eight (MHLG, 1951). There were more, but not that many more, in the pipeline. To some extent these problems were perhaps inevitable. Entirely new towns required major planning and infrastructural work before housing production could begin in earnest. Continuing shortages of construction materials in the early post-war period did not help. The 1947 'dollar crisis' also had the effect of restricting capital investment in all the New Towns from 1948. As if this were not enough, there were other more specific difficulties that delayed particular New Towns, notably Peterlee's problems with the National Coal Board. Local opposition to designations, especially near Manchester, was also important in restricting the programme. Thus the initial scheme never reached the twenty New Towns that had been originally intended.

Criticisms of the New Towns
All this meant that they were open to criticism for not addressing the tremendous post-war housing shortage and being expensive and irrelevant luxuries. In March 1948, for example, Ministry of Town and Country Planning officials were perturbed to receive a letter responding to their new promotional cartoon film, *New Town*, which had just gone on general cinema release. The film (of which Silkin and his officials were very proud) ended with the words 'just you try it':

> I could be a good citizen in one of your New Towns. Send me more details – if you can – and I will endeavour to try it sometime! If you cannot, please omit these words from your next similar production, and save the people who are leg weary and mind weary (from looking for a house); from groaning aloud in the cinemas.
> (Gold and Ward, 1994)

This kind of criticism touched a raw nerve. Powerful and popularly based political arguments were already taking shape against excessively bureaucratic Labour planning failing to provide people with what they wanted. Primarily deployed against the continued shortages and rationing of early post-war Britain, the perceived failings of the New Towns also made them into targets of such arguments. Other aspects of the new physical and land use planning system soon followed.

The Town and Country Planning Act 1947

Why was it necessary?
By 1947 planning powers were available to make serious progress in implementing the strategic planning policies of regional balance and planned decentralization. There were, however, no planning powers to implement metropolitan containment policies such as those central to the 1944 Greater London Plan (Cullingworth, 1975). It was only the government's continued control of building licences (which it used to squeeze the amount of private development) that conferred any real powers to stop development without compensation. No permanent controls existed that would allow this on the scale now being envisaged, without saddling individual planning authorities with heavy claims. Moreover there were still only rather limited powers to undertake planned redevelopment and none to undertake the planned 'remodelling' (or partial changes) of many built-up areas as proposed in, for example, the 1943 County of London Plan. The prospect of effective planning controls had been flagged by the Uthwatt Committee. Yet, as we have seen, the wartime coalition government had been unable to agree on suitable measures. The massive contribution of the Town and Country Planning Act 1947 was essentially to fill this yawning gap (MTCP, 1947, 1948).

Provisions of the Act
(i) Compensation: Much the most politically contentious part of the Act predictably related to compensation and betterment (Schaffer, 1974; McKay and Cox,

Figure 4.8 The New Towns, along with other aspects of planning, became the subject of increasing public criticism by the late 1940s. Allegations of bureaucratic muddle and delay postponed tangible achievements and tarnished the wartime vision.

1979). Silkin and the government sought a different solution to Uthwatt, but with very similar objectives. The Act involved state acquisition of development rights on all land (not just undeveloped) with once and for all payments in compensation drawn from a fund of £300 million. A great deal of Cabinet political consideration had gone into the manner and amount of these payments. The final figure was apparently derived from the Barlow estimate of 1937 development values at £400 million plus £100 million for redevelopment values, scaled down to reflect the sharp fall in property values during the 1940s (Cullingworth, 1975, p. 237).

Very importantly, these payments for loss of development rights were to be handled by a new body, the Central Land Board, entirely divorced from local planning authorities. Where planning authorities were acquiring land compulsorily they remained directly liable for compensating the former owners, however. Grant aid for this was available from central government, as under the 1944 Act, though on more generous terms. The 1947 Act also greatly simplified the procedures for compulsory purchase. Amongst other changes, existing use value was now to be the basis for compensation. This avoided having to determine the value of land before the war, as the 1944 Act had required. This new parity meant that if a local authority was buying slum housing to create a new commercial area, it simply paid the current value as slum housing, even though the proposed use would command a higher price in a free land market. Bomb-damaged property had, of course, to be treated slightly differently.

(ii) Betterment: The central logic of the Act was that if the state had acquired all development rights, then all future increases in land value should be the property of the state, again in the guise of the Central Land Board. Accordingly the legislation as enacted incorporated a 100 per cent betterment tax called the development charge. This has been widely seen as one of the great weaknesses of the 1947 Act. Essentially it meant that there was a contradiction at the heart of the Act. It did not abolish private ownership of development land, as Uthwatt had wanted, but by taking all the increase in value that accrued from development, it removed the profit incentive in the development process (McKay and Cox, 1979; Ambrose, 1986). This contradiction in the betterment scheme contained the seeds of its own destruction, as we will see in the next chapter.

These sections of the Act were not simply a product of a dogmatic (if partial) state socialism on Silkin's part, as sometimes portrayed. The bill had started with a proposal for a charge of only 80 per cent, and the principle of a variable charge was favoured by the Ministry of Town and Country Planning, allowing sensitivity to particular circumstances at different times and in different places (Cullingworth, 1975, p. 202–6). But these proposals failed to find favour with the Treasury, which disliked any agency having fiscal discretion other than itself. Other proposals for a more active land buying and selling role for the new Central Land Board, closer to that envisaged by Uthwatt, were also victims of Treasury pressures. Generally, official and departmental pressures seem to have been as important as Labour ideology in shaping the actual compensation and betterment scheme that was adopted (e.g. Cullingworth, 1975, p. 254). Yet it seems unlikely that a Conservative government would ever have adopted exactly similar measures.

(iii) Development planning: By contrast the actual planning provisions of the 1947 Act were much less contentious. The number of local planning authorities in England and Wales was drastically reduced from 1,441 to 145 by making the counties the planning authorities rather than county districts (Cullingworth, 1986). The position of the county boroughs was unchanged. Each planning authority had to prepare a development plan for its whole area. The plan's proposals were to be based on a thorough survey and analysis, finally adopting something of the ideas of Patrick Geddes into statutory planning practice (Breheny and Batey, 1982). Yet the rationality of analysis that supposedly led from survey to very precise physical and land use proposals was never very clear. In practice many of the wider assumptions and objectives which underlay the proposals were not made explicit in the plans. The development plan itself was intended to be (and was) rather more flexible than the 1932 Act schemes. Paradoxically, later planners also came to see the 1947 Act plans themselves as inflexible.

They were essentially precise physical blueprint plans that specified an end state for land use at the end of the plan period (intended to be twenty years). They were more sophisticated than the 1932 Act schemes partly because they incorporated phasing of programme proposals. In addition they included more detailed small area proposals (e.g. comprehensive development areas (CDAs) or town maps). The CDA, based on the strengthened compulsory purchase powers referred to above, became a major instrument of planned redevelopment, enabling planning authorities to act decisively to implement their planning intentions within the development plan framework. In other ways too these development plans differed from the crude zoning approach apparent in earlier planning schemes. The 1932 Act schemes had in effect given permission to develop land uses or buildings in accordance with their provisions. Now a discretionary element remained with the planning authority, exercised through its development control function.

(iv) Development control: This had existed before 1947, but was only significant when a development proposal contravened a planning scheme's zoning (or occasionally other) provisions. All other development went through 'on the nod'. This changed dramatically after the 1947 Act and gave planning authorities the discretion to refuse or impose conditions on detailed aspects of all development proposals and gave them effective enforcement powers. They were now wholly free from any fears of compensation claims, though were subject to appeal and the minister 'calling in' particular cases. As initially introduced, the scheme was very tight indeed though was relaxed in 1950 to allow more permitted development (for example of farm buildings), not needing development control approval (MHLG, 1951). The distinction between the 'outline' and 'detailed' applications was also introduced at the same time, adding a further refinement to the system.

Apart from this early revision, the 1947 Act development control system has endured with remarkably little change (even in the 1980s). It is though very important to appreciate that the circumstances in which this system was introduced were very different from those today. Until 1954 the development control system was to a large extent shielded by the building licence system which was operated in a way that discouraged large-scale private development. Over 80 per cent of housing built between 1947 and 1954 was public-sector housing and there was also very little commercial development at this time, despite some loosening of the building licence system (McKay and Cox, 1979). In general, development control disputes affecting the public sector were likely to be determined by means other than development control procedures. New Towns, for example, were approved by a ministerial Special Development Order. Planning decisions about local authority housing estates would typically be resolved by inter- or intra-authority discussions. This meant that at the outset the new development controllers were largely dealing with small applications (albeit many of them) from a highly residualized private development sector. This was to change.

(vi) Other provisions: Amongst many other provisions in this mammoth Act, those for historic buildings and industrial development were probably most important. The former greatly strengthened existing powers for dealing with historic buildings, including the introduction of the listing system. The latter introduced the IDC referred to earlier as a permanent control mechanism under the Distribution of Industry Act 1945, replacing the building licence. Issued by the Board of Trade, it was required before any new industrial developments of over 5,000 square feet could be permitted under local planning proposals. It became the formal point of contact between regional policy and town and country planning.

The 1947 Act and the public interest

This Act is justifiably seen as a major landmark in planning history (e.g. Cherry, 1974b, 1988; Donnison and Soto, 1980; Reade, 1987; Hall, 1992). Its operations mark the beginnings of effective and comprehensive physical and land use planning. For the first time it allowed local planning authorities to produce outcomes significantly different from what would have happened without planning. As had already happened in the New Towns, planners everywhere were now able to think of land more as a neutral platform for activities and buildings than as a source of private gain and object of speculation. Accordingly planning conducted in the public interest was henceforth intended to determine the uses to which land was put, rather than private interests expressed through the land market (McAuslan, 1980).

It was an approach which put great responsibilities on the planners themselves. Just as the formulation of the Act had been heavily dependent on expert Civil Servants, so the system in operation depended to a substantial extent on unelected professional experts. It was they who produced the development plan and comprehensive Development Area proposals, and operated the development control system. This raised two important questions. The first was concerned with the appropriate technical qualifications for planners and ensuring that sufficient trained people were available to run the new system. The second, and more fundamental, issue involved establishing a proper and accountable relationship between the professional planner and public. Both were extremely important, given the much greater role that planning was now to have under the 1947 Act and must be considered to complete our review.

Numbers and qualifications of planners

There had been mounting concerns during the war period that there would be insufficient trained planners to operate a new system effectively. In 1946, for example, the Town Planning Institute had only 1,700 members, roughly one-fifth of whom were not-yet-qualified student members, compared to over 1,400 planning authorities. This shortage had been a key consideration in reducing the number of planning authorities. Significant steps were taken in the later war years to introduce accelerated courses for ex-service personnel and to strengthen the more established training routes. Unprecedented numbers joined the profession in the early post-war years. The new planning authorities seem not to have been as seriously hampered by lack of trained staff as their precursors.

Increasingly though the question became the nature of the planning qualification. The traditional emphasis had been on the parent professions of architecture, engineering, surveying and law (Cherry, 1974b). But the newly adopted survey and analytical techniques of the more influential wartime and early post-war plans like those for London and, above all, Max Lock's brilliant Middlesbrough Survey and Plan (Lock, 1946), drew much more heavily on geography, economics and sociology (Hebbert, 1983). The same approach was increasingly encouraged in the 1947 Act development plans. Silkin and his officials became more worried about this mismatch and he appointed a Departmental Committee on the Qualifications of Planners, chaired by Sir George Schuster. The Schuster Report (Schuster Committee, 1950) endorsed the need to widen planning education to reflect the realities of the new system, proposals which were gradually introduced. There were also more controversial proposals to undermine the autonomy of the professional body to make it a 'national institution', which were not adopted.

Public involvement and the 1947 Act system

As enacted, the 1947 Act rested on the presumption that the relationship of professional planners to the public, in whose 'interest' they supposedly planned, would be mediated through the established frameworks of electoral democracy. Councillors, Members of Parliament and ministers would ultimately be responsible for planning actions undertaken by local and central government. They would ultimately act as a check on the fallibility of the professional planners in determining what was in the public interest. This assumption was rather more conservative than that contemplated in wartime Civil Service discussions about formulating a comprehensive planning system. At that time (when there had been, of course, an intense public interest in all matters related to planning, assiduously fostered by a whole range of propaganda activity), attention was given to the question of the direct involvement of the public in planning proposals at a formative stage.

Yet little of this spirit of public involvement was apparent in the Act and the regulations that accompanied it (McAuslan, 1980). The reduction in the number of planning authorities appreciably increased the remoteness of planners. There were also fears that excessive public consultation would result in delays and indecision. And there were also clear fears that the public could not be trusted to reach the correct judgements. In 1949, for example, Silkin himself advised planners:

> I think it is necessary to lead the citizen – guide him. The citizen does not always know exactly what is best; nevertheless we have got to take into account his wishes and his ideas and give him an opportunity of expressing them and give him an opportunity of understanding what is proposed.
>
> (TCPSS, 1949, p. 49)

Conspicuously there was no suggestion here that planning proposals should be derived directly from what the public wanted. Plans were presented to the public largely as a *fait accompli*. The problem was seen as getting public approval after the plan had been prepared by the experts. It was partly because the new planning orthodoxy interpreted the notion of public interest in such a technocentric way, defined by professionals rather than directly by the public, that popular enthusiasm for planning had waned by the late 1940s. But there were other reasons.

Completing the New Orthodoxy and the First Retreats 1947–52

The Expanded Towns

General context

By 1948 planners had the powers that would allow them to make enforceable proposals to fulfil all the major planning objectives for urban regions that had emerged from the plans and reports of the years 1940–6. Yet two further important pieces of planning legislation were passed in 1949 and 1952 that completed the initial phase of post-war legislative action. The first of these, the National Parks and Access to the Countryside Act, did not affect urban areas, though we need to be aware of its existence (Cherry, 1975b). It followed inter-war pressures and the recommendations of three official studies (Scott Committee in 1942, the Dower Report of 1945 (MTCP, 1945) and the Hobhouse Report of 1947 (Hobhouse Committee, 1947)). The Act now provided a legislative basis for the designation of National Parks and Areas of Outstanding Natural Beauty. The second piece of legislation was the Town Development Act 1952.

The Town Development Act 1952

The last of the five key pieces of post-war legislation, this was passed by the Conservative government which replaced Labour in 1951. As with so much else, its origins can be traced to the Greater London Plan of 1944, which proposed the achievement of planned decentralization by the planned expansion of existing towns in the outer metropolitan area, in addition to the entirely new towns. It provided a mechanism for achieving planned decentralization that relied on local government initiative (though with central financial aid) rather than the more centralist instrument of the New Town Development Corporation (Seeley, 1974). Several alternate models of the relationships between the exporting city, the importing district and the county council were provided, allowing for variations in the source of initiative and financing of the town development schemes (often referred to as Expanded Towns or overspill schemes).

The Town Development Act clearly avoided some of the perceived centralist excesses of the New Town mechanism noted above. In that respect it was certainly closer to Ebenezer Howard's localist ideal of the garden city (Ward, 1992a). But it would be wrong to see its sponsorship by the Conservatives primarily in this light. Silkin had always intended a parallel measure (Cullingworth, 1979). The bill had been drafted before the 1951 election and was fully supported by the Labour opposition. Its passage had been delayed for more subtle political reasons. Thus it was obviously going to be very difficult to persuade any local authorities to house people from another area (even with generous financial assistance) while the housing shortage everywhere was as severe as it was immediately after the war. As these problems eased, town expansions were more realistic, though progress under the legislation was always disappointing, especially in the 1950s. Fewer than 10,000 houses had been provided in all the schemes by 1958. This partly reflected particular and arguably inherent difficulties with the Act itself, but it also reflected a growing disenchantment with town and country planning that was becoming more evident by the early 1950s.

The faltering consensus for planning

Waning popular interest

The war years had seen the coincidence of unprecedented public support for planning and a severely weakened property market. Both of these had been crucial elements in securing the political acceptability of a much more far-reaching and ambitious planning system. Yet both these underlying preconditions were themselves changing by the early 1950s. The New Towns were, as we have seen, often irritating to pre-existing populations in their areas while failing to address metropolitan housing needs on any scale. The new 1947 Act system was remote and bureaucratic in its operations. It was also slow: only 22 of the 145 planning authorities managed to submit their development plans within the three years stipulated in the Act. More seriously it too had done practically nothing to accelerate the meeting of housing needs.

These were important weaknesses. In 1945 the voters, working and middle class alike, had voted Labour essentially because they wanted housing, full employment and social security. Planning was to be one of the means of achieving these objectives, but there was scant little interest, especially working-class interest, in planning for its own sake. In fact the Distribution of Industry Act 1945 had done a good deal, especially in the 1945–7 period, to push new factories to the former depressed areas (Parsons, 1986). This, in conjunction with macro-economic management, ensured that jobs remained plentiful everywhere. But the housing shortage was not easily solved and planning now appeared to be part of the problem, rather than the solution. More generally there was disappointingly little evidence of the new post-war Britain that had been promised; the bomb sites remained largely unrebuilt. One of the key reasons for all this lay in the controversial 1947 Act development charge, which had become a major point of conflict between planning and a much strengthened property lobby.

Planning and property interests

Towards the end of the war land values had begun to rise again and property and development interests, though still weak by historical standards, began to become more significant in the early post-war years. Despite all the problems of shortages and building licences, and the limited housebuilding role available to the private sector, there was increased private developer and land market activity in 1945–8. This was before the 1947 Act came into operation, but the Act's new betterment provisions threatened to choke off this activity. The most frequent complaint was that although land was supposed to change hands at existing use value, there was in practice a marked reluctance to conform to this requirement (Cullingworth, 1980). The supply of building land dwindled to virtually nothing. Where it was available, buyers of development land typically found they had to pay the full market value (that is the existing use value plus the value accruing from development) to the seller and the full development charge to the Central Land Board, in effect a 200 per cent betterment tax!

A remarkable array of property and business interest groups and professional bodies rose in opposition to these sections of the Act. During 1950 complaints and proposed amendments were received from the Royal Institution of Chartered Surveyors, the Chartered Auctioneers' and Estate Agents' Institute, the Council of the Law Society, the Federation of British Industries, the Association of British Chambers of Commerce and the Country Landowners Association (MTCP, 1951). The Conservative opposition also committed themselves to drastic reform of the development charge. This was particularly important since they made huge gains at the 1950 general election, leaving Labour struggling with a majority of only five seats. There was clearly a good chance of another election soon, with the Conservatives strongly placed to win. Clearly this encouraged those pressing for radical change. But it also made it less likely that the existing system could be made to work, since owners of potential development land were now even more inclined to hold it back, in the growing expectation that the system would soon be changed. Events soon proved these assumptions correct.

The new Conservative government 1951

We will deal with the actual dismantling of the compensation and betterment provisions of the 1947 Act in the next chapter. The change of government that preceded it was important, however. It reflected a tangible shift away from the 1940s' belief in planning as an overall principle for ordering human affairs. One telling instance of this change came when the Ministry of Local Government and Planning, finally created in 1950 to implement Labour's long-promised merger of housing, local government and town and country planning (Dalton, 1962), was immediately renamed by the Conservatives. Planning was dropped from the title and it became the Ministry of Housing and Local Government, an important symbol of a new emphasis (Sharp, 1969).

Reformulating the consensus for planning

It was at this time that the consensus for planning assumed its more enduring character that characterized it through the next two decades (Donnison and Soto, 1980; Reade, 1987). No longer did it have the mass support of the war and early post-war years. Socially it was actively sustained by the narrower and decidedly middle-class bases of organized pressure groups and professional bodies that had always been central to the advance of planning ideas. Economically it retained the rather passive support of industry. Particularly important though was the new relationship between planning and an increasingly buoyant property sector that was emerging in the early 1950s. After the highly statist 1940s' notion of planning replacing the land market, planners and property interests were now moving into an ambiguously symbiotic relationship. Party political agreement was to be critical in the definition and articulation of this new version of the post-war consensus. What was increasingly evident during the 1950s was that, despite changes in emphasis, planning was the subject of a very wide measure of party political agreement. There were differences, certainly, but the forces of continuity were stronger.

OVERVIEW

The 1939–52 period was, then, a pivotal one in the history of British planning policies. A comprehensive conception of physical and land use planning, rehearsed in the 1930s, was now firmly embedded in a new orthodoxy of government policies that would endure in its essential elements until the late 1970s. Yet already some retreats were imminent. In planning, as in other aspects of government policy, the high tide of state intervention was set to recede, reflecting the emergent mixed economy of the 1950s. Even during the 1940s, the highly interventionist model of the planning system promised at the beginning of the decade was compromised. Reith's desire for an all-powerful planning superministry was sacrificed to existing departmental interests. And Uthwatt's scheme for the permanent subordination of private property interests was weakened as the urban land market recovered from its wartime low point. Bold and definitive as it was in most other respects, Labour's 1947 Act, by placing such heavy reliance on the development charge to assert the dominance of planners in the shaping of urban change, was condemning itself to be an early victim of the revival of private development.

5

ADJUSTMENTS AND NEW AGENDAS: I. THE CHANGING PLANNING SYSTEM 1952–74

INTRODUCTION

The passage of the Town Development Act 1952 by Churchill's Conservative government, with full Labour support, provided the last of the major elements of post-war planning orthodoxy. At its core was the Town and Country Planning Act 1947 and it has become common to refer to the 1945–52 planning legislation as 'the 1947 Act system'. Everything else that followed in British planning, at least until the second half of the 1970s, rested on these impressive legislative foundations. This did not mean that the system was immutable. The most contentious parts of the 1947 Act were soon dropped, allowing the re-emergence of a rather freer market in development land. From about 1960 there was a very marked revival of government interest in planning. A welter of new legislation and planning initiatives gave planners a more positive role, in contrast to their rather more limited position in the 1950s. This new role built on the pattern of state-sector–private-sector interaction that had emerged, almost accidentally, in the 1950s. Planning was now firmly located within a mixed economy and a society increasingly engaged with the pursuit of private affluence. There was certainly no return to the high tide of 1940s' statism.

In this and the following chapter we consider the main elements of the development of planning policies and ideas during these years. This chapter sketches out the broad pattern of social, economic and political change, the overall importance accorded to planning in its widest sense of government intervention and the more specific evolution of the legislative and administrative framework of physical and land use planning. Chapter 6 will then focus in more detail on the evolution of specific strategic and detailed planning ideas and policies. We begin by reviewing the rapidly changing social and economic agenda of the 1950s and 1960s.

POLITICS, THE 'AFFLUENT SOCIETY' AND PLANNING

The economic and social context

The long boom

The quarter century from 1947 was a period of virtually uninterrupted prosperity (Wright, 1979). It was this central economic fact with all its wider impacts on society that shaped the basic demands placed on the new planning system in this period (Hobbs, 1992). The experience was virtually unprecedented. Thus, apart from the odd hiccup, the economy (gross domestic product – GDP) grew steadily from 1951 to 1973 at an average rate of 2.8 per cent per annum. Compared to other advanced industrial countries this was low, and more sophisticated measures underline this differential. But to those experiencing it, it was real enough. The newer manufacturing industries of the inter-war years now began to form the basis of the post-war export economy. The decline of the older industries continued, especially from the late 1950s, but such structural changes no longer occurred in the precipitate manner of the pre-1939 years. Unemployment, the absence of which had by now become a key measure of economic well-being, remained extremely low by inter-war standards (Law, 1981). Thus the 1951 Census recorded 1.8 per cent of workers unemployed (compared to 12.0 per cent in 1931). There was a steady, though modest, upward trend to 2.8 per cent in 1961 and 5.2 per cent in 1971, but this was very low compared to what came after.

The sources of prosperity

There were two basic sources of this virtually unprecedented prosperity. The first was the successful creation of a stable world framework in the late 1940s allowing the growth of international trade and the global economy. The second (and not unrelated) factor was the commitment of all British governments (and indeed all the major advanced countries) to Keynesian-style economic management (Peden, 1991). This willingness to use public spending and intervention did much to iron out cyclical fluctuations in the economy and eased the effects of the decline of older industries. Other factors were also favourable to growth, such as the very high proportion of economically active population relative to dependents (largely because of married women permanently entering the workforce). The greater application of new technologies was also important. And it was new technology, more than anything else, that provided a symbol of economic and social progress.

Social impacts

GDP growth over the two decades from 1951 was almost eight times greater than population growth. With full employment and higher wages than ever before, the social consequences of the long boom were huge (Bogdanor and Skidelsky, 1970; Marwick, 1970; Morgan, 1992). The ending of rationing in 1954 allowed the fortunate majority of the population to become a mass consumer society, completing a shift which had begun in the more prosperous areas during the 1930s. Thus a new and technologically sophisticated consumer good like television had become virtually universal by the early 1970s, itself forming a powerful reinforcement of rising consumer expectations. Numbers of the other main, though more expensive, symbol of the age, the motor car, increased from 2.5 million in 1952 to nearly 12.6 million in 1971 (Plowden, 1973).

Already by 1958 (the year of its publication), J. K. Galbraith's book *The Affluent Society*, a critique of a society (the USA) motivated by the pursuit of private affluence, was describing well-recognized if not so firmly established trends in Britain. The war had fostered a collective sense of society and common interest in the 1940s that disrupted the inter-war tendencies towards home-based suburban consumerism, privacy and the nuclear family noted in Chapter 3. Moreover it had reinforced the more communal way of living that was still well established in most older working-class areas. But the changes of the 1950s and beyond were far more powerful. They dispelled not just the temporary sense of wartime social collectivism but also undermined a good deal of the traditional cohesion of the working class. The young and ambitious particularly began to opt into the consumer society, if only as aspirants. It took a long time for the full effects of these trends to become apparent. But even by 1959 the Conservatives were able to win a handsome electoral victory on the message that 'You've Never Had It So Good' (Morgan, 1992, p. 176).

Demographic changes

This sense of increasingly self-centred well-being and affluence had marked though rather unpredictable effects on the birth rate (Marsh, 1965). Wider knowledge of reliable contraception methods, reinforced by the widespread availability of the Pill in the 1960s, made it much easier to prevent conception. Moreover, as more women undertook paid work which continued after marriage, the already declining desire for larger families diminished further. And the pursuit of private affluence was associated with a growing tendency to view having children in almost the same light as acquiring consumer durables, to be fitted in after the home furnishings, the car, the foreign holiday. Yet against all this the very fact of social well-being itself also seems to have encouraged more couples to have children.

The upshot was a fluctuating birth rate that defied the forecasters (Cullingworth, 1976, p. 32). There was a surge in the early post-war years, a lull in the 1950s, another surge in the late 1950s and early 1960s, before a more sustained downturn from the late 1960s. Combined with continued decline in the death rate, as life expectancy steadily increased, this brought an overall 10.5 per cent population increase between 1951 and 1971, more than might have been expected given pre-war trends. Such overall changes were paralleled by changes in the family. Thus the trend towards the nuclear family, fewer children and the decline of multi-generational families was apparent in a marked rise in the number of one- and two-person households.

Outside the affluent society

But the reality of affluence was not universal, despite a widespread belief that poverty had disappeared (Coates and Silburn, 1970). This belief reflected the absence of serious unemployment and an overestimation of the effects of the post-war welfare state reforms. The truth was that poverty had certainly diminished in absolute terms, but that many people were still being left behind by the new culture of affluence. The welfare state was simply not sufficient to prevent the poverty of families with unskilled or unemployed heads or those dependent on state benefits such as the old, the chronically sick or disabled.

Entirely new social groups were also trying to get into the affluent society. These were immigrant populations from the Caribbean, south Asia and other parts of the so-called 'New Commonwealth'. Between 1950 and 1967 their numbers (including those born in Britain) increased from about 100,000 to just over a million, about 2 per cent of the total population (Daniel, 1968). Those of working age filled the lower-paid jobs that indigenous white workers were no longer prepared to fill in the long boom. Until the Race Relations Act 1965 overt and crude racism was widespread, producing an initial residential concentration of these immigrants in many of the worst slums of the big cities and barring their access to better-paid jobs. Signs of a considerable white resentment of the newcomers became apparent, especially in the Notting Hill riots of 1958 and the anti-immigrant marches ten years later.

Yet discrimination, though still considerable, was generally becoming less open by the late 1960s. This was partly because of legislation but also because the passage of time and the needs of a growing economy were slowly allowing immigrant communities to move from the least attractive parts of the job and housing markets. Thus, even those presently unable to enjoy the fruits of affluence had grounds for believing that the prevailing prosperity would bring them within reach. This sense of optimism, even amongst the least favoured sections of society was an important element reinforcing the general mood of well-being.

Politics and government attitudes to planning: an overview

Political changes

Economic and social changes were therefore structuring a formidable series of pressures for urban change which in turn exerted significant demands on the planning system. The exact nature of the planning response to these basic demands was primarily conditioned by the political process (Morgan, 1992). Britain had elected a Conservative government under Churchill in 1951. The elections of 1955 and 1959 brought progressively increased Tory majorities under Anthony Eden and Harold Macmillan (Pinto-Duschinsky, 1970). The pattern changed in 1964 when Labour, led by Harold Wilson, won a smaller than anticipated majority which was consolidated in the 1966 election. Unexpectedly, the 1970 election returned the Conservatives under Edward Heath, who remained in power until 1974. But despite this highly developed pattern of two-party politics, there were strong elements of policy continuity. Changes of government produced shifts in emphasis but, despite what contemporary commentators often thought they saw, no fundamental changes in direction. To understand this, we need to appreciate how physical and land use planning policies were located within the larger post-war political consensus on economic and social policies.

Politics and state intervention in the 1950s

By the 1950s the Conservatives, with a pragmatism that used to be typical, had absorbed the lessons of their crushing defeat in 1945. The British people, particularly the working class, wanted a government that would ensure full employment, decent housing provision and social services (Hennessy, 1993). On the other hand the widening middle classes did not want an overly interventionist state or the continued austerity of the Attlee years. After many years of shortages and rationing all classes looked forward to greater material prosperity and the variety of goods that only a market system could provide. These kind of considerations had a marked impact on economic and social policies.

Thus the Conservatives quickly jettisoned the last of the austerity controls and Labour's grander aspirations to economic planning. There were few regrets when rationing ended in 1954. The Tory Chancellor, R. A. B. Butler, retained the overall notion of Keynesian-style demand management, however, providing real continuity with his Labour predecessor, Hugh Gaitskell (Brittan, 1971). Contemporary commentators coined the term 'Butskellism' to describe this political agreement in economic management and its ethos permeated many areas of policy. Most importantly the Conservatives committed themselves to sustaining the goal and the reality of virtually full employment. Neither did they embark on any orgy of denationalization remotely comparable with that followed by Conservative governments in the 1980s. There was no significant undermining of the welfare state. And there was surprising enthusiasm too for housing, so that they soon reached their election pledge of 300,000 completions per annum (Macmillan, 1969; McKay and Cox, 1979). Under Harold Macmillan, the first Minister of Housing and Local Government, they even managed to reach an all-time high for public-sector completions in 1953 and 1954, each at just under 262,000 dwellings (UK figures). But, as we noted in the last chapter, this special emphasis on housing was accompanied by an apparent downgrading of the position of physical planning.

Pressures for a new approach

This rather limited role for planning conceived by the Conservatives began to change during the late 1950s and early 1960s. The rising birth rate brought confident, though false, expectations of massive population increases that challenged the rather limited planning approach of the 1950s. There were growing worries too about the scale and nature of economic growth. Would it be sufficient to maintain this growing population in the circumstances of ever increasing affluence to which it was rapidly becoming accustomed? Would the growth occur in the right places and in the right sectors? And above all the nagging doubt – why was Britain's economic growth rate so clearly inferior to that of other comparable countries? Finally there were a growing series of worries about the public problems that the pursuit of private affluence was already creating, most notably the growth of car ownership. Given the growth projections, these problems were set to get much worse.

Macmillan's planning experiment 1960–4

The upshot was a renewed political faith in planning in the widest sense during the early 1960s (Brittan, 1971; Blackaby, 1978). This shift occurred under the premiership of Harold Macmillan, one of the progressive Tory economic planners of the 1930s. The new economic planning was not on the highly statist pattern of the 1940s, but rather an attempt at indicative planning within a mixed economy. The state would indicate goals and use its own resources and activities to give a lead and point directions, but other agencies, particularly private business, would play key roles. One of the key elements of the approach was a close co-operation of government (including local government for sub-national initiatives), private sector and trades unions.

This kind of indicative approach, based largely on French experience, pervaded many of the new planning initiatives of this period (Shanks, 1977). The key agency was the new National Economic Development Council (NEDC) and Office (NEDO) established in 1961 as a forum for discussing problems related to growth and modernization and an agency to give advice (e.g. NEDC, 1963). In the same year a series of important regional planning initiatives were launched, broadly consistent with the same approach. And we can detect elements of the approach in the important proposals for urban planning which also appeared in the early 1960s. We will consider these more fully in later sections.

Labour and the 'technological revolution'

Yet it was Labour, under Harold Wilson, who seemed to exhibit the clearest idea of how planning, in the widest sense, ought to be used. In an important speech at the 1963 Labour Party conference, he spoke of the need for 'conscious, planned, purposive use of scientific progress to provide undreamed of living standards and the possibility of leisure ultimately on an unbelievable scale'. And in the most memorable passage, he called for a New Britain, 'forged in the white heat of a technological revolution' (cited in Hardy, 1991b, p. 63).

The vision was a powerful one. It allowed the different factions of the party, who had spent most of their opposition years fighting each other, to agree on a common programme (Bogdanor, 1970). And it seized on one of the most potent themes within the affluent society. There would be an active state, more active than Macmillan's indicative planning experiments had envisaged. But its main role would not rest on the traditional socialist preoccupations of nationalization and redistribution. Rather it would accept the existence of the mixed economy and take the lead role in planning it, promoting and harnessing technology to secure the planned achievement of a higher level of growth. This would then pay for improved social welfare and public services, while still allowing the private pursuit of all the consumerist rewards of the affluent society.

Planning in the Wilson years 1964–70

On election in 1964 the Labour government initially pursued its new vision with great energy and imagination (Wilson, 1974). Wilson established a new super-ministry, the Department of Economic Affairs (DEA), charged with the task of concentrating on longer-term economic policy (including spatial policy), offsetting the traditional 'short-termism' of the Treasury. The DEA was initially headed by Wilson's deputy, George Brown, a mercurial and rumbustious political figure whose forthright style and fondness for alcohol quickly brought him press notoriety (Brown, 1972). Despite this, his political seniority helped force the pace to make the early Wilson years exciting ones for planning.

Brown's most immediate tangible achievement was the preparation of a national plan (DEA, 1965a), laying out a programme for accelerated economic growth 1965–70. In fact the plan, if it ever had been attainable, was soon blown off course by short-term financial pressures in 1966–7 (Brittan, 1971). Growth rates, though they improved, remained too low to fulfil the promises of the planned technological revolution. But the extent of these failures was not fully appreciated until later. The more immediate significance of the early Wilson years was an enthusiastic endorsement of the principle that planning was essential in a mixed economy to promote growth, deal with its problems and ensure that it brought wide social benefits (Saville, 1988). Overall, therefore, the Wilson years, while they may not have delivered the promised 'technological revolution', did bring a 'second planning revolution' (Hardy, 1991b, p. 63) that stimulated physical and spatial planning activity, as we will consider below.

The last phase of consensus 1970–4

By the early 1970s many of the clever hopes of the previous decade were beginning to expire. There was a growing recognition on all sides that planning alone was insufficient to ensure the achievement of a modernized Britain (Morgan, 1992). Nevertheless much of the planning momentum of the Wilson years continued under Heath's Conservative administration. Despite an initial adherence to the so-called Selsdon programme, involving a vigorous reassertion of free market ideals, the Conservatives ultimately accepted all the main precepts of the post-war consensus about the mixed economy and the role of government. Planning in the widest sense no longer had the high priority that Labour had given it and there were some significant changes in physical and land use planning, but, in contrast to what came after, the Heath government brought no truly radical departures from the previous agenda. It marked the end of a long period of government, coinciding with the long post-war boom, where the elements of consensus were more important than the differences.

THE CHANGING PHYSICAL PLANNING SYSTEM

Conservative adjustments 1952–64

The shifting consensus for physical and land use planning

Within these larger policy continuities, there was also a sizeable degree of political agreement on town and country planning. As we noted in the last chapter the wartime mass consensus which had generated the political momentum in favour of the new planning orthodoxy in 1945–52 soon evaporated. By the early 1950s physical planning was being actively sustained by the support of the narrow band of interests represented by the planning movement. The pressure groups, the professionals and the local and national politicians and Civil Servants directly involved in the management of urban change were now its main protectors. It is the recognition of the narrowness and often weak or partial nature of this support that has led Hardy (1991b, p. 22), following Donnison and Soto (1980, p. 6) to see the post-war story of physical planning as that of a 'fragile consensus'.

Yet such views tend to ignore the growing, if rather ambivalent, accommodation of property and development interests within the planning system from the 1950s (Ravetz, 1980; Reade, 1987). The planning framework of the 1940s, conceived when property interests were extremely weak, had sought to peripheralize them. But as the statist economy of war and austerity was dismantled and a more enduring mixed economy was created, it was inevitable that such interests would become a more important consideration in planning policies. The rapid receding of the high tide of popular wartime pressure for a conception of planning that was potentially competitive with such an involvement of property and development interests obviously smoothed the transition. And the new social engagement with the pursuit of private affluence also made it easier to accept a stronger role for developers of, for example, private housing, within the planning system.

Overall then we can see many of the changes of the 1950s and beyond in terms of this underlying shift, set within the wider context of the emergent affluent society. It marked, of course, a serious retreat from the radicalism of the 1940s. But within the larger framework of policy continuity, it also added a political robustness to the planning consensus that has certainly helped ensure its remarkable resilience. Although the nature of the accommodation of private property was, to say the least, ambiguous and much less explicit than that of the planning movement, it began to take increasingly positive forms during the 1950s and 1960s. And without this growing together, the planning consensus might well have been too fragile to survive in any effective form. We must now consider the specific changes to the planning system.

Housing, planning and Conservative ideology

The Conservative shift in emphasis to housing, initially public-sector housing, was a remarkably successful act of short-term pragmatism (McKay and Cox, 1979). It also had a longer-term and more ideological dimension. At its heart was the concept of the 'property-owning democracy', first introduced into Conservative policy debates by Eden in 1946 in an attempt to find an idea to counteract Labour's supremacy in social policy matters (Pinto-Duschinsky, 1970). Ideologically it rested on the argument that social democracy was to be achieved by widening social access to private property. It was an argument that the emergent affluent society of the 1950s was eager to hear. It contrasted sharply with the socialist idea of abolishing or heavily controlling private property and relying on state ownership and control.

There was no suggestion that a property-owning democracy would not require physical planning, as has consistently proved to be the case (see for example the section on containment in the next chapter). But it would clearly be rather different in emphasis to the unreformed 1947 Act system. In fact the greatly enhanced powers with which town planning entered the 1950s, especially with the continuance of the building licensing system, were easy to represent as the very antithesis of this property-owning democracy. This allowed the Conservatives to adopt a medium-term strategy which gave greater encouragement (and of course greater profits) to the private developers. As the new Minister, Harold Macmillan, remarked in 1952, 'the people whom the Government must help are those who do things: the developers, the people who create wealth, whether they are humble or exalted' (Marriott, 1969, p. 15).

And its housing policies, although in the short term boosting public-sector building, involved a longer-term move towards the private sector and the encouragement of owner occupation. Thus the private sector grew from just 15 per cent of all housing completions in 1952 to 60 per cent in 1961, an absolute increase from under 37,000 to over 180,000 (UK figures).

Mr Pilgrim's suicide

A more specific episode also brought the land provisions of the planning system into great disrepute in 1954 (Cullingworth, 1980). This was the unfortunate case of Mr Pilgrim who, having purchased land in 1950 at market value, failed to make a claim against the compensation fund set up under the 1947 Act. When the local council then compulsorily purchased the land at existing use value, he lost a great deal of money and killed himself. The case was widely reported and Churchill, normally indifferent to most planning matters, took a great personal interest. Macmillan later recalled being summoned to see the great man one morning:

> When I got there, I was ushered upstairs and found Churchill in bed, finishing a substantial breakfast, soon to be followed by the inevitable cigar. He was wearing the famous Chinese dressing gown, and his favourite budgerigar was perched on his head. The bed was strewn with newspapers. But there was no benevolence in his manner. Instead he was in a fierce and angry mood, and poured out a flood of accusation and reproach. 'Why have you done this man to death – you and your minions?'
>
> (Macmillan, 1969, p. 428)

Despite this air of tragic comedy, the Prime Minister's political instincts were entirely sound. The issue was raising real fears amongst many ordinary Conservatives (and others) about the vaguely totalitarian character of the planning system, hounding the 'little man'.

Town and Country Planning Acts 1953 and 1954

By this time though the amendment of the 1947 legislation had already begun (MHLG, 1952a; Cullingworth, 1980). Indeed the marked upturn in private housebuilding essentially reflected the impacts of the planning legislation of 1953 and 1954, together with the ending of the building licensing system in 1954. The two planning Acts were exclusively concerned with the financial sections of the 1947 Act and involved no other changes. The 1953 Act abolished the development charge though neither Act contained any alternative proposals to collect betterment. Between them the two Acts also changed the provisions on compensation for loss of development rights. This was now to be paid on a rather more limited basis than under the 1947 Act, spreading the public expenditure implications over a longer period. Under the original scheme this would not theoretically have been a problem because compensation could have been funded from the development charge income.

Finally the 1954 Act redefined the valuation basis for land transfers. In contrast to the 1947 Act's reliance on existing use values, by now regarded as quite unworkable, the 1954 Act used existing use plus development value as the basis for land sales to private purchasers. Specific provisions safeguarded people in exactly the same position as the unfortunate Mr Pilgrim. Yet in a more general sense the new system increased the risk of individual losses of this kind. This was because a different system operated when land was purchased by public agencies. Here the 1947 principle of transfer at existing use value was retained. The 'dual market' thereby created was much criticized by property and professional interests.

Town and Country Planning Act 1959

There was nothing to rival the Pilgrim scandal, but the operation of this dual land market undoubtedly created cases of individual hardship for small property owners. The system was roundly condemned by the official Report of the Franks Committee (1957) on Administrative Tribunals and Inquiries. Accordingly the 1959 Act corrected this anomaly, restoring full market value as the normal basis for all land transfers. It also contained other provisions, most notably to improve the quality of information given to the public and especially to property owners about planning applications. The need for this arose because there had been cases of developers who evaded paying market prices for property by concealing their planning intentions from the original owners of the sites they were assembling.

Policy hesitations in the early 1960s

This amendment was itself a confirmation of the general direction in which planning was moving in the late 1950s and early 1960s (Cullingworth, 1980). With the end of building licensing, the planning legislation of the 1950s had created a much more favourable climate for private property development of all kinds. Meanwhile the generally buoyant economy and mood of business confidence accelerated development pressures. More specific government initiatives of modernization gave further encouragement. The most notable was the long-delayed motorway programme, first promised in 1946, begun tentatively in 1953 and finally started in earnest in 1957 (Starkie, 1982; Charlesworth, 1984). The first short section (of what ultimately became the M6) was opened in 1958 as a bypass for Preston. The first long stretch, seventy-two miles of the M1, opened the following year and the creation of a national system was by then well under way.

Reflecting these and other more general growth pressures, the intensity of demand for development land caused land values to rise sharply, especially in the south east. This effect was intensified by the rather tight approach to the allocation of development land that had developed in the previous decade. We will examine the sources of this tendency in more detail in a later section, but for the moment we can note that it was creating pressures for amendments to planning law, especially in relation to betterment. The Conservatives were rather resistant to this, arguing that greater land allocations would address the problem (in ways to be considered more fully in Chapter 6). The Chancellor, Selwyn Lloyd, did, however, introduce a modest 'speculative gains tax' on property development in the 1962 Budget.

Town and Country Planning Act 1963

The final legislative effort before the 1964 election (apart from a consolidating 1962 Act) closed one of the most serious loopholes used by commercial property developers in London (Marriott, 1969; Cullingworth, 1980). The Town and Country Planning Act 1963 modified the permitted development rules under the 1947 Act, which allowed a 10 per cent increase in the volume of a building in existing use. In other words an old office could be demolished and redeveloped as a 10 per cent larger office without being subject to planning consent, other than on detailed design matters. Yet, given the lower ceiling heights that were possible in modern buildings, 10 per cent cubic increase could easily mean 40 per cent extra floorspace. If all the older office space in central London were redeveloped with such gains, the extra floorspace alone would have been roughly equivalent to the total increase in office floorspace in the same area over the years 1951–60!

Moreover the developers of buildings like the Hilton Hotel and the West London Air Terminal were increasingly pressing, with some success, to exercise these rights in brand new buildings, immediately building them 10 per cent bigger than the intended planning permission. There was no way that the planning authority, the LCC, could stop any of this without facing extremely heavy compensation claims for revoking permitted development rights. Accordingly the 1963 Act substituted only 10 per cent increase of floorspace in development rights, to apply only to buildings erected before the 1947 Act came into operation. The new legislation was effective so far as it went, though open-plan layouts meant that a good commercial architect could still do a lot with a 10 per cent floorspace increase. And on its own it represented no solution to the wider problem of the growth of office employment in London. In keeping with Macmillan's economic planning experiment of the early 1960s, however, this question had become part of the wider review being undertaken of regional growth problems in the south east (MHLG, 1963). We will examine this more fully in the section of Chapter 6 dealing with regional policies.

The development plan system

During the early 1950s there had been widespread talk within the planning movement of a '1932 mentality' (Beaufoy, 1952, p. 3). The fear was of planning being sacrificed to mindless Conservativism and going back to the ineffectual system of the pre-war years. Yet the changes of the 1950s did not, with hindsight, justify this pessimism. The decade saw the completion and approval of virtually all the statutory development plans. As a group of documents they exhibited little of the panache and innovation of the non-statutory plans of the previous decade. Thomas Sharp, who had himself prepared several of these earlier plans, pulled no punches in 1957 when he commented, 'Far too few [development] plans contain any positive proposals that can excite interest . . . they are presented in documents of such insupportable dreariness . . . specially designed to produce the maximum amount of bewilderment and boredom' (Sharp, 1957, p. 135).

Their main problem, however, and it was becoming all too apparent by the early 1960s, was that they had simply not allocated enough land for residential development (Hall *et al.*, 1973, Vol. II). Although the Conservatives had encouraged a particularly 'tight' approach in the 1950s, to be examined in the section on containment in Chapter 6, the root problem was the faulty population forecasting referred to earlier. But however dreary and overly restrictive the plans might be, at least they were there. For the first time Britain had a fully comprehensive development planning framework, providing a basis for controlling development proposals.

Development control and the private developer

Despite the shock of a private development boom, the other staple of local planning activity, development control, also remained intact following its reform in 1950 under Labour. By mid-1964 the system was handling some 400,000 applications a year, 84 per cent of which were approved (Cullingworth, 1980). On the whole the system was felt to work well, though the number of appeals on the refusals was growing from under 4,500 in 1953 to over 12,000 in 1963, bringing more delays. There was, however, ministerial and official reluctance to go further in enlarging permitted development rights, especially since pressures were growing for greater involvement in planning decisions by an increasingly critical public.

Inevitably the development controller's relationship with the developer was to some extent that of gamekeeper and poacher, as the amendments in the 1959 and 1963 Acts rather suggest. But there were clear signs too of the emergent accommodation of the private developer within the planning process through development control. Each party became accustomed to the ways in which their actions could potentially assist the other, in return for concessions of various kinds. Already by the early 1960s crude forms of 'planning gain' (not yet called this) were in use and other partnership arrangements were becoming more common in the promotion of planned commercial redevelopments. As with other strategic policy details, we will examine these more fully in the next chapter. But they are important to note here as an indicator of how much town and country planning now reflected the norms of the mixed economy.

Labour's new agenda 1964–70

The 'technological revolution' and physical planning

Labour's overall strategy for planned economic growth based on harnessing technology was important for physical and land use planning in three main ways.

Figure 5.1 *Development plans prepared under the 1947 Act were essentially detailed physical blueprints for change. This map, part of the Leicester City Development Plan, shows how the intended use and density of every piece of land was specified, together with areas for comprehensive development.*

Firstly it set a general tone for government actions that brought all forms of planning activity into greater prominence. Secondly it fixed the general political parameters for planning policies and actions. Above all there was to be greater reliance on private–public co-operation within the mixed economy than on the heavily state-dominated planning approaches of the 1940s. Finally it thoroughly imbued the practice of physical planning with the prevailing optimistic ethos about technology and its capacity to address society's needs in a democratic and accountable way. None of these things was entirely new of course. All had emerged during the Conservative 'planning experiment' of the previous few years. But Wilson, especially during his first administration, managed to give the whole approach a coherence and positive sense of direction that had not been apparent under the Conservatives.

A Ministry of Land and Planning?

Yet even in the first days of his administration he was forced to compromise his ideas. Wilson had intended to create a Ministry of Land and Planning (MLP) to expedite the supply of land needed to implement Labour's regional and housing proposals (Wilson, 1974; Crossman, 1975). His first Minister of Housing and Local Government, Richard Crossman, had readily accepted this scheme. But Crossman's long-established and rather formidable Permanent Secretary, Dame Evelyn Sharp, fought an intense and successful four-day Whitehall battle to stop MHLG's planning powers being lost. The proposal was quickly reduced to a rather small Ministry of Land and Natural Resources (MLNR). Most planning responsibilities remained with MHLG, though MLNR carried through the Land Commission legislation before it was abolished.

It is difficult to assess the significance of this early disruption. Looking back, Wilson himself blamed the subsequent problems of the Land Commission at least partly on his failure to insist on this change. MHLG was, he felt, too committed to the local planning system to push it to release enough land for development. This Civil Service conspiracy theory may well be a convenient rationalization, however. Wilson never intended MLP to be a super-ministry that would have had the 'clout' to force through the planning and land changes that were needed to do what was intended. Yet the episode showed how much institutional inertia locked planning into a particular governmental setting that was perhaps less appropriate for the new approach to planning that was being canvassed in the 1960s. (The 1965 battle between MHLG and DEA over regional planning, reported in Chapter 6, rather reinforces this point.)

Control of Office and Industrial Development Act 1965

Labour's first legislative effort, although rather specific in intention, demonstrated a desire to take a stronger line than its predecessors. In November 1964 George Brown had introduced his so-called 'Brown Ban' on further office development in London. The 1965 Act introduced a permanent mechanism, the office development permit (ODP), directly comparable to the industrial development certificate which applied to factory building. But despite Brown's initiative, ODP control was exercised by MHLG (unlike IDC control which was handled by the Board of Trade). The Act added some teeth to the advisory approach of the Conservatives, evident in the Location of Offices Bureau set up by them in 1963. Yet there was no serious opposition to it, even amongst the developers themselves. Marriott (1969, p. 22) has even described it as the 'crowning gift' to the office developers by calling a halt to what was by then becoming a speculative overprovision of offices, thereby enhancing the profitability of the final schemes.

Land Commission Act 1967

Such an effect had not, of course, been Brown's major intention, though it doubtless contributed something to the widespread political support the proposals received. Labour's second attempt, twenty years after its first, to address the question of land values and betterment was, however, a conscious expression of their moderate intentions towards private landed and development interests (Hall *et al.*, 1973, Vol. II; Cullingworth, 1980). The Land Commission had two major aims: to tax betterment and to accelerate the supply of land for building development, especially for housing. In their pre-election proposals it sounded very much like an implementation of the wartime Uthwatt recommendations, but the Act and even more the practice were more modest (MLNR, 1965). Thus the betterment levy on development values was introduced at a level of 40 per cent, well below Uthwatt's suggested 75 per cent, the 1944 White Paper's 80 per cent or the 1947 Act's 100 per cent. The intention was to leave a considerable incentive in private hands, a sharp contrast to the 1947 development charge. An increase to 50 per cent was promised but it never occurred. The second aim was the most controversial since it included extensive compulsory purchase powers. It also included powers for subsequent disposal to developers, with discretion to offer concessionary terms where residential development was involved. There was a particular proviso that the Commission could ensure that owner-occupiers rather than developers secured these concessions.

Labour was therefore clearly following the parameters of its larger strategy in these proposals. It leaned very heavily in the direction of private landed interests and the mixed economy. It also clearly embodied the notion of promoting owner-occupation and the property-owning democracy. In a real sense therefore Labour showed that it too was pragmatic enough to learn from its political opponents. Unfortunately this did not make the Land Commission work any better. The main cause of the shortage of development land was not that landowners in general were wilfully holding it back (though this happened in some cases) but that development plans had not allocated enough. This was not helped by local authority mistrust of this new central agency. Until this problem was resolved (as Labour believed it could be by reforming local government and introducing a new planning system) it was difficult for the Land Commission to do very much. Moreover the immediate Conservative commitment to abolish it at the first possible opportunity did not make its work any easier. Events gave it even less time than the 1947 Act's Central Land Board to prove itself.

The Planning Advisory Group (PAG) Report 1965

The problems of the development plan system had in fact already been addressed by a group of officials, local authority and private practice representatives. This group had been established by the last Conservative Minister of Housing and Local Government, Keith Joseph. His Labour successor, Richard Crossman, an ebullient and intellectually heavyweight enthusiast for Wilson's modernizing vision, quickly recognized its members as potential heroes of the planned technological revolution and judged them 'stunningly able and successful' (Crossman, 1975, p. 184). The key item on the PAG agenda was the inflexibility of the system in meeting the needs for greatly increased development land allocations consequent on population and economic growth, especially for housing. It had also become increasingly clear that it was going to prove practically impossible to adapt the 1947 Act development plans to all the new ideas and proposals about urban and regional planning that were then emerging.

The PAG Report (PAG, 1965), *The Future of Development Plans*, made an important distinction between the policy or strategic questions, which would require ministerial approval, and tactical issues, which could be left for local decision. PAG envisaged several new kinds of plans. 'Urban plans' (for towns of over 50,000 population) and 'county plans' would deal with broad patterns of development and redevelopment and handle land use and transport in an integrated way. Each would identify 'action areas' requiring comprehensive planning action over ten or more years. In addition detailed 'local plans' could be prepared, most significantly for the action areas. Finally the Group entered fully into all the exciting proposals for national and regional economic planning that were just emerging from the DEA, and looked for strong linkages between local physical plans and these higher-level economic proposals.

Overall the Group was reinventing the planning process to fashion something that was more explicitly rational and responsive to change. It stressed the policy basis of plans and sought to avoid the excessive elaboration of physical and land use detail that dominated the blueprint approach of 1947 Act development plans. Some planners clearly regretted the retreat from precision, finding a voice in one of the more irritable (though eloquent) champions of the 1940s' planning revolution, Thomas Sharp (1966). Yet Sharp was by now well known as a quirky oppositionist and carried few with him (Stansfield, 1981). Most of the profession, who were rather younger than Sharp, were excited by the new proposals. The PAG proposals were amongst the most important conceptual developments of the 1960s. And, in sharp contrast to comparable earlier conceptual innovations, which had their origins in the ideas of key individuals of the planning movement, the PAG Report showed how new ideas were now generated entirely within official agencies and fed directly into policies. The distinction between ideas and policies was becoming more blurred.

Town and Country Planning Act 1968

The PAG proposals formed the basis of the 1968 Act, the first major reform of the planning elements of the 1947 Act system (Bor, 1974; Bruton and Nicholson,

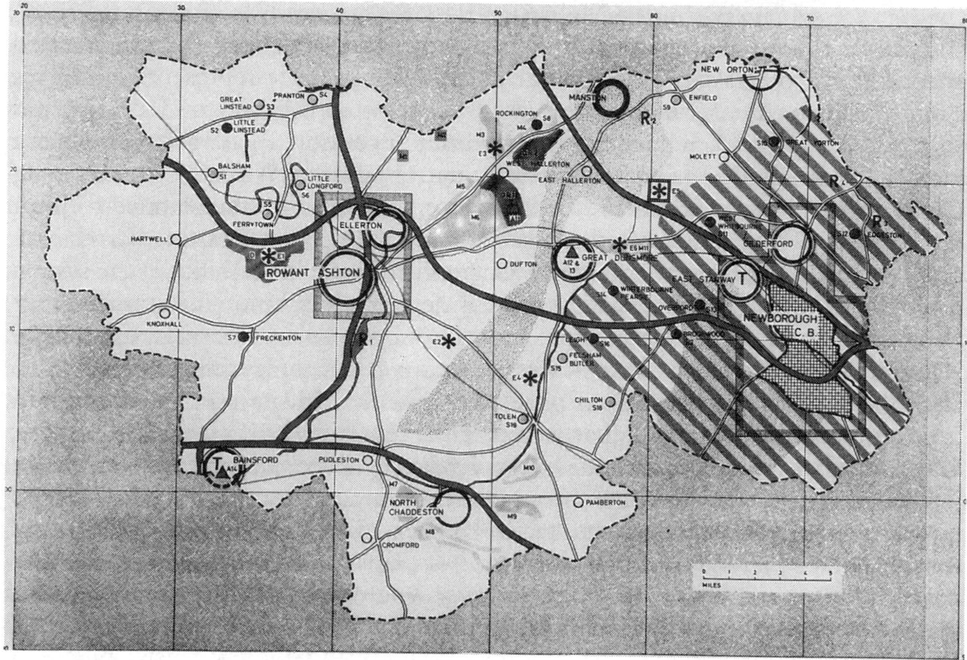

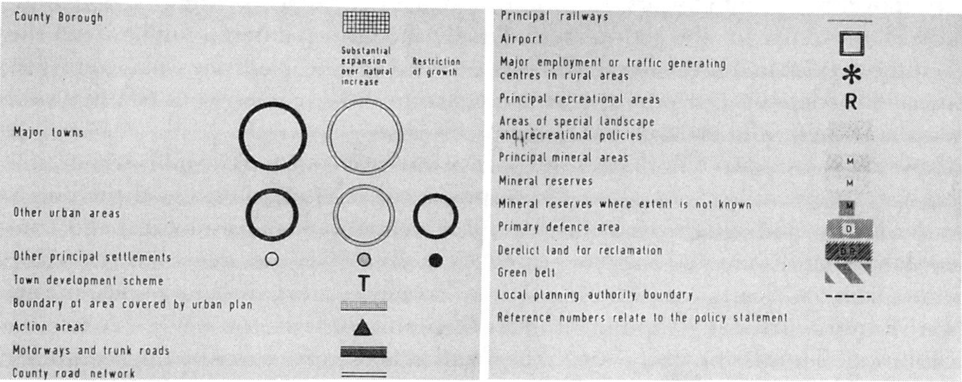

Figure 5.2 *The PAG Report of 1965 proposed radical changes to the planning process. In place of the blueprint plans of the 1947 Act (see Figure 5.1) there were to be rather generalized structure plans, intended to show only broad land use and transportation patterns and proposals for growth/restraint, etc. Where detailed plans were necessary, these would be elaborated in various kinds of local plans.*

1987). It was introduced by Anthony Greenwood, the second of Labour's Ministers of Housing and Local Government (and, incidentally, son of Arthur Greenwood, who had played an important role in advancing planning policies during the 1929–31 government). Greenwood was a rather quieter figure than Crossman, less given to independent thinking or impetuous action. Accordingly the new Act contained few surprises. It was an attempt to overcome the inflexibility and cumbersome procedures of the 1947 Act system, largely on lines spelled out by PAG. It also sought to provide more opportunities for participation by individual citizens in the planning process, an aim which did not sit easily with attempts to speed up decisions. Finally it was supposed to strengthen the ability of the planning system to stimulate the creation of a good environment, rather than merely stop what was bad.

There was little to object to in what were, according to the dominant currents of thinking at the time, such worthy intentions. Now that planning legislation had been divested of its controversial land values elements, little political heat was generated in the passage of the 1968 Act. The Conservatives would almost certainly have passed a very similar measure at that time. There were, however, some justifiable worries about adopting a new planning system without first completing the local government reforms that were then starting to be considered (and which we will discuss in a later section) (MHLG and WO, 1967). Already in 1966, Greenwood was proposing delaying full implementation of the Act, instead inviting groups of existing local authorities within sub-regional areas to begin working together voluntarily in ways that anticipated, without of course knowing, the post-reform boundaries. It was a messy solution, fiercely opposed by Crossman (1976, p. 199), at least. But it was, in effect, what happened and the sub-regional studies, especially the earlier ones begun in 1966, came to be seen as methodological dress rehearsals for the new planning process, to be considered more fully below.

Structure plans and local plans

True to the spirit of PAG, two new types of plans were introduced in the 1968 legislation – structure plans, which embodied strategic policies, and three varieties of more detailed local plans – for districts, action areas (replacing the CDAs of the 1947 Act) and subjects (such as recreation or landscaping). These plans would be imbued with the whole ethos of the planned mixed economy, offering positive guidance to private developers with the clear intention of marrying profitable development with good planning. The new Act also expedited planning procedures, chiefly by strengthening the position of planning inspectors. There was also provision for greater public participation, though the exact forms this might take were still unclear.

The Skeffington Report 1969

The 1947 Act system had not been encouraging to public involvement, even to the extent of denying information about planning proposals in their formative stages. By the late 1960s, however, there was a growing agreement of the need to change this, reinforced by much wider pressures calling for a dispersal of power within society (Morgan, 1992). The principle of participation was being endorsed both in planning and more widely (Hill, 1970). Student protests throughout the West in 1968 gave particular prominence to the idea. In the same year Tony Benn, Labour's Minister of Technology, put forward the idea of the participatory democracy within the vision of the new technologically based future, where citizens considered public affairs on television and expressed opinions by pressing

***Figure 5.3** In contrast to the 1947 system, which put the public into a passive role, citizen participation was to be integral to the new planning system proposed by PAG and implemented by the 1968 Act. The 1969 Skeffington Report spelled out the ways in which this was to be achieved.*

buttons in their own homes. It was against this general background that the Committee on Public Participation in Planning, chaired by Arthur Skeffington, a junior minister at the MHLG, was established. A more specific concern grew directly out of the 1968 Act system. Since the new local plans, uniquely did not need central approval, there was understandable MHLG concern that the planning process was properly conducted. In this light, involvement of the public at large could be a potential check on an approach that was too narrowly based.

Yet the Skeffington Report (Skeffington Committee, 1969), *People and Planning*, was a disappointing and cautious document. This was partly because the Committee's terms of reference had limited it to looking only at planning procedures, not at local affairs in general. The Report outlined the rudiments of a participation system for planning involving more information and involvement at the preparation stage. And it recognized the need to ensure that participation extended beyond the organized pressure groups. Yet it did not consider the fundamental question of how much this should involve a real shift of power to shape local planning proposals from its established guardians, the local politicians and planners. The failure to deal with this ensured that official public participation in planning has never fulfilled the promise of a real sense of involvement in local affairs. But it has certainly increased the scope for more organized interest groups to influence the planning process. A growing concern of such groups in the late 1960s was limiting the process of change. Here too there was important legislation.

The Civic Amenities Act 1967

This is an Act that superficially appears to fall outside the main story of the development of the planning system over these years. Very unusually, it was passed as a private member's bill, introduced by a former Conservative Minister of Housing and Local Government, Duncan Sandys (Punter, 1985). As we will see in the next chapter, his major achievement as Minister had been the encouragement of green belts. He had also played a key role in the establishment of the Civic Trust, whose influence was strongly apparent in this legislation. The Act's main concern, with the protection of the historic environment, also appears somewhat out of step with the dominant planning themes of growth and technologically based modernization.

Yet such distinctions were more apparent than real. Had Sandys not brought forward his bill, the Labour government would certainly have come up with its own proposals. In the event the 1967 Act quickly assumed major significance within the 1968 Act system (and was consolidated with it in 1971) because it addressed a question that was of real importance as the pace of urban change quickened: what was to be done to handle the pressures for change in urban areas which were valued in their unchanged state? Essentially the Act extended planning's concern with the historic environment from the level of individual buildings and monuments to wider areas. Specifically it introduced the conservation area, where a stricter regime of planning regulation could operate and measures for enhancement could be implemented. Authorities speedily took up this new instrument of policy and by mid-1974 there were over 3,000 conservation areas throughout Britain.

Planners in the 1960s

By the end of the 1960s therefore there had been a tremendous change in the powers and duties of town planners. Even more significant was the change in the actual practice and methodology of planning, particularly urban planning. As Hall remarked in an instructive piece of hyperbole, 'a planner trained in 1930 would still have felt reasonably at home in the planning office of 1960. The same cannot be said for the planner trained in 1960 who finds himself in the planning office in 1970' (Hall, 1973, p. 49).

Already by 1965 there were sufficient signs of the new directions to generate a major professional crisis of identity for the town planner, manifest in serious tensions within the Town Planning Institute (Cherry, 1974b). Increasingly, as the extent of planning expertise widened, the professional ideal of the generalist town planner, synthesizing a range of specialist skills, was becoming untenable. The new directions of the 1960s relied more on the collaboration of specialists within a team approach.

Yet, the reformers of the Town Planning Institute, in common with Hall, overestimated the spread of these new approaches during the 1960s. In fact their full impact did not affect all planners until after 1972, when local government reform created a new institutional framework for planning. But the shift certainly originated in the 1960s as a direct consequence of two major changes that gave planners higher status within urban local authorities and brought a spread of more sophisticated planning methodologies.

New city planning departments

In the 1950s the planning profession had been dominated by county planning officers, a consequence of the enhanced role of the counties under the 1947 Act. City planners had been the poor relations. Apart from the LCC and occasional exceptions like Coventry, planners had done little more than dent the traditional dominance of the city engineers and other departments in the big provincial cities before 1960. Now, led by Newcastle, Liverpool and Leicester in the first years of the 1960s, this began to change decisively (Cullingworth, 1970, pp. 123–42). New full-status municipal planning departments were created. In Newcastle, an extraordinarily charismatic and visionary Labour council leader, T. Dan Smith, who himself played an important, if ultimately corrupt, role in the 1960s' planning revolution (Smith, 1970), brought in Wilfred Burns from Coventry to plan the city's renewal (Burns, 1967). Liverpool and Leicester soon followed, appointing Walter Bor and Konrad Smigielski. A new pattern had emerged which gave planners a pivotal role in co-ordinating municipal intervention in rapidly changing cities.

New planning methodologies

These new planning departments took many important methodological initiatives. Planning research and intelligence were improved, increasingly using the computers which the big local authorities were acquiring to handle their payrolls. Yet it was a slightly different group of planners who made the most important methodological advances during this period. From 1966 the promise of the PAG Report was elaborated in the series of centrally encouraged sub-regional planning studies that we referred to earlier (Glasson, 1978). Leicester and Leicestershire; Coventry, Solihull and Warwickshire; Nottinghamshire and Derbyshire; and South Hampshire formed the first group. More followed after the 1968 Act became law, though by this stage their remit was more specific: to prepare actual structure plans. The studies pioneered the structure planning process, using sophisticated modelling techniques to generate clear policy choices (Cross and Bristow, 1983). Their concerns were with the general spatial patterns of land use, transport and development, a clear break with the detail of the blueprint 1947 Act development plans.

Systems planning

As interest in the new approaches spread in the early 1970s, the Geddes-inspired model of survey–analysis–plan that had underpinned the 1947 Act development plans was supplanted by a goals- and objectives-led approach (Marshall and Masser, 1982). Instead of beginning with a survey, the new process began with broad policy goals and more specific objectives, with survey and analytical work, often involving computer modelling, used to generate choices that could be evaluated against the original intentions. It all mirrored the vogue for indicative planning in national economic management, which established strategic targets to be achieved by a combinations of public and private actions.

PAG had marked the beginning of this retreat from the Geddes formula. The increasingly sophisticated elaborations of it now derived their growing impetus from advanced US space and defence systems planning. The systems approach, which in some respects revived the notion of planning as a synthesizing discipline, was then translated into a physical planning context by McLoughlin (1969), Chadwick (1971) and others. Here truly was a rational and scientific approach to planning that was entirely appropriate to the new and exciting technological age. Or so it seemed. What was perhaps more significant in the longer term was that, as we have noted for the PAG report, most conceptual innovation in planning was now fully incorporated within and driven by the policy process. Where new ideas derived from outside, they had precious few connections with the planning movement in the traditional sense and they were soon locked into policy concerns. New more radical ideas about the city and society were beginning to emerge, as we will see in Chapter 7. But they grew mainly outside the planning movement and their advocates often perceived planners as part of the problem, rather being any longer a path to real reform.

More Conservative adjustments 1970–4

The Department of the Environment

Meanwhile the inter-party agreement on most physical planning issues remained very strong. The change of government in 1970 produced some immediate shifts including, as in 1951, a change in the central department (McKay and Cox, 1979). The new Prime Minister, Edward Heath, created a new super-ministry, the Department of the Environment (DOE), with overlordship over housing, construction, local government, planning, water supply and transport. In many respects it was a continuation of the process begun by Wilson in 1969 when a new super-ministry called the Department of Local Government and Regional Planning had been created with federal powers over MHLG and the Ministry of Transport. The new DOE brought a tighter structure, abolishing the older ministries and establishing a single concentration of most of the key ministries related to land use planning (while still excluding industrial development control). In the circumstances there were no worries of the kind voiced in 1951. It was, after all, a strengthening of the central apparatus.

The end of the Land Commission

By contrast, the Land Commission Dissolution Act 1971, whose objects were self-explanatory, was a clear break with Labour's land value policies. Even in this, though, there was a stronger element of agreement than at first sight appeared. Labour had by this time tacitly accepted that the Commission had failed, for the reasons discussed earlier, and only went through the motions of opposition. Moreover the Conservatives were not going back to 1951, they were picking up things from where they had left them in 1964. Accordingly they replaced the 40 per cent betterment levy with a 32 per cent capital gains tax on land value increases. There was an important difference of principle in Labour's desire to tax betterment in a clearly different way to other forms of capital. But taken overall the new proposals were evidence of a really rather modest distinction in the attitudes of the different parties to the balance of private and public interests within the planning system.

The Town and Country Planning Acts 1971 and 1972

Nor was there anything very dramatically different in the two planning Acts of these years. The 1971 Act was a consolidatory measure, introducing no new powers, though significant amendments were made the following year. In many ways these were a response to the experience of the Greater London Development Plan (GLDP) Inquiry in 1970–2. The GLDP was not actually a structure plan (though it was later deemed as such) but it was seen as an important rehearsal of the process. Yet the length of inquiry and the 28,000 objections to the published plan stunned the DOE into seeking some short cuts (Ardill, 1974, pp. 69–73). The most important introduced the 'examination in public' procedure, expediting the consideration of structure plans. It allowed the DOE to shape the agenda more closely than was possible in a traditional public inquiry. There have been many fears that this attempt to expedite the structure planning process has brought a denial of legitimate objections and a weakening of public participation.

The Land Compensation Act 1973

Although the general issue of compensation for loss of development rights, so prominent in the 1940s, had now been settled once and for all, the question of compensation for environmental deterioration produced by planning or related actions was of growing importance. The massive scale of urban redevelopment, which we will examine in greater detail in the next chapter, often created serious environmental problems for neighbouring areas. Despite important moves during the 1960s to break down the traditional separation of transport and town planning, roads were much the most serious problem (Starkie, 1982). From about 1965 national trunk motorway routes were increasingly being constructed within major built-up city areas, for example in London, Leeds, Glasgow and Birmingham. In turn they were complemented by new, locally initiated (but largely centrally funded) urban motorways. These new road systems within built-up areas involved major demolition of buildings and disturbance to adjoining areas when they were being built and serious problems of noise and fumes when they were in use. Not surprisingly therefore the 1970s saw a growth of anti-urban roads protest campaigns. A notable and widely publicized early instance came in 1970 when the new elevated section of Westway in London was opened. Houses whose upper storeys were only a few metres from the viaduct carried a huge and direct message: 'GET US OUT OF THIS HELL' (Aldous, 1972). In addition the 1968 Act planning system, by creating the possibility of rather non-specific identifications of action areas of major change in structure plans, was felt to run the risk of intensifying planning 'blight' (that is uncertainty). The 1973 Act broke new ground in admitting rights to compensation in the case of environmental deterioration produced by roads, airports, etc., and extended the compensation provisions on blight.

Local government reform 1972–4

Much the most fundamental change under Heath's premiership was the reform of local government. Virtually every government since 1945 had toyed with the idea of reform. From the planning point of view the separation of the urban county boroughs, usually rather tightly developed within their boundaries by the 1960s, and the surrounding counties was a general problem. In the main conurbations, particularly London, the absence of any overall local government body for metropolitan planning was a serious problem. Yet reform was a huge and politically rather thankless task. The Conservatives had finally reordered London's government in 1963, creating a two-tier system with a new Greater London Council and thirty-two London boroughs (Herbert Commission, 1960). But it was not until 1966 that the Royal Commission on Local Government (in England) was announced under the chairmanship of Lord Redcliffe-Maud. A parallel body (the Wheatley Commission) fulfilled a similar function in Scotland. Proposals for Wales were already being considered.

The Redcliffe-Maud Report (Redcliffe-Maud Commission, 1969a, 1969b) recommended a two-tier London-type metropolitan system for the three main provincial conurbations around Birmingham, Manchester and Liverpool, but otherwise a unitary, one-tier system. Labour had broadly accepted this and was preparing legislation when it lost office. But the Conservatives, while broadly implementing the Redcliffe-Maud two-tier concept for an extended list of metropolitan areas, also preferred a solution based on a rather different two-tier split elsewhere. This was also the basis for Wales with a rather more varied, but overwhelmingly two-tier, pattern for Scotland (Wheatley Commission, 1969). Moreover, in England particularly, they frequently defined the boundaries of the lower-tier authorities in ways that perpetuated many earlier problems. Overall there were many doubts from the outset about how well the new system would meet the requirements of planning.

Physical planning under the new authorities

The impact of the reforms was particularly marked in planning, because the strategic/tactical divide of the PAG Report was now split between the new counties and districts. This severely compromised the integrity of the whole 1968 Act system which had rested on the assumption of unitary authorities. It made it impractical to assume that local plans, produced by the districts, could simply be the tactical elaboration of the counties' strategic structure plans, since there was greater likelihood of real conflicts between plans produced by two different authorities. Development control was to be largely a district matter unless wider county issues were involved, though it contained similar potential for inter-authority conflict.

Planners in the new authorities

Despite these serious problems, local government reform served to extend the 1960s' mood of optimism amongst planners. From their point of view it had one wonderful consequence. As a result of the reforms there were now 504 local planning authorities throughout Britain, 455 in England and Wales compared to just 145 after the 1947 Act (Cullingworth, 1976). Quite simply it created many more planning jobs. Demand greatly outstripped the immediate supply, but major expansions in planning education accelerated this during the 1970s. There was, it is true, some uncertainty as to the exact policy roles that planners were fulfilling in the new authorities. The old certainties of a heavily land-use-based approach, though remaining reasonably intact in staple activities like development control, were now becoming blurred in other aspects of planners' work, especially in the main urban areas. The new 1960s' style of strategic policy planning, although it had first appeared as an underpinning to PAG, now seemed to pervade all aspects of the policy process in the new local authorities (Maud Report, 1967; Bains Report, 1972). It was no longer the special prerogative of the planner. And in the initial inter-departmental jockeying for position that accompanied local government reorganization, the relative importance of the planners varied. But everywhere there were more of them than there had been in the early 1950s.

OVERVIEW

The most important point to emphasize is how much the changes in the physical and land use planning system reflected the larger shift away from the highly state-directed society of the 1940s. It had become something that, though still centralized by international standards, exactly mirrored the mixed economy, based on public-sector–private-sector interaction. There was an associated shift from the promotion of collective welfare goals in the 1940s to facilitating the pursuit of private affuence. In this chapter we have noted its impact particularly in the field of housing, with associated effects for planning. It will feature strongly in the more specific discussions of the next chapter.

By changing with the prevailing mood of post-war Britain, physical planning became part of the 'post-war consensus'. It was never totally immune to changes in government, but the elements of bipartisan continuity were always stronger. Thus 1960 appears as a more significant turning point than 1964 or 1970, reflecting a general recognition of social and economic changes rather than electoral shifts. As we have seen, physical planning acquired a high political profile in the 1960s. Yet it never regained the mass commitment that had been glimpsed in the war years. This had laid the essential political basis for the translation of planning ideas into effective policies. By the 1960s, however, the whole conceptual dimension, including the generation of planning ideas, was firmly incorporated within the planning policy process. We can trace more of the implications of this in the next chapter.

6

Adjustments and New Agendas: II. Strategic Policies 1952–74

Introduction

By the early 1950s four major strategic physical planning policies of decentralization, redevelopment, containment and regional balance had been adopted to deal with the problems of urban Britain. The thinking on which they were based had grown out of the experiences of the inter-war years, as we saw in Chapter 3. And they had become government-sanctioned policies largely because of the changed political priorities of the later 1930s and especially the 1940s, as we saw in Chapter 4. They remained centrally important throughout the following quarter century, implemented through the evolving planning system outlined in the previous two chapters. Yet, as we will see, their importance relative to each other varied significantly over time, reflecting many of the wider shifts we identified in the previous chapter.

Thus the strong 1940s' commitment to decentralization and regional balance shifted in the 1950s to a stronger emphasis on containment and redevelopment. The new priorities of the 1960s brought a continued and strengthening commitment to redevelopment, a re-emphasis of regional balance and decentralization and a downgrading of the role of containment. And within these larger policies, there was also a great deal of more detailed policy innovation. Thus new ideas were being applied to the planning of commercial and public-sector redevelopment areas, new towns and other large green-field planned areas, green belts and the urban fringe, traffic management and elsewhere. Again the pace of policy innovation reflects the larger picture, with the 1960s standing out as a period of particular importance. Overall, then, we have a picture that complements that of the previous chapter, of shifting emphases against a background of strong continuity. We begin with the policy area that wrought the most dramatic changes on the towns and cities of Britain over these years, planned redevelopment (Ravetz, 1980; Esher, 1983).

PLANNING AND REDEVELOPMENT

The central areas

Redevelopments pre-1954

In the previous chapter we saw how the reforms to the 1947 Act and the end of building licences in 1953–4 brought increasing development activity in central areas, especially London. Before that time there had been some rebuilding, especially in the blitzed cities, many of which had lost large parts of their central areas. The demand for shops was very high in large cities like Plymouth or Southampton so that it was easy to pre-let and thereby get the necessary building licence (Marriott, 1969). Moreover few individual retailers or local small-scale developers were willing to tackle the scale of development that was needed. It was therefore in these kind of locations that the developers who were to dominate commercial redevelopment over the following decades cut their teeth, especially Ravenseft Properties.

Early public–private partnership for shopping redevelopment

Although few people realized it at the time, such schemes constituted an important organizational innovation in planned redevelopment. In contrast to pre-1939 development practice, where public–private partnerships were virtually unknown, the developers were now working closely with local authorities. This was because local authorities in the larger towns and cities were either already the landowners of redevelopment areas, or in a position soon to become so, as well as being planning authorities. The compulsory purchase powers of the 1944 and 1947 Acts had eased land assembly problems in designated redevelopment or comprehensive development areas (Hasegawa, 1992). (By the 1950s the CDA mechanism had largely superseded the 1944 Act's redevelopment areas.) Typically the local authority would remain ground landlord, with the developers as lessees and middlemen, undertaking the development and negotiating rentals with the retailers.

In some cases local authorities themselves might undertake initial development to prime the pump. This happened in Coventry, for example, where there was some private developer reluctance to fall in with the rather radical planned proposals (Gregory, 1973; Mason and Tiratsoo, 1990). It was here that the first British traffic-free central shopping precinct development appeared in the mid-1950s. The scheme gave modified effect to Gibson's famous wartime plan, providing a British practical example to rival that of the earlier Lijnbaan pedestrianized area in Rotterdam. And after initial developer and retailer caution, this notion of pedestrianized precincts was soon accepted so that private developers increasingly took on such schemes. This growing developer willingness to produce more innovative planned designs mirrored the narrowing opportunities for purely municipal developments, where planning priorities could be paramount. The Conservative government became increasingly reluctant to allow completely municipal central area developments during the 1950s.

Speculative development and partnership

The rather developer-friendly system created in 1954 encouraged the emergence of a more speculative approach, especially for office buildings (Marriott, 1969). After fifteen years which had seen very little development of shops, offices or other central commercial buildings, there was a strong unfulfilled demand, particularly as consumption of goods and services soared following the end of rationing in 1954. The multiple retailing chains whose inter-war advance had been checked by austerity resumed their growth. Similarly there was an expansion of financial services and demand for new headquarters buildings for major companies which stimulated office development in London. Meanwhile the range of potential investment funds was also widened as pension funds, which were beginning to become much more significant, were allowed to invest in property from 1955 (Whitehouse, 1964). Overall therefore the element of financial risk for speculative developers in these early years was rather low. And although there were a growing number of purely speculative schemes, usually individual office blocks, the larger schemes retained a high degree of partnership (and risk-sharing) with local authorities.

Examples of office developer/local authority partnerships

(i) Planning-led: The nature of the partnership could vary considerably. Two notable London office schemes in the late 1950s/early 1960s indicate something of the range (Marriott, 1969). A more planning-led approach was apparent in the development of the six office towers lining London Wall, in the southern part of the Barbican CDA in the City of London, based on municipal ownership of a huge blitzed area. There had been much preliminary planning and design work, part of a hugely ambitious mixed-use redevelopment scheme of over sixty acres on the fringe of the existing commercial area. In particular the traffic planning was innovative, with a pedestrian deck above the road level, reminiscent of some of Le Corbusier's inter-war schemes. From 1957 the rights to develop individual blocks were then sold by tender to developers. Because the location was then seen as rather risky, the bids were not high but in the event the schemes were hugely profitable.

(ii) Developer-led: Further west the creation of the Euston Centre over the late 1950s and 1960s indicates how the partnership could work when local authorities were not landowners. Having already secured outline permission for office development on part of the site for a client, the West End estate agent Joe Levy and his partner, Robert Clark, conceived a plan to redevelop a thirteen-acre area adjoining a major new road scheme. The LCC planners and valuers had realized that to build the road would require revoking the permission Levy already held, a hugely expensive operation. Instead therefore they co-operated with him in his surreptitious assembly of the whole site, involving some 315 individual purchases, mainly in 1956–60. He meanwhile co-operated to the extent of giving them parts of the site needed for the road (worth perhaps £2 million) and also land to rehouse displaced population with established tenancy rights. This was done on the basis that the planners would allow more intensive development on the remaining site to compensate for the area lost. This was the first really big example of another important dimension of public–private interaction – planning gain – though the term was not yet used. Smaller examples were present in many other office schemes of the time (Elkin, 1974).

New planning concerns in the 1960s

As we noted in the previous chapter, the quickening pace of this redevelopment had rather overtaken the planning system by 1960. We have seen how important,

Figure 6.1 *The Euston Centre (background, right) was one of the first major examples of 'planning gain'. The developers, Robert Clark (left) and Joe Levy, surreptitiously acquired a large rundown area for commercial development. Without disclosing what was happening, LCC planners connived in the proposals and the developers gave them the land needed for the widening of Euston Road, which otherwise would have been prohibitively expensive.*

though often incremental, organizational innovations in the 1950s drew planners and developers into closer union. And we saw in the last chapter how modifications of the planning system under the 1959 and 1963 Acts were largely directed against specific unacceptable practices of commercial developers. But neither of these trends constituted effective policy-making for planned central area redevelopment. The sheer pace of commercial development pressure, especially in areas where there was little bomb damage (and normally, therefore, rather less muncipally owned land) threatened to swamp aspirations for a planned approach. The danger was that without a more comprehensive approach and the funds to implement it, planners would become too dependent on the work of developers to be able to control or shape their operations adequately.

By 1962 the planners were trying to regain the initiative. MHLG with the Ministry of Transport was officially encouraging local authorities large and small to renew their centres in partnership with private developers (MHLG and MT, 1962). It recognized the limitations of the 1947 Act system in enabling the development of a coherent planning approach. The statutory town maps and CDA maps were too limited to allow an approach that was sensitive enough, particularly for districts and non-county boroughs which were not actually planning authorities. Accordingly they were advised to develop non-statutory central area plans that would allow a more detailed approach to evolve. The main elements were familiar enough, based on the experiences of innovative blitzed cities like Coventry (and the new town centres).

The problem of the motor car

A key factor that underlined the dangers of a less-planned approach was the increasing problem of the motor vehicle. Following the end of rationing the number of cars particularly began to increase rapidly, from 3.5 million in 1955 to 5.5 million in 1960 (Plowden, 1973). This trend, allied with all the other forecasts for population and economic growth, pointed to huge future increases. And the problem was acquiring great public and political saliency. More than anything else, it brought home Galbraith's message about the pursuit of private affluence creating public problems, in this case traffic congestion and deterioration of the urban environment. This submission to the House of Commons All-Party Roads Study Group in May 1960 by one of the rising stars of 1960s' planning, Konrad Smigielski, typified contemporary fears:

> London is faced with a disaster on a far bigger scale than the bombing during the last war. The threatening heavy black cloud hanging over the metropolis has nothing to do with the danger of nuclear war. It is the result of our own faults, of our inability to think ahead and to plan ahead.
>
> There is already sufficient evidence that the economic and physical disintegration of London has set in under the onslaught of motorisation. Of course the results will not be noticeable overnight, but they will be pretty soon, certainly in our lifetime, and possibly within the next ten or fifteen years.
>
> (Smigielski, 1960, p. 208)

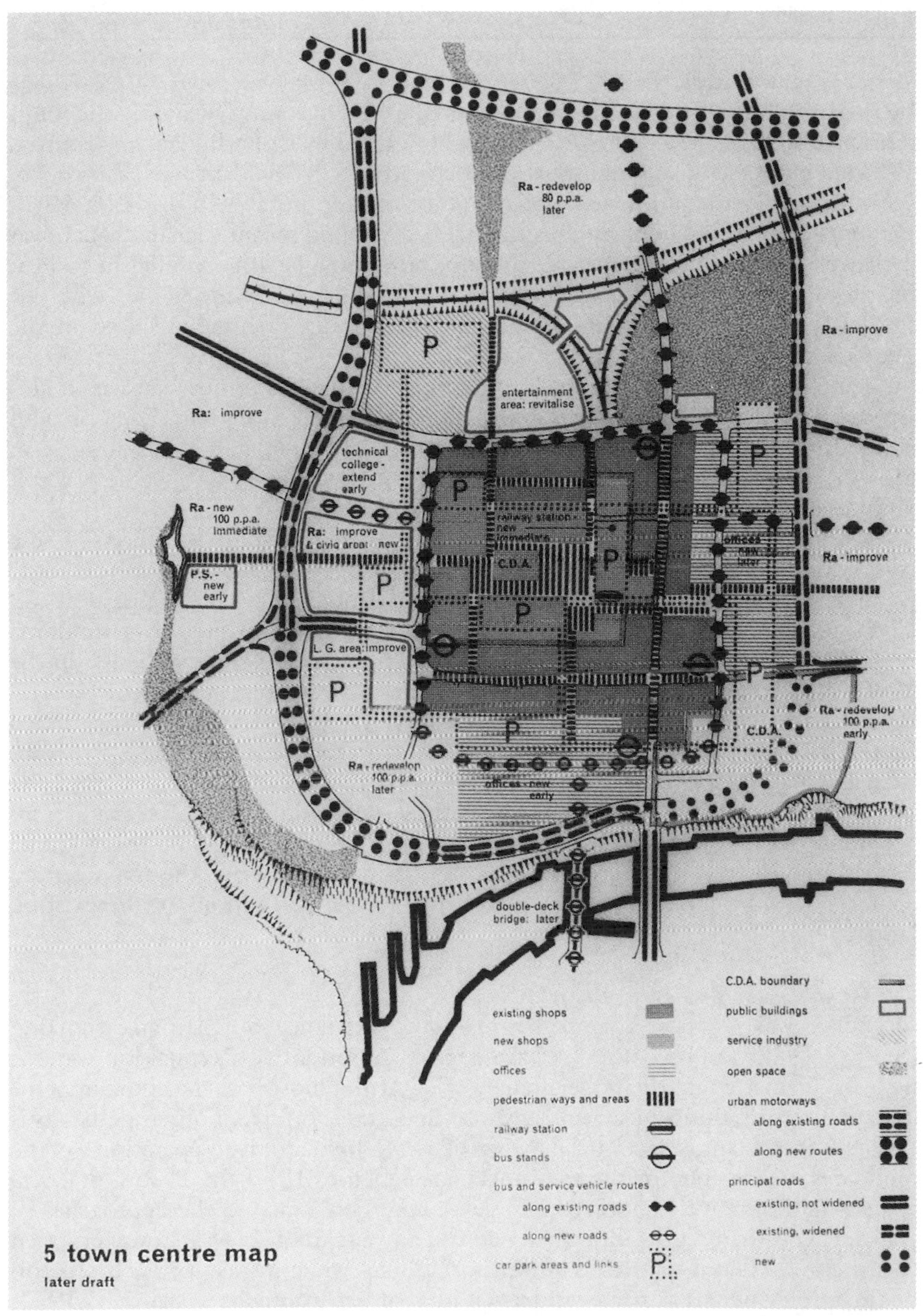

Figure 6.2 By the early 1960s central government was giving strong encouragement to central area renewal, using CDA powers in partnership with private developers. The approach highlighted the technical limitations of the statutory CDA plans, however, and a more sophisticated non-statutory planning process was being encouraged, as shown in this 1962 publication by the MHLG and the Ministry of Transport.

Traffic in Towns Report 1963

In 1961 the Ministry of Transport appointed a study group to examine the long-term problems of traffic in towns. The group was led by Colin Buchanan (Bruton, 1981), a particularly able transport planner and ex-Civil Servant who had shown great interest in the problems of planning for the car, notably in his book *Mixed Blessing* (1957). This book and his general background meant that Buchanan was widely respected as an independent figure, bringing a healthy scepticism to bear on the Ministry's enthusiasms for roads. His report (MT, 1963) vies with the PAG Report as the most important official planning document issued in the 1960s. Despite Buchanan's earlier scepticism, it was essentially a pro-car report, in the sense that it did not seriously consider options that did not use the car. It also left the reader in no doubt that cars would not go away (though in common with many forecasts of the period, it significantly overestimated their long-term growth). But it pointed out that the capacity of towns to absorb traffic was not infinite, however much redevelopment was undertaken.

At a more practical level, the report also indicated a range of options. These could accommodate different traffic capacities in ways that would, in the study group's view, be reasonably tolerable in avoiding the twin evils of congestion and environmental deterioration. In effect Buchanan was returning to a modified version of the precinct principle which Alker Tripp (1942) had elaborated in the 1940s, providing for the designation of 'environmental areas' from which extraneous traffic would be excluded. Yet, whereas Tripp had envisaged specialized single land use precincts, Buchanan's environmental areas were mixed use. Pedestrian circulation systems were to be developed that were segregated from the major traffic routes. The exact form of these environmental areas would vary according to the degree of restructuring and the required traffic capacity. The most drastic involved the creation of a new living deck above the main central routeways. More modest variants merely involved modifications of the existing fabric.

The impact of the Buchanan Report

Overall it was a report that had a profound influence on planning thinking (Starkie, 1982) (and we note, yet again, how important conceptual innovations were occurring within the policy framework). At the most basic level the impetus it gave towards thinking about land use and transport planning together was crucially important, especially in view of the administrative separation of the ministries responsible for the two policy areas before 1970. In a more practical sense it authoritatively endorsed and gave more coherence to the approaches of vehicle–pedestrian segregation that were already apparent. It was, however, open to criticism in that it tended to ignore accessibility when it was not car based and because it exhibited a naive understanding of environment (Plowden, 1972; Ravetz, 1980). In particular it underestimated the intrusive effects of noise and fumes of traffic on the proposed environmental areas.

Planned shopping redevelopments in the 1960s and 1970s

Yet by adding planning arguments to the existing commercial pressures for change, Buchanan further encouraged the central redevelopment boom. Thus as

Figure 6.3 The Buchanan Report of 1963 gave further encouragement to the integration of planning and transport concerns. Although the report posed a series of options, it was largely understood and promoted as being pro car. The illustration shows how the needs of drivers and pedestrians are supposedly reconciled by recourse to multi-level segregation in a redeveloped modernist cityscape. Yet the environmental impacts of traffic noise and fumes seem to have been ignored.

more local authorities initiated inner roads schemes, they typically found they were automatically creating important redevelopment opportunities. The pace of activity quickened markedly in the early 1960s (Marriott, 1969). Thus the numbers of CDAs being considered by MHLG rose from fifteen in 1959 to seventy in 1963 (though not all of these were central schemes). The pace of shopping redevelopment quickened and traffic-free precinct developments, now endorsed by both Buchanan and the main retailers, became the norm. In 1964 the first modern British enclosed shopping precinct was opened at the Bull Ring in Birmingham, developed by Laing on municipally owned land in conjunction with the city's major inner ring-road project begun a few years earlier (Sutcliffe and Smith, 1974). Yet this and another early enclosed centre at the Elephant and Castle in London (opened in 1965) were not very successful, largely because of overambitious designs and poor locations.

The loss of confidence was only temporary, however, with growing awareness of successful examples in North America (Darlow, 1972). An important factor was the attitude of the major retailers, whose market position was strengthened further following the abolition of resale price maintenance in 1964, allowing them to undercut smaller independent shops (Wright, 1979, p. 80). Such retailing giants were becoming more interested in the new centres and their position as 'anchor' tenants was a key factor in schemes that were successful in retailing terms. Thus by 1966–7, Arndale, one of the most successful of the 1960s' shopping developers, was engaged on a large number of enclosed schemes, most notably the St Sepulchre Gate scheme at Doncaster (390,000 square feet) (APT, c.1966). Within a few years it was actively developing much bigger schemes in concert with local authorities, especially in Luton (755,000 square feet, opened in 1972) and Manchester (1,200,000 square feet, opened in 1976). Cities like Leeds, Nottingham and Newcastle also had similar large schemes in hand by the late 1960s and 1970s, with a variety of developers.

Planned office redevelopments in the 1960s and 1970s

By contrast the amount of office redevelopment underwent rather more fluctuation (Cowan *et al.*, 1969) and (Moor, 1979). After the boom of the late 1950s and early 1960s, a short-term oversupply and the impact of the 1964 Brown Ban and subsequent 1965 Act (discussed in the previous chapter) brought a comparative lull in building (if not profits) in London and, as the ban was extended in 1965–6, the midlands. Activity picked up again in the early 1970s as ODPs (and interest rates) were relaxed. Thus the five years between 1967 and 1972 saw a 21 per cent increase in office space in Britain, with very high demand in London especially encouraging yet more schemes (Jenkins, 1975). There had also been a significant growth of office building in the provincial cities during the middle and later 1960s. In part this reflected the impact of regional policies intended to encourage office dispersal from London, especially of government

Figure 6.4 *The Birmingham Bull Ring, developed by the Laing Development Company in conjunction with the City Council, was Britain's first enclosed shopping precinct, opened in 1964. Notice the close integration of the city's new inner ring-road with the proposals.*

offices. Local desires to encourage a new service employment base in the older industrial cities were also important. City councils like Newcastle typically had to act as developers themselves in the early 1960s, to replace obsolete buildings and create new space for dispersed government offices (NPD, 1973). Private developer interest grew during the decade, however, and partnership schemes were typical by the early 1970s.

Changing nature of the public–private partnership

By the late 1960s and 1970s the actual form of these partnerships had evolved still further, blurring even more the distinction between public and private actions. On the one hand the requirement of developers to disclose their planning intentions for sites to property owners, introduced in 1959, together with the increasingly public nature of the whole planning process in the 1960s, had limited the possibilities for large-scale private land assembly on the Euston Centre pattern. Indeed Joe Levy found this when he tried to repeat the formula on a neighbouring site at Tolmers Square in the early 1970s (Wates, 1976). As one developer commented in 1973, 'Too many schemes have been fouled up in the past when the public hears about them' (Ambrose and Colenutt, 1975, p. 65).

On the other hand the sheer pace of redevelopment had triggered tremendous activity in central area land markets that meant that a private developer might hold a quite large individual parcel of land. While these would not in themselves make up a completely viable redevelopment area, they would allow developers to approach local authorities from a position of strength and with a definite proposal in mind. The local authority could then use its powers to assemble the rest of the site to allow a jointly planned scheme to proceed. The Doncaster Arndale, one of the earliest really successful covered shopping precincts, was a very early example of this kind of approach (APT, c.1966; Marriott, 1969). The logic of this, then, increasingly drew local authorities and developers together into joint commercial development ventures, whereby they shared the equity of the final scheme. Coventry and Newcastle had increasingly struck such deals for both shopping and office developments by the early 1970s (Gregory, 1973; Galley, 1973).

Overall assessment

The period was a very important one for the renewal of British central areas, when planning and commercial pressures came together as never before. Although there were obvious planning and design innovations, most notably precinct developments and the universal reliance on modernist designs, it was the organizational innovation of public–private partnership that was of more fundamental significance. This was one of the most important symptoms of how the planning system was changing to reflect the new mixed-economy consensus. And it was something that was fully endorsed by Labour as well as Conservative politicians at local and central levels. Thus developers like Sam Chippendale of Arndale were adept at gaining the confidence of Labour councils, especially in northern towns and cities (APT, c.1966; Marriott, 1969). Another property developer, Hammerson, employed Frank Price, a Birmingham Labour leader with much experience of negotiating partnership deals, for similar reasons.

In turn Labour ministers were happy to be seen encouraging this rebuilding of the old industrial areas. Such ventures were an important symbolic and material expression of their wider programme, creating ultra-modern city centres that extended the technologically advanced affluent society to the heartlands of Labour support. Indeed both they and the Conservatives were anxious to extend the idea of close co-operation and partnership into other spheres of planning activity, especially the implementation of the housing programme. Richard Crossman was particularly keen on this idea, even to the extent of considering recruiting as his 'friend', Harry Hyams, perhaps the most notorious of the 1960s' developers, during the autumn of 1964 (Crossman, 1975, p. 48). The idea was pursued more formally by the Heath government in its 1971–2 Working Group on Local Authority/Private Enterprise Partnership Schemes (DOE, 1972). In fact though, a somewhat different form of private–public relationships had evolved in the other arena of planned redevelopment, the slum areas of the inner city.

The inner city

Resuming clearance

In Chapter 3 we noted the origins of mass slum clearance and redevelopment as political, social, economic and design pressures coalesced behind the notion of a clean sweep approach during the 1930s. As we saw then, a great deal of work had been undertaken in many cities to define both the nature of the problem and the extent and location of their slums. War had worsened the situation, yet it was not until the Housing Act 1956 that slum clearance and redevelopment began again to become priorities. Despite the hiatus, inter-war experience remained important, bequeathing a perception of inner-city problems that was overwhelmingly dominated by the need to replace slum housing and modernize the city (Ravetz, 1980). The war years had emphasized the case for comprehensive planning, involving the redevelopment not just of housing, but all other land uses as well, including industry. In contrast to the extensive use of the CDA approach in the city centres, however, only a few authorities applied this in the inner city (Burns, 1963). It was more common simply to use Housing Act powers. The LCC and Birmingham had used the 1944 and 1947 Planning Acts to initiate comprehensive redevelopment proposals for inner-city areas. But such initiatives were very unusual and even here housing priorities dominated the redevelopment agenda.

The scale of the slum problem

Given the extent of housing obsolescence, such dominance is easy to understand. The problem that had been identified in the inter-war years had barely been touched by the truncated 'Great Crusade' of the 1930s, discussed in Chapter 3. Moreover housing obsolescence had increased in extent during the 1940s and 1950s because of bomb damage and an absence of investment to maintain or improve older housing. In addition the rising expectations engendered by the 1930s' housing boom, the war years and the new emergent affluent society of the 1950s had a very marked effect. Yet it took some time for governments to recognize all this. The initial assessment of unfitness in 1955 relied on local surveys, as had happened in the inter-war period. As then, many (though fewer) authorities found it more convenient simply to find unfitness only to the extent that they could deal with it in agreed clearance programmes (Gibson and Langstaff, 1982). Accordingly the figure it produced – 853,000 unfit dwellings (in England and Wales) – was a gross underestimate. By the 1960s, when there was a greater willingness to face reality, the same figure actually rose to 1.8 million based on a nationally organized survey in 1967. This was despite the demolition of 644,000 dwellings between 1955 and 1966. Accordingly the rate of clearance remained at a high level, though it began to tail off from the late 1960s.

Physical and social modernization

In all approximately 1.165 million dwellings were demolished between 1955 and 1974. This involved the moving of some 3.1 million people and profound disruption and dispersal of established working-class communities (Young and Willmott, 1957). Yet throughout much of this period these devastating changes, which destroyed much of the social and cultural fabric of life for the poorer working class, were accepted without any serious question (MHLG, 1970a, 1970b). The slum-dwellers themselves offered little organized opposition. For those shaping policies, including planners, the social effects were seen as inevitable, even desirable. They were part of a wider process of social and physical modernization that echoed the imperatives of the affluent society and its characteristic optimism about material progress. Thus Wilfred Burns, Newcastle's first City Planning Officer and one of the most influential planners of the period, enthused about the beneficial effects of destroying established communities:

> this is a good thing when we are dealing with people who have no initiative or civic pride. The task, surely, is to break up such groupings even though the people seem to be satisfied with their miserable environment and seem to enjoy an extrovert social life in their own locality.
>
> (Burns, 1963, pp. 94–5)

The clearance process

Clearance typically began with the declaration of unfitness and compulsory purchase either of an individual house or, more often, a wider clearance area (English, Madigan and Norman, 1976). This part of the process was controlled by the public health inspectors, though the town planners and other officials were heavily involved in determining the scale of the intended redevelopment. This was important because it was rare for large areas suitable for redevelopment to be entirely unfit. To compensate for this, housing legislation allowed the removal of sound or well-maintained buildings that would otherwise prevent a more comprehensive approach to rebuilding.

The redevelopment process

Unlike the city centres, the cleared inner-city sites were invariably developed by the local authority (Burns, 1963; Esher, 1983). This put a great priority on their administrative structures and expertise. In fact the exact departmental division of

Figures 6.5 and 6.6 *Netherthorpe, Sheffield, in 1956 and 1961, showing the breathtaking physical impact of planned redevelopment. The potency of the modernist vision is abundantly clear as the smokey unplanned disorder of the industrial era is supplanted by the clean, bold rationality of the planned, modern city. The consequences of these changes for the communities concerned can only be imagined.*

responsibilities varied by authority and over time. Housing architects, engineers and town planners were always centrally involved, however, overseeing the detailed housing designs, road and infrastructural considerations and overall land use. Consistent with the strengthening of planning's position within the big city authorities in the 1960s, town planners (such as Burns) became more significant in shaping later redevelopment schemes.

One of the most fundamental characteristics of redevelopment was that it invariably involved a reduction from pre-existing densities. This was partly because the new homes were larger than the old slums, reflecting the elimination of overcrowding, the provision of bathrooms, better kitchens, etc. It also reflected an improvement in external space standards, especially the increased provision of open space, car-parking spaces and more spacious arrangement of communal facilities such as schools. Such improvements were further encouraged by the Parker-Morris Report, *Homes for Today and Tomorrow* (Parker Morris Committee, 1961), which defined new, higher standards, though their adoption remained voluntary. It followed therefore that rehousing required rather more land than the redevelopment site alone and it was here that planned redevelopment impinged on other strategic policies, especially containment and planned decentralization (Cullingworth, 1960). By 1960 most of the major cities had nearly exhausted the supply of building land within their boundaries. This had wider implications, as we will see below. It also underpinned pressures to redevelop at as high a density as possible during the 1950s and 1960s, to minimize the impact on green belts.

The push for high rise

In turn this was linked to the increasing use of progressively higher flats, either on their own, or in the mixed-development schemes which had been recommended in the 1940s (Bullock, 1987; Horsey, 1988) and endorsed in MHLG's booklet *Flats and Houses* (MHLG, 1958). In fact the density increases on the higher schemes were often more apparent than real, because of the need for wider spacing of blocks. Yet as demographic pressures increased in the late 1950s and 1960s the image became more important than the reality. There was a coincidence of a strong central government push, reinforced by architectural opinion, and the near impossibility of securing enough housing land because of tight green-belt policies. In addition, the ease with which high rise could be incorporated into the parallel drive for more industrialized building methods (and the parallel involvement of major building groups) were also of crucial importance (Dunleavy, 1981). As Keith Joseph, the responsible minister in 1962–4, later regretted: 'I suppose that I was genuinely convinced that I had a new answer. It was prefabrication and, Heaven help me, high blocks' (1973, cited in Dunleavy, 1981, p. 170). Not that Joseph was the only advocate of these solutions. They reflected the essential 1960s' vision of modernization based on technological progress. At the local level they expressed a sense of forward-looking renewal that was extremely important for the older cities, paralleling the redevelopment of their city centres.

The rise of high rise

Between 1956 and 1967 central government housing subsidies gave progressively higher grants for higher blocks, offsetting their greater real costs. Accordingly flats of all kinds rose from 23 per cent of all tenders approved for public-sector dwellings in 1953 to 55 per cent in 1964 and remained near that level through into the 1970s (England and Wales figures) (Cooney, 1974). Most of these were under five storeys, but within that total there was a marked increase in the number of flats in higher blocks. Thus flats over five storeys increased from under 7 per cent in 1953–9 to nearly 26 per cent in 1964, before falling below 10 per cent in 1970. Very high blocks of over fifteen storeys similarly rose from under 1 per cent in 1953–9 to over 10 per cent in both 1965 and 1966, before falling to less than 2 per cent in 1970. Apart from some limited use in suburban and New Town locations, most such high blocks were concentrated in the inner redevelopment areas.

Generically there were three basic types: the point (or tower) block, the slab block and 'streets in the sky'. In their design conception, all came from the

***Figure** 6.7 Two leaders of the 1960s' planning 'revolution': T. Dan Smith, the charismatic Labour leader of Newcastle City Council, and Dame Evelyn Sharp, the forceful Permanent Secretary of the Ministry of Housing and Local Government, admire a model of the tower blocks proposed for the Scotswood Road area of Newcastle in 1961. Unlike most promoters of such schemes, Dan Smith actually lived in a similar block, after his fall from grace.*

modern movement in architecture (Johnson-Marshall, 1966; Esher, 1983). The point blocks were derived from various continental precedents, including French inter-war schemes like Drancy-la-Muette, though post-war Swedish schemes were also very influential. The slab blocks came from Le Corbusier's ideas, especially the famous Marseilles block, Unite d'Habitation, built 1946–52 (Marmot, 1982). Both types appeared during the 1950s, especially at the LCC's Alton East and West estates (1952–9) at Roehampton, south-west London (actually a suburban rather than a redevelopment site). Finally the linear multi-storey 'streets in the sky' deck-access blocks were another Corbusian variant, pioneered in Sheffield's Park Hill/Hyde Park estates from 1957 (HDCCS, 1962).

The role of the planner

Throughout there was conspicuous opposition to this flatbuilding boom from important (though, in the 1950s, rather weakened) sections of the town planning movement. The TCPA, guardian of the garden city tradition, was resolute in its opposition (Hardy, 1991b). Osborn, its ageing leader, kept his own blacklist of modernist architects who designed high rise, but themselves lived in low-rise houses with gardens. Occasionally individual chief planners also publicly distanced themselves from their authority's housing policies, for example Walter Bor in Liverpool (Muchnick, 1970). Yet it should also be recognized that, even if they had doubts, few planning officers had the status within their local authorities to challenge effectively the actions of more powerful departments until the later 1960s (Dunleavy, 1981). And the simple truth was that many planners were happy to acquiesce in high rise. It suited the concerns of the employers of many of them, the county councils. Moreover a significant proportion of the more urban-oriented planners were as much seduced by the modernist imagery as their architectural colleagues.

Figure 6.8 *'Streets in the sky'. One of the main flatted alternatives to the tower block was the deck-access medium-rise block. The form was pioneered by the Park Hill estate in Sheffield, visible in the middle foreground, developed from 1957. It was joined within a few years by the neighbouring Hyde Park estate (immediate foreground).*

The fall of high rise

By the late 1960s, however, the changes in planning methodology and greater awareness of research discussed in the last chapter were reinforcing a widening scepticism about high flats (Cooney, 1974; Dunleavy, 1981). There was a mounting critique of their social impacts, especially for families with children. This was paralleled by a growing concern about the high cost of housing subsidies needed to offset the extra costs of higher towers. The switch began with the critical subsidy change in 1967. An event in the following year completed it. In May 1968 a gas explosion on the eighteenth floor of Ronan Point, a twenty-two-storey block in the London Borough of Newham, blew out the side walls and caused the progressive collapse of one corner of the block (MHLG, 1968). Remarkably only five people were killed, but the impact on wider awareness of the problems of high-rise redevelopment was tremendous. In particular it underlined the low structural quality of many of the blocks and was effectively the beginning of the end of high-rise housing.

Medium-rise deck-access ('streets in the sky') schemes, usually built by industrialized systems, continued to be built for a few more years. Yet they soon began to experience social, design and constructional problems at least as severe as those affecting many high-rise blocks. This was certainly the experience of Hulme in Manchester, where the 'Crescents' (Hulme V), designed by the architect of Sheffield's Park Hill scheme, did almost as much for 'streets in the sky' as Ronan Point did for tower blocks. By the late 1960s and 1970s there was a very marked reversion to lower-rise high-density schemes, such as St Ann's in Nottingham (Coates and Silburn, 1980), or predominantly low-rise mixed developments such as Byker in Newcastle (NPD, 1973). And by this stage the traditional emphasis on the redevelopment of the inner-city slums was itself changing.

System building and package deals

Before considering the shift towards improvement policies, we should note the particular variant of the mixed economy that evolved in inner-city planned redevelopment. Instead of the partnerships that characterized central area redevelopments, it was the package deal that became the usual model. This meant that local authorities would often purchase a complete scheme 'off the peg' from the contractor. This was important because only a handful of extremely large firms had the capacity to offer industrialized systems, which required extensive off-site prefabrication facilities and expensive on-site handling equipment (Dunleavy, 1981; Russell, 1981; Finnemore, 1989). Normal competitive tendering procedures, whereby contractors submitted secret bids to build schemes that were entirely designed by the local authority, were simply not possible. Central government also discouraged local authorities from switching between systems, rather they should each establish a close relationship with one firm.

All this had the effect of diminishing the significance and control of the local authority as developer relative to the builder as contractor. The process actually had more in common with central area partnership deals than with the traditional model of competitive tendering. There was great emphasis on persuading a few key decision-makers, usually officials or committee chairs, to adopt particular systems. The professional sensibilities of the municipal architects were usually

Figure 6.9 *The effects of the Ronan Point explosion, May 1968. The twenty-two-storey Ronan Point and several similar council blocks were built in Canning Town, London, by the Taylor Woodrow Anglian Company, using a Scandinavian industrialized building system. The system had no structural frame or even fixings, and one floor simply rested on the wall panels of the one below, rendering it particularly vulnerable to 'progressive collapse'. The disaster brought the whole drive to build high into great disrepute.*

assuaged by leaving a few detailed design aspects to their discretion. And a whole public relations machine was created to oil the wheels and, on occasions, grease palms. Some commentators at least have portrayed this as one of the decisive forces shaping the boom in high-rise redevelopment: 'There was no human need for the tower blocks or most of the industrialized building systems of recent years. But a commercial demand was organized by political manoeuvring and high pressure salesmanship, helped along by corruption, regardless of human needs and consequences' (McEwan, 1974, cited Dunleavy, 1981, p. 121).

Notable figures in this extremely grubby corner of the mixed economy were John Poulson, a Yorkshire architect, and T. Dan Smith, who had started the 1960s as the dynamic Labour leader of Newcastle City Council (Smith, 1970). As a public relations consultant Smith later became an important intermediary between Poulson, the big construction companies and local authorities, especially Labour ones. Poulson, Smith and others were later jailed for corruption (Ravetz, 1980, p. 177). Another notable fall was that of Alan Maudsley, the Birmingham city architect. Both cases revealed wide networks of dubious practice. It was clear that several figures much closer to the centre of power, most prominently the Conservative Cabinet Minister Reginald Maudling, had been fortunate to escape prosecution (Morgan, 1992, p. 343).

The emergence of improvement

Such revelations during the early 1970s seemed to confirm a growing sense of public and political revulsion against the comprehensive redevelopment approach. By this time clearance was running well ahead of replacement building, and there were growing allegations of many basically sound dwellings being unnecessarily removed. Moreover there was growing resentment at the draconian character of the clearance process and its disruptive effect on local communities. Young and Willmott had warned about this in 1957, but there was little organized opposition until the later 1960s, when some areas, such as Millfield in Sunderland (Dennis, 1970), began successful challenges to redevelopment proposals. (We should add though that some of the last major redevelopment schemes, notably Byker, were handled with great sensitivity, and only took place after residents had rejected the option of improvement (NPD, 1973; Ravetz, 1980).) In fact there had been mechanisms and grants to facilitate the improvement of older housing since the Housing Act 1949. The Housing Act 1964 further introduced the policy instrument of the improvement area. Yet such policies had little real impact; only 5,600 dwellings were improved under the 1964 Act by 1969.

Housing Act 1969

The 1969 Act marked a watershed, coming as it did after several influential studies and pilot schemes (most famously at Deeplish, Rochdale) (MHLG, 1966). It also reflected something of the mounting reaction to redevelopment, though it was actually presented as a parallel, rather than a replacement approach (MHLG and WO, 1968). Its primary significance was that it created a strong area instrument for comprehensive improvement, the general improvement area (GIA), allowing more direct local authority action than before. The Act was a great success in that grants for improvements grew from 222,000 in 1965–9 to 1,009,000 in 1969–74 (Gibson and Langstaff, 1982). This success was qualified, however, because the major policy innovation, the GIAs, accounted for less than 10 per cent of this total. And the social impacts of the policy were doubtful in that the grants were used least by the poorest and rather reinforced a process of gentrification of some older areas (Ferris, 1972). But little was made of this at the time so that improvement began to look like a serious alternative to clearance, particularly as central funding for housing began to be more constrained. Thus only 42,000 dwellings were demolished in 1974 compared to 71,000 in 1968.

The position in 1974

The 1969 Act was a Labour measure, but the marked advance of improvement and retreat from clearance actually occurred under the Conservatives. Like much else though, this was a shift based on consensus rather than party politics (McKay and Cox, 1979). By 1974 both parties, as we will see in the next chapter, were fully agreed that improvement should be strengthened further. The early 1970s were therefore the beginning of the end for large-scale planned inner-city redevelopment on the model developed in the 1930s and applied so consistently and energetically from the 1950s. This was not without consequences for the second major strand of strategic planning policy, another product of the 1930s: planned containment.

PLANNED CONTAINMENT

'Land-saving' and green belts in the 1950s

Conservative 'land-saving' and housing policies

One of the most striking features of Conservative planning policy in the early 1950s was a marked concern to restrict the conversion of land to urban uses (e.g. Block, 1954). This was, of course, something which local planning authorities were only able to do following the 1947 Act's compensation provisions, which relieved them of the heavy burden of paying landowners not to develop. Very significantly the Conservative reforms in 1953–4 did not alter this and, indeed, actually reduced the burden of such compensation on central government (Cullingworth, 1980). Yet 'land-saving' was not an automatic consequence of these planning powers; it was a policy they were used to implement.

Contemporary critics such as Osborn of the TCPA attributed it to the particular strengths of the agricultural lobby in the Conservative party (Hughes, 1971, p. 213; Hebbert, 1981). It was certainly bolstered by contemporary declining population projections and had the weighty backing of Stamp's Land Utilisation Survey of the 1930s and the Scott Report of 1942. Osborn (1959, p. 34) had actually satirized the approach in his poem 'Jerusalem Replanned':

And did freestanding family homes
Invade A2 and A5 fields
And was Jerusalem builded here
Reducing agricultural yields?

But its influence was palpable by the early 1950s. In 1952 the MHLG publication *The Density of Residential Areas* (MHLG, 1952b) made many detailed proposals for tighter layouts to save land. The Ministry pushed this general message more directly in public-sector housebuilding. We have already noted how in 1956 the new housing subsidies began to give strong encouragement to flatbuilding, with progressively bigger subsidies to higher flats.

'Land-saving' and planning issues

'Land-saving' was apparent too in the residential land allocations of the county development plans for areas adjacent to the big cities. Although some smaller cities (such as Oxford) secured boundary extensions at this time, the big cities were refused either extensions or permissions to build housing on their peripheries (Hall *et al.*, 1973, Vol. I; Elson, 1986). Sheffield, for example, had a proposed major extension disallowed in 1952. The most famous instance was MHLG's repeated refusal to allow Birmingham to build on land it owned in neighbouring Wythall, Worcestershire, following a series of important planning inquiries culminating in 1959 (Long, 1962). Nor, as we will see, were there new designations

Figure 6.10 *In the first decade after 1945 many towns and cities continued to meet their housing needs in low-density peripheral estates, intensifying the pressures for a stronger approach to urban containment. Greenhill, Sheffield, developed in the early 1950s, was one such scheme and an early British example of the Radburn layout (see Figure 3.4).*

under the New Towns Act 1946, except in the rather specific circumstances of central Scotland. Even the Conservatives' own mechanism of planned decentralization, the Town Development Act, was little used. Manchester, for example, found its ambitions to build in Cheshire thwarted, following another famous 'test case' inquiry, concerning a proposed overspill scheme at Lymm, in 1958.

Wider support for 'land-saving'

Although the government's stance was making itself unpopular in the big cities and in some planning circles, these views were by no means general. The farmers and rural/outer suburban populations of the county areas, who almost invariably supported the Conservatives, were not generally anxious to see large council housing estates for the big cities, bringing largely Labour-voting populations to their areas. There were too the aesthetic objections to sprawl, which received support with the publication of 'Outrage', the famous special number of *Architectural Review*, in June 1955 (Punter, 1985). The tone throughout was apocalyptic and doom-laden especially as its author, Ian Nairn, fulminated on the evils of 'Subtopia': 'This thing of terror, which will get you up sweating at night when you begin to realize its true proportions . . . the universal suburbanization not merely of the country or the town, but of town-and-country – the whole land surface. Suburbia becomes Utopia. Utopia becomes suburbia' (Nairn, 1955, pp. 365–6).

And planning, because of its traditional emphasis on urban development based on the house and garden, was roundly condemned for speeding rather than stopping Subtopia's advance. Overall, 'Outrage' was a forceful attack that reinforced many existing public and political concerns about the spread of towns. In fact the central issues it raised were already being addressed in a new and more positive policy initiative that was to put mere 'land-saving' on to an altogether higher plane.

The Green Belts circular

In April 1955 Duncan Sandys, Macmillan's successor as Minister of HLG, issued the famous Green Belts circular, 42/55 (MHLG, 1955). Senior Civil Servants were rather worried at the consequences of giving such positive policy endorsement to a planning concept that was potentially so negative and inflexible (Mandelker, 1962). But despite their opposition, Sandys knew he was on to a good thing politically and did not allow them to deflect him. In effect the circular extended the green-belt principle which had been accepted for Greater London in 1946, following the 1944 Abercrombie Plan, and provided for green belts to be incorporated in development plans (Thomas, 1970; Munton, 1983). Now the floodgates were opened for county areas around provincial towns and cities to bring forward proposals (Elson, 1986). Some basic ground rules were laid down, stating that green belts were to check the growth of large built-up areas, prevent coalescence of neighbouring urban areas and/or preserve the special character of particular towns. And, in deference to his officials' fears, Sandys warned that he would be strict in applying these criteria. This did not stop several areas which did not fit these criteria trying their luck, however.

Green-belt proposals

Between 1955 and 1960 no fewer than sixty-nine sketch plans of preliminary proposals had been submitted, many of them covering different parts of what ultimately became the same green belt. By 1963, 5,585 square miles of England were subject to green-belt policies. Over a third of this was accounted for by the original London green belt (which was the only fully approved scheme) and its interim (that is not finally approved) extensions (Thomas, 1970). All the main English conurbations had interim schemes: Birmingham–Coventry, West Yorkshire, Tyneside–Wearside, Manchester–Merseyside and South Hampshire. The remainder comprised fairly large non-conurbation cities like Bristol–Bath, Stoke and Nottingham–Derby and historic centres like Oxford, York and Cambridge. Several other English bids were completely unsuccessful. Salisbury, Swindon, Scarborough, Teeside, Flintshire, East Sussex and most of the County Durham proposals were all rejected. Meanwhile green belts for Edinburgh and Aberdeen were designated from the mid-1950s (and for Falkirk–Grangemouth, Dundee and Prestwick Airport in the 1960s) (Skinner, 1976). Yet the 1946 Clyde Valley Plan's proposals for a green belt around Scotland's largest city, Glasgow, were only partially incorporated in development plans. This hinted at the perceptibly more sceptical Scottish attitude to the green-belt concept that was so uncritically adopted in England.

Refining green-belt policies

The consideration of the first round of English green-belt proposals (and further MHLG policy pronouncements, especially a 1957 circular) refined the notion of what a green belt was actually supposed to be. It became clear that formal green belts were not intended to apply to smaller towns (unless they were of special character) and were unnecessary where growth pressures were not felt to be strong. This did not mean such areas could not operate policies of restraint and containment through normal planning policies, merely that they did not need the extra reinforcement of a formal green belt. There was also a clear reluctance to sanction green belts in areas where there was uncertainty about future development, to avoid the need for frequent reviews. With the same consideration in mind the MHLG was also anxious that green belts should allow for some increases in population in the urban areas they surrounded. In addition to land already allocated for development in development plans or town maps, the 1957 circular had introduced the notion of 'white land', within the green-belt inner boundary, but not actually part of it. This could be made available for development at some future date, though until the need arose it would remain in an undeveloped state (MHLG, 1957).

Overall assessment of the early green belts

Such refinements were essentially pushed by MHLG officials who were worried about the long-term implications of such a potentially negative planning instrument (Elson, 1986). Yet they failed to dent the rather simple conception of the green belt held by their ministers and many others, especially in England. Thus Henry Brooke, Sandys' successor as Minister of Housing and Local Government, commented in 1960:

> The very essence of a Green Belt is that it is a stopper. It may not all be very beautiful and it may not all be very green, but without it the town would never stop, and that is the case for preserving the circles of land around the town.
>
> (Heap, 1961, p. 18)

This view of the green belt as a simple, permanent and readily understandable restriction on the spread of towns was (and remains) its great strength (MHLG, 1962; Mandelker, 1962). But it was increasingly questionable whether such a highly specific expression of containment was appropriate in the 1960s, against the background of accelerating growth forecasts.

Declining emphasis on green belts in the 1960s

Doubts in the early 1960s

It was therefore ironic that no sooner had Sandys and Brooke established green belts as a key part of planning policies, than their successors began to experience serious doubts. The population forecasts of the early 1960s caused the whole process of green-belt approval to be placed in abeyance in 1962. In Greater London particularly a serious crisis of housing land supply was being recognized, manifest in shortages and high costs of development land (Cullingworth, 1980). Yet, as we noted in the last chapter, the government was reluctant to intervene in the land market in any decisive way to expedite supply or stabilize values. The then Minister, Keith Joseph, saw the answer in simply urging local planning authorities to allocate more land. But this expedient sat very uneasily with strong commitment to the green belt.

In 1963 he tried to open up further the issue for his fellow ministers. Yet, even in the secret confines of ministerial discussion, the issue had to be raised gingerly. His desire to 'clear the air' had to be prefaced by words that confirmed just how quickly the green belt had become a sacred cow of planning policy: 'Nobody intends to destroy the Green Belt. On the other hand, the pressure for housing land within easy reach of London is tremendous' (Cullingworth, 1980, p. 235).

The results of these rather tentative deliberations were rather limited. A White Paper – *London: Employment, Housing, Land* – invited local authorities in the home counties to come up with the necessary land voluntarily (MHLG, 1963). It came as no great surprise when they conspicuously failed to do so.

Green belts out of favour 1964–70

Labour was more inclined to agree with MHLG official advice that more development land needed to be allocated in the green belts. In quick succession, Richard Crossman, their first Minister of HLG, granted requests to build in green-belt areas in Kent (New Ash Green), Birmingham (Chelmsley Wood) and Sheffield (Stannington). As he confided to his diary in December 1964:

> I'm making these three decisions quite deliberately because I've decided, if rigidly interpreted, a green belt can be the strangulation of a city. With so many people to house we can't put them all in New Towns thirty miles away on the other side of the green belt. We have to find places nearer the cities to house them in, even if this means that in some cases we shall trespass into the green belts or turn a green belt into four or five green fingers. I know this will cause me my first major row but I'm

> pleased about it. I've decided to do it and I think it's good ground on which to fight.
>
> (Crossman, 1975, p. 87)

Yet, as he subsequently realized, great caution was still necessary in addressing such issues, which were capable of generating heat even on his own side. Accordingly his 1965 policy statement began with a ritual reaffirmation of commitment to the purposes of green belts (which were now extended to embrace recreation for urban populations). He suspended judgement on the provisional green-belt proposals until the major questions about the need for and supply of development land had been resolved, however. Behind this decision was a widening professional and official opinion that the policy as then defined was out of date and too static for the growth pressures facing Britain at that time (Elson, 1986). Certainly the trend of central decisions throughout the late 1960s saw a continuation of the pattern Crossman had set. Around Birmingham, for example, further peripheral development was permitted on appeal at Moundesley, Hawkesley and Pirton in 1968–9.

Green belts and other strategic policies

In fact the green-belt issue had become tied up with several other dimensions of strategic planning policies. We will consider these more fully in later sections, but for present purposes we need to be aware of how they impinged on the issue of green belts. It is, for example, important to realize that a whole series of regional and later sub-regional studies were under consideration during the middle and later 1960s. The justification for Crossman's decision to suspend consideration of green-belt proposals was because he was awaiting their reports, largely concerned with the location of new development. The kind of growth-corridor and green-wedge solutions that Crossman was referring to in his diary above might well have been favoured, radically altering the form of green reservations. And both green belts and regional planning were linked to the increased emphasis on both planned redevelopment of the inner city and planned decentralization throughout the 1960s. Clearly the amount of rehousing needed outside the inner city and the form, character and locations of any New Towns would have a major bearing on green belts that had been proposed in the 1950s.

Green belts and local government reform

Another key imponderable was local government reform. In many respects the proposals for green belts were an expression of the political separation of town and country within the local government system. It is significant perhaps that it was in West Yorkshire, where much of the 'countryside' had a particularly urban character, that the most workable green belt was produced from the point of view of development land availability. The more typical pattern was of amenity-conscious, Conservative-voting, urban-fringe counties seeking to block incursions from the Labour cities. A radical reformulation of local government that created unitary authorities embracing both town and country would certainly reduce political polarization, generating a new form of fringe management. As noted in the previous chapter, the publication of the Redcliffe-Maud proposals in 1969 endorsed such a solution and it was also assumed in the Town and Country Planning Act 1968.

A recreational dimension

Meanwhile in his 1965 policy statement, Crossman (following Labour thinking since the inter-war period) had asserted an urban recreational role for green belts that had been neglected in the 1950s' circulars. In the long term this was perhaps the main significance of this period for containment policy. The Countryside Act 1968 reflected the new emphasis by introducing country parks (Cullingworth, 1976; Elson, 1986). These were areas of countryside reserved specifically for urban recreational needs, within easy travelling distance of city populations. Many were in either formal green belts or urban-fringe areas where ordinary planning or land ownership controls were preventing building. Many of the early parks had already been used for recreation, but grant aid and more formal status improved what was being provided. Smaller picnic sites could also be created, and again many were in formal or informal green belts. By 1974 there were 111 country parks in England and Wales, covering 15,000 hectares, and 141 picnic sites.

Green-belt revival 1970–4

Conservative enthusiasm

Peter Walker, the youthful and first ever Secretary of State for the Environment, indicated an intention to accelerate green-belt approvals, signalling a real revival of enthusiasm for the policy of containment under the Conservatives (Elson, 1986). Moreover he implemented a pattern of local government reform that perpetuated much of the pre-existing urban–rural separation, retaining the essential institutional basis of green belts. There was more to this than mere party politics. Growth pressures were slackening as population and economic growth forecasts were revised downwards, making the more static notion of green belts seem more appropriate. Uncertainties about regional growth strategies that had dominated the 1960s also ended as the radical options were discarded, allowing traditional conceptions of the encircling green belt (as opposed to green wedges, for example) to remain dominant.

The position in 1974

Not for the first or last time, Conservative ministers soon found that firm commitments to green belts, however comforting to their supporters in urban-fringe areas, could prove embarrassing. By 1972–3 it was clear that more housing land was needed than simple population forecasts suggested. This was largely because a series of reflationary budgets, aimed at bolstering an increasingly ailing economy, had triggered a massive private housing and property boom (Ambrose and Colenutt, 1975). Moreover it was becoming abundantly clear that as household size declined, housing needs could still remain high even if population increase was slowing. All this meant searching for more land in or near the green belts and wholesale release of 'white land' for building was being contemplated by 1973–4. In the event, therefore, Walker and his successor Geoffrey Rippon were unable to grant the speedy approvals of green belts. By 1974 therefore the position had barely changed from 1962 and less than half the total area of green belt was actually approved. For something that had become such an important component of strategic planning policies, this was indeed a paradox. The indecision over containment over these years had been closely linked to the heightened concern for wider regional issues. It is to these we now turn.

Regional Balance

Regional policy in the 1950s

The relaxation of regional policy

The Distribution of Industry Act 1945 had been used very strictly in the years to 1947 but was then applied in an increasingly relaxed way (McCrone, 1969; Parsons, 1986). The 'Butskellite' macro-economic policies of both Labour and the Conservatives after 1950, relying on a Keynesian-type demand management approach, allowed the reliance on physical controls (like IDCs) to be reduced. The fact was that unemployment had not re-emerged as a problem in post-war Britain and, although there were some regions which suffered more, unemployment everywhere was very low by historical standards. In 1951 the worst British regions, Scotland, Wales and the North, had unemployment rates of only 3.5 per cent, 3.5 per cent and 3.0 per cent respectively. Northern Ireland, at 6.6 per cent, was appreciably worse, though still relatively modest by the standards of the 1930s or 1980s (Law, 1981). These generally low levels did not, however, reflect the emergence of new regional economies. Although the war economy and post-war regional policies had brought new factories to many of the pre-war depressed areas, there had been a marked revival in employment in the older industries on which these regions remained heavily dependent.

But regional policy had not been intended to remodel regional economies. It was primarily a response to regional unemployment and, in the circumstances of the 1950s, the Conservatives felt able to relax regional policy. Expenditure fell from an average £8.13 million per annum in the last three years of Labour to a mean annual figure of only £4.76 million in the first eight years of Conservative rule. The object of policy was no longer the application of serious obstacles to new industrial development in the buoyant regions so much as the diversion of a few industries to the worst-off regions. The wider objective of regional balance was very much in the background.

Regional policy revival from 1958

From 1958, however, older industries began to experience serious problems and responded by shedding labour. In response the Conservatives were forced to abandon their hands-off approach. There was new legislation in the Distribution of Industry (Industrial Finance) Act 1958 and the Local Employment Act 1960 (McCrone, 1969; McKay and Cox, 1979). The first signalled a marked increase in spending on regional policies, from £3.6 million in 1958–9 to £11.8 million in

1960–1 and allowed assistance to smaller blackspot areas outside the development areas. The second extended this principle and actually abolished the development areas, replacing them with development districts, smaller areas to be defined solely according to unemployment criteria, essentially those where unemployment was roughly twice the national average or 4.5 per cent. This move was heavily criticized by planning commentators who saw it as the culmination of a decade of retreat from regional policy. Certainly its focus was less on the balanced distribution of industry and more on the avoidance of local unemployment. It seemed to be, in Cullingworth's phrase, 'first-aid', falling well short of being a comprehensive approach to regional planning (cited in Parsons, 1986, p. 177).

From regional policy to regional planning 1960–4

Regional policy 1960–4

Despite such criticisms, the 1960 Act coincided with a much more vigorous approach to regional policy under Reginald Maudling and later Edward Heath as successive interventionist Conservative Presidents of the Board of Trade. Thus Maudling embarked on a major initiative to push new expansion in all the main firms within the motor industry into the depressed areas, especially Scotland and Merseyside. In February 1960 the *Guardian* described it, with only slight exaggeration, as 'the largest piece of planned industrial dispersal that has ever been attempted in Britain' (cited in Parsons, 1986, p. 146). (Wartime dispersal had of course been more significant.) And regional policy spending was further extended under the Local Employment Act 1963. Important too was the increasing concern to encourage more office development away from London, manifest in the creation of the advisory Location of Offices Bureau in 1963, and noted in the last chapter.

Sources of the shift

This broad policy shift in traditional regional policy was essentially a reflection of continuing high regional unemployment (by the standards then applied). In turn, of course, this highlighted the importance of the all-party, consensual commitment to the maintenance of full employment by government policy. The ghost of the 1930s still haunted the decision-making of the 1960s, especially for the Conservatives who had been so successfully labelled as the party of unemployment in the inter-war years. But despite increasing regional policy expenditure, regional unemployment was appreciably higher in 1961 than it had been a decade earlier. The north, Wales and Scotland stood at 3.8, 4.2 and 4.5 per cent, respectively, with Northern Ireland at 9.4 per cent, despite major legislation extending assistance there as early as 1954 (Law, 1981).

There were, however, other important underlying factors. We have already noted the growing early 1960s' concern with accelerating population growth and the need to ensure it was matched by economic growth (e.g. NEDC, 1963). In such circumstances it was imperative that the older regions were properly integrated in the national economy, to avoid excessive congestion and growth pressures in more buoyant regions. As we have seen, MHLG was simultaneously struggling with the management of growth pressures around London, particularly the specific problems of office development and the green belt (MHLG, 1963). It was beginning to dawn on government (and wide sections of business) that a more active regional policy could play a key role in avoiding the mounting wage, land and general congestion costs of the south east and the midlands. This involved moving beyond administering regional first-aid to the worst areas, but looking seriously at the capacity of the regions to attract and accommodate growth. All of which was pointing beyond regional policy to regional planning.

Regional planning frameworks 1960–4

One of the main difficulties in moving to the more comprehensive notion of regional planning was the deeply entrenched departmentalism at central government level. This kept physical planning, controlled by the MHLG and largely applied through local government, separate from spatial economic policy, applied through the Board of Trade. The further separation of the Ministry of Transport, busily engaged on the national motorway programme by the early 1960s (Charlesworth, 1984), did not help either. As in other aspects of their planning policies, the Conservatives began to address this question but failed to act decisively.

The opening of regional offices by the key central departments was a gesture towards fostering a regional approach (and dispersing offices from London) (Cross, 1970). Yet it did practically nothing to transcend the departmentalism of Whitehall, where all the real power remained. In Scotland the creation of the Scottish Development Department (SDD) in 1962 went furthest down the road to integration. And in early 1963 the government took the extraordinary step of appointing a special Minister for the North East, in the person of Lord Hailsham (also Minister for Science) (Hailsham, 1975). A Cabinet Minister and political heavyweight, Hailsham greatly advanced the case for regional planning in England's main problem region (Bulmer, 1978). Moreover he knew how to get the local politicians of this overwhelmingly Labour region, particularly the key figure, T. Dan Smith of Newcastle, on his side (Smith, 1970). His accidental substitution of a flat cap for his usual bowler during an early visit attracted some ridicule as a clumsy attempt to identify with working-class north-easterners, but it gave regional problems a popular prominence they had not had since the 1930s. Hailsham had only a temporary remit, however, inherited by the last of the Conservative Presidents of the Board of Trade, Edward Heath. Rather grandly titled the Secretary of State for Industry, Trade and Regional Development, Heath let himself be known as 'Minister for the Regions' and seemed to symbolize the beginnings of a full regional commitment during the last months of Conservative rule.

Regional planning initiatives

Meanwhile the several agencies with regional planning responsibilities had been taking important initiatives during the early 1960s. In 1961 MHLG initiated the

South East Study 1961–1981, an analysis of growth pressures in the largest and most prosperous region. The report (MHLG, 1964) highlighted the huge need for housing that was building up largely as a result of a forecast natural increase of some 2.5 million within the region itself and a further 1 million by migration. It advocated massive increases in development plan land allocations and major new town and city designations, together with additions to existing New and Expanded Towns.

Furthermore, in 1963 new centrally sponsored regional plans had also been prepared for central Scotland (SDD, 1963) and the north east (at Hailsham's instigation) (DITRD, 1963). The plans identified major growth areas within the regions and looked for a concentration of public and private investment in such areas. Such proposals marked a clear break from the 'blackspot' approach of the development districts because they involved focusing on the most likely areas for growth within the regions. The approach, which included proposals for new towns and major road investments, was partly based on the then fashionable concept of the growth pole, a rather dubious though influential spatial adaptation of the ideas of the French economist Francois Perroux (Parsons, 1986). Yet it shared an important characteristic with the blackspot approach, the principle of selective, concentrated assistance. It was, therefore, an approach which depended on more careful planning.

From regional planning to regional policy 1964–74

The Department of Economic Affairs and Regional Planning

As elsewhere in planning, these regional aspirations were given great initial encouragement by Labour in 1964–5. The creation of DEA, noted in the previous chapter, and the National Plan were intended to link closely with regional proposals (DEA, 1965a). Under George Brown, DEA instigated the ODP, the new national mechanism for office control, under the 1965 Act discussed in the last chapter. More importantly from the point of view of regional planning, Brown (1972) created a new framework, of Regional Economic Planning Councils (REPC) and Boards (REPB) (Smith, 1970; Martins, 1986; Pearce, 1989). The Councils comprised collections of regional worthies, including local politicians, businessmen, trades unionists, academics, etc. The Boards were the Civil Servants seconded from various departments to work on particular regions. Together this regional apparatus was charged with producing regional economic plans that would relate national economic planning objectives and local physical planning objectives. Brown himself saw them as something that was the embryo of a new form of regional government that he hoped would emerge as a by-product of local government and wider constitutional reform.

Weaknesses of the new regional machinery

The new machinery was seriously compromised from the outset. Despite some sympathy with Brown's planning objectives the MHLG, with the tacit support of the Treasury, managed to emasculate the DEA proposals by the time they were announced in December 1964. Richard Crossman savoured his Ministry's victory in his diary:

> I had been continually negotiating on the draft. It had come across to us with the word 'economic' added in longhand before the word 'planning' as a concession to me, so that it read not just 'regional planning' but 'regional economic planning', ha, ha, ha! I also wrote into the Statement the explicit assurance that nothing in this scheme would affect the existing powers of the local authorities with regard to planning. It seemed to go down fairly well in the House, mainly because it will take some time for people to discover how inadequate and meaningless it has now been made.
>
> (Crossman, 1975, p. 93)

Brown himself took a more hopeful view:

> I reckoned that by agreeing to have the word 'Economic' in their title we could leave enough ambiguity in the situation for our scheme to go ahead without too much fuss. My hope was that, as time went on, the connection between planning for economic development of an area and planning for its physical development would become so obvious that the two would naturally fuse.
>
> (Brown, 1972, pp. 102–3)

In the event Crossman's assessment was the more apt. This MHLG–DEA demarcation dispute was only part of it, because the regional bodies also found themselves with no decisive say in the exercise of regional economic policy (McKay and Cox, 1979). IDC controls remained with the Board of Trade (briefly shifting, in 1969–70, to the Ministry of Technology), while the new ODP was exercised by the MHLG. Even the DEA's National Plan, which included much regional trumpeting, was drawn up without any truly regional input. The Regional Councils and Boards thus came to occupy the no-man's land between the central and local centres of power and authority. By 1967 prominent planning figures like T. Dan Smith, then Chairman of the Northern REPC, and David Eversley, former member of the West Midlands Council were going public on their frustration and disillusion (Parsons, 1986). And after George Brown ceased to be Secretary of State, what central commitment to the regional planning machinery that existed soon evaporated. In 1969 DEA was abolished and the short-lived overlord Department of Local Government and Regional Planning assumed control of the regional planning apparatus. In 1970 it passed to the new Department of the Environment.

Regional economic studies

The most tangible achievement of this period was the production of regional economic studies (in effect, plans) (listed Cullingworth, 1976, pp. 277–8). DEA itself produced two studies, for the west midlands (DEA, 1965b) and the north west (DEA, 1965c). Thereafter the pace, such as it was, was set (in England) by the REPCs, all of whom produced documents for their regions between 1966 and 1968 (e.g. NEPC, 1966). Three of the REPCs also produced sub-regional studies, most notably for the Halifax, Huddersfield and Doncaster areas by the Yorkshire and Humberside REPC. Several other notable sub-regional studies were also initiated in the late 1960s by central agencies for estuarine zones that were felt to have major growth potential at Humberside, Severnside and Tayside. Scotland meanwhile continued the pattern established pre-1964, of its own more direct form of economic planning. It was joined in this respect by Wales, a change which reflected the establishment in 1964 of the Welsh Office as a permanent 'regional' department of central government, on the model of the Scottish Office. The Scottish and Welsh proposals accordingly formed a more authoritative basis for action than those in the REPC reports, which, though producing a few suggestions that were taken up, were largely ignored.

Regional economic policy 1966–70

Instead, as regional unemployment began to move upwards from 1966, the emphasis swung back to a strengthened regional policy, firmly controlled from Whitehall (McCrone, 1969; Jay, 1980; McKay and Cox, 1979). The Industrial Development Act 1966 reintroduced the Development Areas and strengthened available grants. The new areas covered 40 per cent of Britain's area and accounted for 20 per cent of the population, most of them Labour voting. This was much less spatially discriminating than either the blackspot or the growth-pole

Figure 6.11 *T. Dan Smith as Chairman of the Northern Economic Planning Council, launching their regional study, Challenge of the Changing North, in 1966. The (unfulfilled) hope was that the Councils would be able to link regional and national economic concerns with physical and land use planning.*

approach to regional assistance. There was also less discrimination about what kinds of firms were aided. Rather than particularly encouraging the more profitable firms by assistance paid as tax allowances, the emphasis shifted to more labour-intensive manufacturers, encouraged through bigger grants and the regional employment premium of the new selective employment tax, introduced in 1967. Also in 1967 the Special Development Areas were introduced, giving extra assistance to areas of coal-mining closures. Finally the Local Employment Act 1970 offered a lower level of assistance to newly created Intermediate Areas (the so-called 'grey' areas) which did not justify Development Area status but were significantly less prosperous than the most favoured regions. This followed the recommendations of the important Hunt Committee (1969) on the Intermediate Areas.

Factors in the resurgence of regional policy

This shift was largely driven by rises in unemployment. This traditional concern began to assume priority over the growth-oriented concerns as the optimism of the early 1960s began to fade. This was intensified by the political priorities of the Wilson government. For Labour the broad approach was more politically acceptable than the more selective strategy of blackspots and growth points that it had inherited from the Conservatives. Quite simply it avoided the political damage of refusing or reducing aid to areas which were predominantly Labour voting. Thus by 1970 (an election year) the government was unwilling to tackle the issue of descheduling some of the development areas, as the Hunt Committee had recommended. Industry meanwhile was quite content to accept the ever-increasing grants to move into areas with generally lower labour costs, objecting only (in evidence to the Hunt Committee) to the accompanying restrictions on the freedom to expand elsewhere.

Regional policies 1970–4

Nor did the incoming Conservative government resume the type of selective regional planning approach that Heath himself had espoused in 1963–4. Unemployment was much higher than it had been a decade earlier and from 1970 it began to move inexorably upwards. By 1971 the most favoured region, the south east, had unemployment of 4.1 per cent, worse than the least favoured British region in 1951. The three worst British regions – Scotland, Wales and the north – recorded figures of 7.4 per cent, 6.9 per cent and 6.9 per cent respectively (Law, 1981). Accordingly, despite an ostensible electoral commitment to less interventionist industrial policies and a distinctly hard-faced approach to regional policy in 1970–1, the Heath government made its famous 'U-turn' (McKay and Cox, 1979). The Industry Act 1972 further extended regional policy aid, particularly by widening the Intermediate Areas. More assistance was also being given for the clearance of derelict land. There were some elements of the earlier Conservative approach of selectivity in the type and availability of grants. But there was no resurgence of regional planning. The REPCs continued their work, some to the extent of issuing further reports, but by now they were completely sidelined. Despite the initial rhetoric of the Conservative programme, it was all essentially an

Figure 6.12 *By the late 1960s the moves for regional economic planning were being subordinated to a more indiscriminate pattern of regional assistance, reflecting a gradually worsening economic climate.*

extension of the approach of the late 1960s. Overall, it was a story with parallels in the final policy area, planned decentralization.

Planned Decentralization

A limited approach in the 1950s

Conservative doubts

In the early 1950s the position of the New Towns was by no means secure. Only 3,126 dwellings had been built in the twelve designated areas in England and Wales by the end of 1951 (Cullingworth, 1979; Aldridge, 1979). The Board of Trade was reluctant to grant IDCs for the London New Towns, arguing that they undermined regional policy. Nor were the New Towns being given favoured treatment by supply ministries, an important factor when major building materials were still subject to rationing. Not least, the Treasury was worried by the large deficits on the programme. Given their very dubious progress, it was hardly surprising that the Conservative government seriously considered scrapping the programme. Macmillan himself admitted privately in September 1952 that 'I do not think it was a very good idea in the immediate post war conditions' (Cullingworth, 1979, p. 555).

This was with the gift of hindsight; the Conservatives had not opposed the New Town programme under Labour. And the fact was that, despite early disappointment over poor progress, the New Towns remained a basically popular idea. Clearly it would be politically unwise to jettison the programme, given the Conservatives' very slender majority at the 1951 election. And Macmillan decided that it was better than the alternatives, especially around London. Without the New Towns, 'the LCC will be striving to fill the gap – which will save neither public money, nor investment: it will just make a nasty mess' (*ibid.*, p. 118).

The New Town programme in the 1950s

These arguments carried the day. The programme continued to be pressed in 1953–4, especially by the Treasury, until it began to dawn on them just how much profit would soon begin to flow in their direction (Heim, 1990). But during 1952 Macmillan was able to get agreement for a policy whereby the existing New Towns were continued, but there would be no new designations and any further planned decentralization would be handled through the Town Development Act 1952 (Block, 1954). Accordingly the proposals for a New Town at Congleton to accept Manchester overspill was dropped, much to the relief of Cheshire County Council, who trenchantly opposed both it and an earlier proposal at Mobberley (Lee, 1963). (It was this that sowed the seeds of the pressure to develop at Lymm, referred to earlier.) There was, however, one apparent contradiction of this general policy line (Cullingworth, 1979). A further New Town was actually designated, in 1955, at Cumbernauld in Dunbartonshire. The anomaly reflected the extreme seriousness of Glasgow's rehousing problems, particularly the higher than average additional land requirements when the city's extremely high-density slums were redeveloped

(Osborn and Whittick, 1977). Moreover there was no Scottish legislation equivalent to the Town Development Act 1952 until the Housing and Town Development Act 1957, so that local authority overspill schemes were not an option.

Generally the development of the existing New Towns was accelerated, and they benefited from the relaxation of IDC availability to establish sound industrial bases during the 1950s. By the end of the decade, progress had been sufficiently good to raise the question as to what would happen to the assets of the development corporations when the towns were completed. The Commission for the New Towns was the favoured device, created under the New Towns Act 1959. Its essential concern was to prevent the local authorities getting their hands on New Town assets. Initially the Commission was fiercely opposed by Labour (though they did nothing when in office after 1964). Hemel Hempstead and Crawley became the first to be handed over to the new Commission, in 1962.

Social and economic development

The New Towns of the 1950s exactly caught the aspirations of younger, inner-city, skilled and semi-skilled working-class families for a better life (Aldridge, 1979). Formal (and, even more importantly, informal) industrial selection schemes ensured that migrants from the older city core areas had the kinds of skills which were in demand in New Town industries. Housing allocation was therefore directly linked to having a suitable job or the good prospect of getting one. This meant that the London New Towns particularly were an industrialist's dream. They combined modern, well-serviced locations, land and premises that were extremely good value for money with (almost literally) hand-picked workforces living in good-quality and inexpensive housing.

Many of the promised community facilities also began to appear during the 1950s and the New Town centres were well advanced by the middle years of the decade. Most were developed directly by the development corporations (with increasing Treasury approval) (Cullingworth, 1979). Private developers operated in six, however. A single company, Ravenseft, alone developed four (Marriott, 1969). It had established its reputation redeveloping the centres of blitzed cities, as noted earlier. The appearance of these convenient modern centres of consumption (earlier than their equivalents in most older cities) was, of course, a foretaste of wider changes in British society. For those who were able to move there, they were an important means of opting into the affluent society, providing a setting for a new, and more materialistic, lifestyle.

Planning innovations

The New Towns were also one of the major sources of planning innovation in the 1950s and consequently received many visitors from Britain and overseas. There was little they did that was entirely new, but they provided an opportunity to adopt and develop a series of good ideas that had been around for some time (Osborn and Whittick, 1977). A real innovation came with shopping precinct development. Stevenage vied with Coventry in producing the first pedestrianized central shopping precinct in Britain (Stephenson, 1992). In addition the New

Figure 6.13 *The early New Towns, especially around London, were often in the vanguard of wider social changes, mirrored to some extent in their physical planning. The photograph shows a neighbourhood shopping centre in Basildon c.1961. We can detect the increasing motorization of society and the growing trend to vehicle/pedestrian segregation. Pervading everything is the sense of hitherto unknown working-class affluence.*

Towns were at the forefront of neighbourhood and social development ideas. There was certainly an increasing sociological scepticism about the wider validity of the concept as a physical means of creating social cohesion (Glass, 1948). But by the 1950s, the New Town planners were generally adopting an essentially pragmatic approach. They used the neighbourhood idea largely to provide a useful physical framework for the provision of essential social facilities in reasonable proximity to housing. The closely related idea of Radburn residential layouts also began to appear from the mid-1950s, most comprehensively at Cumbernauld, which was the first New Town plan to reflect increasing car ownership. And there were many other more minor instances where New Towns articulated the best practice of British planning in the 1950s.

Other dimensions of planned decentralization

New Towns were the main, but not the only, dimension of planned decentralization during this period. Little was achieved under the Town Development Act 1952 before the 1960s. By 1960 fewer than 10,000 dwellings had been provided through local overspill agreements. Swindon, where the requirements of planned decentralization from London was matched with a strong local desire for growth and diversification, was the only really successful example on a scale comparable to the New Town programme (Harloe, 1975). Other attempts to use the legislation on a bold scale foundered, for example at Lymm, Cheshire, noted above. By 1957 the LCC had become so frustrated with the constraints of the 1952 Act and central unwillingness to designate further New Towns that it embarked on its own project for a New Town at Hook, Hampshire (LCC, 1961). It fell following intense local opposition to the specific location, but Hampshire counterproposed a scheme for major town expansion at Basingstoke under the 1952 Act, which made an important contribution in the 1960s and beyond (Dunning, 1963).

A good deal of the 'planned decentralization' of the 1950s was actually just a modified version of old-fashioned municipal suburbanization in areas where the intention to develop was established before green-belt proposals were introduced. Manchester, for example, met many of its housing needs by completing its massive Wythenshawe satellite town begun between the wars (Deakin, 1989). Liverpool did the same at Speke and its larger post-war satellite town at Kirkby (Lawton and Pooley, 1986). And, despite Cumbernauld, Glasgow produced huge peripheral estates such as Drumchapel and Pollok (Smith and Farmer, 1985). In London, the LCC was engaged in the building of large 'out-county' estates at Debden, Elstree, Harold Hill and elsewhere (TPI, 1956; Young and Garside, 1982). The main problem with this approach was that, by 1960, these reserves of housing land were running out. Meanwhile the inner-city redevelopment programmes were moving into even higher gear. This meant that, even with the trend to high-density redevelopment noted above, ever larger numbers of people would be displaced from inner city areas and need rehousing elsewhere. As

Figure 6.14 *The New and Expanded Towns were prime locations for manufacturing development. This shows the first of Swindon's industrial estates in the early 1960s. Already the presence of on-street parking suggests a growth of car-based travel to work that went beyond what was anticipated by planners, though the two pedestrians, significantly, are women.*

if this were not enough, it was at this point that higher population growth forecasts were indicating a continuing high demand for additional (rather than just replacement) housing. It was this which produced the remarkable renaissance in planned decentralization policy from 1960.

A policy renewed 1960–74

New Town designations 1961–4

By 1960 it was clear to Conservative ministers that their planning policies would soon collapse under the weight of their own contradictions without some new initiatives. The provincial cities of Birmingham, Glasgow, Liverpool and Manchester were patently unable to resolve their needs for housing without some central assistance (Osborn and Whittick, 1977; Aldridge, 1979; Cullingworth, 1979). Accordingly New Towns were designated for Birmingham at Dawley (1963) and Redditch (1964) (Anstis, 1985); for Glasgow at Livingston (1962); for Liverpool at Skelmersdale (1961) and Runcorn (1964). Manchester was once more left out, partly because past battles had by now completely soured relations with neighbouring Cheshire. Schemes for Risley, near Warrington, and Leyland-Chorley were being considered, though neither was as straightforward as those for other cities.

In addition to this fairly conventional use of the New Town, for big-city overspill, these years were marked by an explicit recognition of the advantages of the New Town in regional development. This was most marked in Lord Hailsham's 1963 proposals for the north east, noted earlier, which included a specific proposal for Washington as a growth point/overspill New Town south of Tyneside. In fact there had been important local steps towards planned decentralization on the north side of the Tyne with the inception of Killingworth (1960) and Cramlington (1963) (Byrne, 1989) as large 1952 Act town developments to take overspill from north Tyneside. The government was attracted to the arguments for merging New Town and regional growth point policies, however, because that allowed more direct central control over growth policy and Washington was duly designated in 1964 (Hole, Adderson and Pountney, 1979; Holley, 1983). There were similar though less directive intentions in Scottish proposals of the same year for a fifth Scottish New Town at Irvine.

New Town designations 1964–70

The issue of further New Towns for the south east came to the fore with the publication of the South East Study in 1964, noted above (Cullingworth, 1979). Without being specific as to the mechanisms for development, it proposed major new cities for Bletchley (Buckinghamshire), south Hampshire and Newbury (Berkshire), new towns for Stansted (Essex) and Ashford (Kent) and large expansion of the existing large towns of Ipswich, Northampton, Peterborough and Swindon, all of which were actually just outside the south east itself. Yet Labour ministers were profoundly sceptical of the impact of such massive investment on the most prosperous region relative to the less favoured regions. But it was at just this time that the population forecasts were suggesting the biggest increases, which gave the study considerable credibility. On this basis a MHLG–DEA coalition of Crossman and Brown carried the day against the scepticism of the Board of Trade and the Scottish and Welsh Offices.

The Bletchley new city, renamed Milton Keynes, was designated in 1967 (Bendixson and Platt, 1992). Peterborough was also designated in 1967 (Bendixson, 1988) and Northampton (Barty-King, 1985) the following year. Both followed the New Town legal framework, though with the intention that their development corporations would operate more in partnership with local government. South Hampshire's expansion was left to the local authorities. Ipswich got to the stage of a public inquiry in 1969 but was then dropped. Agricultural opposition was a major factor, though by that stage the population projections were dropping, casting doubts on the validity of the study's assump-

Figure 6.15 *The new city of Milton Keynes epitomized much of the spirit of 1960s' planning. Its American-inspired grid road layout reflected the presumption of a far higher car-ownership than in earlier New Towns (and lent itself to easy expansion should growth pressures continue).*

tions of just five years earlier. This also allowed the remaining proposals to be dropped or severely trimmed. Thus Newbury and Stansted were dropped, Swindon continued as a large 1952 Act Expanded Town and Ashford became a small one. There were additions to the targets of some of the earlier New Towns to compensate for these changes.

These important developments in the planned decentralization of the south east were paralleled by provincial initiatives (Aldridge, 1979; Cullingworth, 1979). Irvine, recommended in the 1963 Central Scotland Report, was designated in 1966. In 1967 a small New Town was designated at Newtown, to secure rural regeneration in mid-Wales. It represented the cautious implementation of a proposal for a much larger development based on Caersws. The following year Manchester finally got a New Town at Warrington, where the Risley proposal was incorporated in a wider scheme for renewing an older industrial town. Also in 1968 Dawley New Town area was greatly expanded to become Telford, following the general trend by incorporating several existing small towns. Finally, in 1970, the long-promised Central Lancashire New Town was designated to serve Manchester overspill and regional development objectives. Much of the delay had been because of the need to address the fears of north-east Lancashire that they would be left behind by the new growth point based on Preston/Leyland/Chorley.

New Town designations 1970–4

Although the peak of growth expectations had by now passed, government commitment to New Town policies was undiminished by the return of the Conservatives in 1970. Yet, following Ipswich, there were more signs that the New Town solution might not be as relevant as it had seemed a decade earlier (Aldridge, 1979). Thus in 1972 a draft designation order was laid for Llantrisant New Town in south Wales, to serve a regional regeneration function, as recommended in a study published in 1969. A public inquiry in 1973 brought fierce opposition from the nearby local authorities, worried about their own viability, however, and the proposal was dropped. In Scotland, yet another Glasgow overspill/regional growth point New Town was designated at Stonehouse in 1973. Significantly, the new Strathclyde Regional Council (the post-reform upper-tier 'county' authority for the Glasgow area) was soon criticizing its irrelevance to Glasgow's needs. Moreover no action was taken on the major planned population and economic expansions mooted for Severnside, Humberside and Tayside at the high point of mid-1960s' growth optimism. Perhaps the last breath of this optimism came in a further new city proposal that was announced to accompany the decision (subsequently reversed) to develop Maplin (Essex) as London's third major airport in 1973 (Hall, 1980). But though it was not immediately apparent at the time a new agenda was already in the making.

Later New Town plans

This later wave of New Towns showed many important differences to the earlier New Towns, except Cumbernauld (Osborn and Whittick, 1977). Most obvious were the higher population targets. All the 1955–64 New Towns had population targets of 65,000–90,000 at designation; most were 70,000. Thereafter, with the exception of Newtown, which had a target of only 11,500, the target figures moved up even more dramatically. Irvine, the first of the post-1966 group, had a target of 116,000, but Peterborough (175,000), Warrington (205,000), Northampton and Telford (220,000), Milton Keynes (250,000) and Central Lancashire (500,000) soon breached this. The targets of the 1955–64 New Towns were further increased (typically by 10,000–20,000) in the later 1960s, reflecting the pressure of upward population projections.

All the later New Towns were planned with a much stronger sense of the motor car than earlier schemes, though there were some variations in how planners interpreted this. A high degree of vehicle–pedestrian segregation was now the norm. Runcorn was unusual in having a strong public transport orientation, based around a figure-of-eight busway that was essentially a closed adaptation of the Soviet and MARS linear city idea of the 1930s (see Chapter 3). In most cases, though, planners absorbed the main message of *Traffic in Towns* (MT, 1963) and aimed to give the greatest possible freedom to car use. This was nowhere more so than at Milton Keynes (Bendixson and Platt, 1992), which aspired to become a little piece of California in north Buckinghamshire. Early ideas for a public-transport-based city were quickly discarded in favour of a Los Angeles-type extensible road grid (though with very British roundabouts at the junctions). It was a solution which had already been rehearsed by the same planners at Washington, though for the historian it is the plan for Milton Keynes that remains the most perfect expression of the essence of the 1960s' 'planning revolution', a hugely optimistic scheme for planned affluence.

Detailed planning aspects

Reflecting these higher targets, most of these New Towns were planned at significantly higher densities than earlier schemes. This encouraged a much stronger sense of design and architectural innovation, in contrast to the fairly conventional architecture of the earlier New Towns. The short brick-built terraces of pitched-roof dwellings characteristic of, for example, Stevenage or Hemel Hempstead gave way to tighter and more visually exciting groupings of flat or monopitch-roofed dwellings, often built by non-traditional means in Cumbernauld, Livingston, Runcorn or Washington. (Though this is not to say that they were any more popular than the earlier schemes.) In fact the innovative modern architecture of the later New Towns was partly compromised by the increasing emphasis on owner-occupation and private development in the later New Towns (Aldridge, 1979). Reflecting the manner of their growth, the New Towns contained many of the ambitious young working-class families who were most anxious to take advantage of the bipartisan political push for home ownership that had dominated government housing policies since the 1950s. Amongst other things, this suggested that the later New Towns ought to have a more equal balance of public and private housing development which, conveniently enough, also reduced the costs of New Town development to the Treasury. In 1966 the

Figure 6.16 *The 1960s' New Towns were often more adventurous in detailed design aspects than their predecessors. This shows the Donwell area of Washington, with full vehicle segregation from bicycle and pedestrian movement, careful landscaping, site planning and grouping of buildings.*

target of a fifty-fifty tenure split was adopted as policy for all the New Towns. In fact the southern and some of the midland New Towns attracted a good deal of private development, usually of rather more conventional design than public-sector schemes.

Other lifestyle changes were reflected in neighbourhood or district planning. Although it had become less fashionable to refer to neighbourhoods by the 1960s, the concept was very deeply embedded in New Town planning and was essentially adapted and often renamed (e.g. as a district) to reflect social changes. One of the most obvious changes came in the field of local shopping. The earlier New Towns would typically include parades of small neighbourhood shops. Later New Town non-central shopping was more likely to be based around a large supermarket, with a much wider catchment area than the traditional neighbourhood, reflecting widening car ownership and the increasing prominence of larger retailers in food retailing. The higher densities of the later New Towns also allowed a greater reliance of many housing areas on central area shops.

Other aspects of planned decentralization 1960–74

The New Towns invariably dominate any discussion of decentralization, planned or otherwise. In fact, however, town developments, though they never rivalled New Towns in the provision of planned decentralization, took on a new lease of life in the 1960s (Herington, 1984, 1989). Schemes such as Basingstoke, Killingworth, Cramlington, mentioned above, and several smaller schemes for Birmingham and Liverpool such as Tamworth, Daventry, Ellesmere Port and Winsford at least showed the potential of this procedure during the 1960s. The larger schemes particularly took on something of the character of the New Towns, though they were rarely able to generate the same degree of professional interest. Meanwhile cities such as Sheffield, Leeds, Nottingham, Leicester, Bristol, Cardiff (and, possible growth areas notwithstanding, Hull and Southampton–Portsmouth) continued to order their own 'planned decentralization' largely through peripheral town extension or *ad hoc* arrangements with adjoining areas (Gordon, 1986). Even cities that did secure New Towns, like Birmingham, also continued to undertake these kinds of schemes, most notably at Chelmsley Wood, noted earlier (Sutcliffe and Smith, 1974).

'Unplanned' decentralization

Finally we should note the growing significance of 'unplanned' decentralization, whereby largely private development occurred outside New Town, town development or other substantially planning-initiated schemes. (They were of course subject to local planning controls.) This became more important as the volume and relative importance of private building grew during the 1960s and early 1970s. Private housing completions (in the UK) were running at an annual average of a little over 200,000 in the ten years 1964–73, representing 54 per cent of total completions (McKay and Cox, 1979). Virtually all this development took place in urban-fringe locations. As a very rough comparison, New and Expanded Towns accounted for only 3.7 per cent of new public and private housing built in England and Wales between 1945 and 1969, though the proportions were higher in Scotland (and in metropolitan areas with well-established planned decentralization schemes, especially Glasgow and Liverpool) (Hall *et al.*, 1973, Vol. II).

The larger truth was that large, new, private, suburban or semi-suburban areas were rapidly developing. These were mainly grouped around smaller towns in outer metropolitan areas within or beyond green belts, or as peripheral developments around small and medium-sized free-standing towns and cities without green belts. In a loose way their development was being controlled by the planning system, but the strategic relationships of home, work and services were not planned in any overall sense. In some places a real coherence was maintained that exceeded that even of the New Towns (for example at Luton). And there had been a couple of attempts at private New Towns, in Kent (Cullingworth, 1979; Hebbert, 1992). The first was a 25,000 population scheme proposed by Costain, the major housebuilders, for the Isle of Grain in 1956, which was given short shrift by MHLG. As noted earlier, the much smaller 6,000 population scheme at New Ash Green was approved in 1964, though it never achieved the promised coherence.

In fact the cumulative outcome of this 'unplanned' decentralization was a great increase in commuting and, though it is more difficult to quantify, journeys to services (Hall, 1963; Hall *et al.*, 1973). Some differential shift in employment from the core big city areas to outer-city areas was already apparent by the 1960s, though only in the New and Expanded Towns was this occurring in a coherent relationship with housing development. Retailing, especially food and non-luxury items, was showing a strong tendency to shift to reflect population decentralization. As yet this was on a modest scale. The planned suburban shopping centre was the most typical manifestation, for example at Cowley in Oxford, opened in 1963 (DTEDC, 1968). But there were occasional glimpses of more dramatic North American-style decentralized retailing, particularly the 1964 application for an out-of-town regional shopping centre on the American model at Haydock Park in Lancashire (DTEDC, 1971). The refusal of this application, following a public inquiry, combined with the tremendous encouragement given by planning to central area redevelopment had important long-term consequences for the nature and extent of 'unplanned' decentralization in Britain.

PLANNING 1952–74: CONCLUSIONS

The 1952–74 period and especially the 1960s were, then, tremendously important ones for planning. As this chapter has emphasized, the concerns were essentially those of urban and regional modernization and the management of a growth process within a mixed economy. The shifts in emphasis closely reflected the changing social and economic agenda, particularly the changes in population projections and the extent of optimism about economic growth. More even than the last chapter, which showed some party differences in attitudes of the planning system, the theme has been one of party political continuity. Wider changes were mediated, not so much through party politics, as through the internal departmental politics of ministers and Civil Servants. The tensions between the Prime Minister, MHLG and DEA over the creation of what was potentially a radical new planning machinery in 1964–5 is a clear indication of this.

Yet while departmentalism was a crucial element in circumstances of high party political continuity, other elements were important. The growing significance of private development interests, especially in central and inner-city redevelopment policy, though also in the extent and nature of 'unplanned' and even planned decentralization, was a fundamental element in understanding the general trajectory of urban planning policies over these years. Industrialists remained broadly committed to interventionist regional policies which basically paid them to locate in areas with a cheap labour supply. Agricultural interests and middle-class suburbanites were powerful forces shaping containment policies. Meanwhile urban and regional renewal was broadly seen as a benefit to working-class interests, particularly by their elected leaders though, as we have seen in practice, some of these benefits were, to say the least, dubious. Overall therefore there was a broad, not just a party political, consensus supporting these major strands of planning policy over these years.

The exact character of these policies and their evolution over time was, however, determined at the departmental and technical level by Civil Servants and professional planners. Again we have seen evidence of the tremendous fertility of planning ideas, especially in the 1960s, yet all these ideas arose within the institutional frameworks of the policy process or (in redevelopment partnerships, planning gain and package deals) in its interface with the private development industry. It was all a long way from the tradition of radical and independent thought from which planning had grown. At best the planning movement offered a critique of policies, but it was no longer a great source of new ideas for planning.

7

REMAKING PLANNING SINCE 1974: I. THE CHANGING SYSTEM

INTRODUCTION

The last two decades have witnessed dramatic and unprecedented changes in the policies and practices of town planning. The whole notion of planning was deflected to a political trajectory much less favourable than that which it had followed in the previous three and a half decades. Even the very conception of planning as a coherent policy approach was threatened. Although the policy roles of planners are now certainly more varied and diverse than ever before, their strategic ability to shape or even influence the processes of urban change is much diminished. The very notion of a comprehensive approach to urban and regional planning policy, pursued with a genuine (even if not always total) political and administrative conviction from the 1940s to the 1970s, has now become fragmented into a series of piecemeal and disjointed initiatives.

The most common explanation offered for this dramatic remaking of planning can be summed up in one word: Thatcherism. Certainly we will not dissent from the view that the radical right-wing pro-market Conservative governments led by Mrs Thatcher from 1979 to 1990 played a central role in this redirection. But we will also show how the end of the long post-war boom in 1973–4, the worldwide retreat from Keynesian economic management and the onset before 1979 of a prolonged period of economic restructuring and high unemployment were also of fundamental importance. It was these pressures and the responses to them by the left as well as the right wing which broke the political consensus about the mixed economy and the welfare state on which post-war urban and regional planning policies had rested.

More controversially, we will also argue that the British planning movement must carry some responsibility for its own weaknesses. Many planners had supported or acquiesced in profoundly unpopular planning solutions during the post-war years, undermining public support. With few exceptions, they had also allowed themselves to lose the strong independent tradition of radical thought out of which their activity had originally grown. By tying planning thought so closely to the policy process after 1945, planners had in effect lost their independent ability to reinvent their activity in ways that were more appropriate to the changing circumstances of the 1970s and beyond. It is particularly significant that the new environmental ideas of the 1970s and 1980s did not spring from the planning movement (as they might well have done in the 1900s or even the 1930s). They were, moreover, conceived in ways that saw planning as part of the problem rather than of the solution. This intellectual vacuum at its heart made the whole conception of planning very vulnerable to political shifts such as that which occurred after 1979. Before we can fully appreciate these problems, however, we must begin by examining the hugely important economic, social and political underpinning to the specific changes within urban planning. Planning was remade as part of the wider creation of a new political economy.

The Making of a New Political Economy

The economic and social context

The end of the long boom

The economy took a serious turn for the worse in 1973–4. Unwise budgetary stimulation of demand, the sudden quadrupling in oil prices imposed by the Arab-dominated oil producers cartel, OPEC, and mounting industrial unrest in Britain precipitated serious economic crisis (Blackaby, 1978; Morgan, 1992). In retrospect we can see that these events signalled the end of the long post war boom and the certain expectation of economic growth and rising living standards. It was, though, more than a series of one-off shocks that brought the boom to an end (Ball, Gray and McDowell, 1989). Underlying everything was a chronic structural weakness in British manufacturing within a changing world economic system. The destabilization of the international economy in the 1970s heralded the disintegration of the Keynesian world system of monetary and economic management created in the 1940s. Within this wider instability, Britain was also struggling to find a new post-imperial economic role in the world. The sheltered markets of its former colonies, important still in the long post-war boom, were now gone for ever.

Such changes highlighted Britain's profound inability to compete effectively in markets whose expansion was now less certain and which were now far more open to international competition. The writing had been on the wall in the 1960s and earlier in the form of the UK's relatively poor growth performance. Despite all the rhetoric of planned modernization, neither Macmillan's nor Wilson's economic planning experiments of the 1960s had tackled this central problem. Economic policies of the Labour governments from 1974 to 1979 and the Conservative governments since then were to be dominated by short-term management of the consequences of these weaknesses (Morgan, 1992) and the search for alternative models of modernization, without any enduring success.

De-industrialization

One of the most glaring symptoms of Britain's economic woes was the serious and sustained contraction in manufacturing (Townroe and Martin, 1992). Between 1974 and 1991 its relative share of the UK's GDP, already declining, fell from 30.2 per cent to 21.0 per cent. The decline in manufacturing employment was even more dramatic. Between 1971 and 1988 there was a net loss of almost 2.9 million jobs in British manufacturing, taking it from 36.4 per cent of total employment to 23.1 per cent. This loss affected virtually all sections of manufacturing, but was particularly marked in traditionally dominant activities such as metal working, engineering, vehicles and textiles, often with devastating effects for the towns and cities which had depended on them. There was, meanwhile, an even larger growth in services (though not sufficient to prevent a slight employment decline overall). Service employment grew by nearly 3.5 million, from 52.6 per cent of total employment in 1971 to 68.8 per cent in 1988. Yet there were marked variations in where old jobs went and new jobs appeared. To the traditional regional unevenness between the economically weaker north and west and stronger south and east was added a newer divide, between declining urban cores and buoyant outer cities and rural areas (Lewis and Townsend, 1989; Balchin, 1990). This underpinned a new spatial economic planning agenda in these years, emphasizing the urban more than the regional dimension. It also exacerbated problems of unemployment.

Rising unemployment

One of the most obvious manifestations of these shifts came in the abandonment of the hitherto sacrosanct post-war political commitment to ensuring full employment. In truth it was a shift which was effectively forced on governments faced with the greatest wholesale economic restructuring since the 1930s. We can, though, be certain that the application of more traditional policies would have produced lower unemployment in the 1980s, at least in the short to medium term. UK unemployment, measured in Department of Employment figures (which inherently understated the true dimensions of the problem) rose from an average 3.5 per cent in 1971 to 6.2 per cent in 1977 (Law, 1981). And, despite significant modifications in the method of compilation which further reduced the recorded rate, these figures rose even higher in the following decade. From 8.1 per cent in 1981, the figure climbed to 11.1 per cent in 1986. It then fell to 5.8 per cent in 1990 before moving sharply upwards again to 8.1 per cent in 1991. By early 1993 over 3 million workers were unemployed, even on flawed official figures. The real figure was clearly much higher.

. . . And rising affluence

For all the misery and uncertainty that unemployment, even short-term unemployment, created, the overwhelming majority of workers were in work. Moreover many of them continued to experience improvements in their standards of living, especially in the inter-crisis periods of economic growth of the later 1970s and 1983–9. The pursuit of affluence was very much alive and well, even within a profoundly troubled economy. Thus home ownership rose from 52.7 per cent of all British households in 1974 to 67 per cent in 1990. And, despite serious oil price increases in the 1970s, car ownership in Britain rose from nearly 12.6 million in 1971 to 20.25 million in 1991. Nearly a quarter of all households had regular use of two or more cars. The same trends could be found repeated across the whole range of consumer items as a wide section of the population felt itself better off than ever before. The story of the 1970s and especially the 1980s was at least as much one of easy credit and conspicuous consumption as of unemployment and economic restructuring.

The 'culture of contentment'

The co-existence of all this affluence with high unemployment (and the increased poverty that went with it) created an altogether more divided and fragmented society than at any other time since the 1930s and perhaps earlier. For all its faults, the 'affluent society' that J. K. Galbraith had written about in the 1950s had been a way of life that virtually all could realistically aspire to in 1960s' Britain. His 1992 book, *The Culture of Contentment*, described a society where the gap between the lifestyle of the contented and affluent majority and the poor minority had become unbridgeable. Once again he drew largely on the example of America, but we can see clear parallels in the Britain of the 1980s and 1990s. The legacy of social cohesion from the war years that had shaped the post-war politics of social welfare was all but exhausted (Deakin, 1987). It was true that opinion polls consistently showed a wide social commitment to interventionist welfare policies, especially in health and education (Crewe, 1988; Rentoul, 1989). But actual voting patterns suggested a widespread social unwillingness to pay more taxes to sustain policies of income redistribution. If it was not yet literally true that there was, in Margaret Thatcher's memorable 1987 phrase, 'no such thing as society' (Benton, 1987), the concept was certainly seriously weakened over these years.

The realities of discontentment

In line with all this, the symptoms of social fragmentation and breakdown multiplied (Ball, Gray and McDowell, 1989). The annual divorce rate more than doubled over these years, often pushing single mothers into poverty. Homelessness soared and every city experienced a very visible increase in numbers of people 'sleeping rough', a problem which had virtually disappeared in the welfare-oriented 1940s. Crime, especially violent crime, rose alarmingly, a reflection in part of the frustrated material aspirations of those outside 'contented' society. Racist attacks became more common and more blatant. There were also more instances of serious collective disorder and violence, notably at football matches, in industrial disputes and in the increasingly frequent riots which began to occur, mainly in inner-city areas, from 1980 onwards. None of these things had been entirely unknown before this time, but by the early 1990s they had become a familiar part of British life. The traditional, if rather romanticized, post-war notion of Britain as a disciplined and compassionate society was seriously eroded. It was a shift which was both reflected and influenced by political changes.

Politics, state intervention and the demise of economic planning

Political changes

The two general elections in 1974 saw the Conservatives lose, but did not give any decisive mandate to Labour (Childs, 1992). The first election brought Harold Wilson back to power as head of a minority government. The second gave him 39.2 per cent of the vote and a tiny parliamentary majority. This disappeared by 1976 and Labour (by now led by James Callaghan) ruled until 1979 as a minority government, dependent on Liberal and fringe-party support. The 1979 general election broke the political impasse, giving a decisive majority to the Conservatives under Margaret Thatcher, with 43.9 per cent of the popular vote. Her stridently successful leadership in the Falklands War, the serious divisions within Labour and the creation of the new Social Democratic Party allied with the Liberals ensured that her parliamentary majority was handsomely increased in the 1983 election (though the Conservative share of the vote actually diminished to 42.8 per cent). Better economic conditions saw her parliamentary majority (and popular vote) barely dented in 1987. Thereafter the Conservatives began to look much weaker. But after ridding themselves in 1990 of the (by then) hugely unpopular Mrs Thatcher, the Conservatives under John Major unexpectedly won the 1992 election with a majority of twenty-one seats.

The whole period since 1974 has therefore been dominated politically by the Conservatives. They have used their power to reshape the agenda of state intervention, with widespread implications for planning in the broadest sense.

Politics and state economic planning 1974–5

The prelude to all this was a period of policy uncertainty. Labour in the 1970s was being pulled in two directions (Coates, 1980; Holmes, 1985). Following its 1970 election defeat, the party had moved decisively to the left. There was a rejection of the mixed-economy planning of the 1964–70 Wilson governments and a reassertion of a more directive state economic planning approach. These concerns soon became apparent with the formation of the National Enterprise Board in 1975. This was originally intended to extend nationalization into profitable sections of industry to facilitate more effective government economic planning. Provision was also made for 'planning agreements' with private companies, committing them to agreed growth and other objectives (Ganz, 1977).

In practice though, such hopes soon fell by the wayside. As the economy worsened in the middle 1970s, the government found itself drawn into supporting loss-making industries whose very future was threatened. This 'lame ducks' approach had its origins before 1974, following the Heath government's famous 'U-turn'. Now the Labour government found itself taking over aircraft and shipbuilding and the only large British-owned section of the motor industry. Far from being considered acts of economic planning, these takeovers were products of desperate necessity.

The origins of monetarism 1975–9

This was all the more striking because the whole style of macro-economic management was also changing by 1975 (Peden, 1991). In the face of mounting unemployment and rising inflation, the government was turning aside from traditional Keynesian remedies in favour of more deflationary approaches (Healey, 1990). As Callaghan told the 1976 Labour conference, 'You cannot now, if you ever could, spend your way out of a recession' (cited in Morgan, 1992, p. 382). The Environment Secretary, Anthony Crosland, a hitherto committed Keynesian, had already instituted tight controls on local authorities' expenditure. 'The party,' he had told them in May 1975, 'is over' (cited in Morgan, 1992, p. 378). Many town planners and other local government officials saw this as signalling a clear shift from former priorities.

Mounting crisis in late 1976 necessitated a loan from the International Monetary Fund. The price of this was further heavy cuts in public expenditure in 1977–9 (and, incidentally, the first privatization, of most state holdings in British Petroleum). Although in the event the cuts were not as severe as at first appeared necessary, a new orthodoxy of macro-economic policy, based on monetarism and cutting state expenditure, had already been fashioned by 1979. But while Labour had moved in this direction out of perceived necessity, the Thatcher government was to pursue it with missionary zeal.

Thatcherite political economy

The philosophy of what was soon dubbed 'Thatcherism' essentially involved the wholesale re-establishment of market processes and reassertion of individual or private responsibilities in all spheres of economic and social life (Hall and Jacques, 1983; Skidelsky, 1988; Gamble, 1988). State expenditure was accordingly to be cut and government responsibilities passed into private hands. Although it took some time for critics to recognize this, it represented a serious attempt to modernize Britain through the supposed purifying effects of unfettered and widespread capitalism. Instead of the mixed economy, where the state supposedly sapped enterprise by excessive taxation and dependency on state benefits, a new 'enterprise culture' was to be fostered.

The philosophy which underlay it was not particularly new. In many respects it derived from eighteenth- and nineteenth-century political economists such as Adam Smith or Samuel Smiles, whose ideas of individual self-help and enterprise retained a strong appeal amongst rank-and-file Conservatives. Yet it marked a clear contrast with what had become orthodox post-war Conservatism, which (as we saw in Chapter 5) placed more emphasis on the need for government intervention, social welfare and a high degree of political consensus. Ideologically though, the 'Thatcherite' approach had been gaining ground for some time, even before Mrs Thatcher became party leader. It had, for example, been apparent in the Selsdon Programme which had formed the basis of the 1970 Conservative manifesto. Heath, however, had quickly allowed himself to be deflected from it. Mrs Thatcher made no such 'U-turn'.

'Rolling back the frontiers of the state' 1979–?

In fact, she applied this philosophy with a single-mindedness that was very unusual in a prime minister. She did not shrink from using state power in an authoritarian way to assert the legitimacy of market processes, as urban planners soon found (Gamble, 1988; Thornley, 1991). Alternative centres of power and authority, such as local authorities or trades unions, which threatened to frustrate the achievement of her strategy, were emasculated (Young, 1989). A climate of public expenditure restraint was instituted, far more severe than that under Labour. There was little attempt to use government spending to shield the economy or the workforce from the impacts of the severe recession of the early 1980s. Unemployment, as we have seen, rose alarmingly, but this was explained as part of the re-establishment of market processes. The shake-out of labour was justified as a necessary precondition for creating a leaner and more competitive economy. Small business formation was encouraged as a way of creating new sources of wealth.

Gradually, and especially when her position had become more firmly entrenched after 1983, extensive privatization was introduced (Riddell, 1989; Ball, Gray and McDowell, 1989). Telecommunications, airlines, gas, electricity, water, steel, aerospace, car making, shipbuilding and others were returned to the private sector. The profitable public utility sales particularly encouraged a threefold increase in popular share ownership in ten years, to about 22 per cent of adults in 1989. Moreover there was no immediate prospect of these changes being undone by a future Labour government as the opposition tacitly accepted much of what had been done. By 1993 plans were in hand for the privatization of rail and coal, undoing the last of the Attlee government's nationalizations. In other spheres too, the state was pushed back, notably in 'right to buy' housing legislation. This allowed council tenants to buy their homes, a major element in the jump in owner-occupation from 57 to 67 per cent over the 1980s. And there were many other examples of such policies; few areas of the public services remained untouched.

'Ever closer union'?

There was meanwhile one hugely important area of policy which largely transcended the traditional divisions of party politics (Nugent, 1993). This was the matter of Europe. Edward Heath had finally secured British membership of the European Economic Community in 1973, though continued membership was only confirmed following a referendum in 1975. Despite this, British participation was normally less than wholehearted. The reasons for this were complex, but they reflected Britain's traditions as a world power and trading nation operating on a much wider stage than Europe (Morgan, 1992). Its links with the USA and other parts of the English-speaking world remained strong. Its suspicions of other major European nations, particularly the French and Germans, remained profound. This kind of thinking was reinforced by Thatcherism, which emphasized the notion of Britain as a strong and independent nation state, still able to play a world and quasi-imperial role (such as saving the Falklands for democracy).

There was, moreover, resentment at the strongly interventionist tendencies of the Community, which were particularly galling to a national government which was committed to cutting state intervention within its own borders. The notion of a single market within Europe was approved of, but attempts to harmonize European currencies and perhaps move towards a single new currency met strong opposition. So too did the attempts to develop a stronger social dimension, for example in reinforcing workers' rights or providing more regional assistance. The growing environmental dimension of European policy was met with a rather sulky and halfhearted compliance within the British government (McCormick, 1993). Yet by the early 1990s few could deny the logic of some kind of closer union growing out of ever stronger trading links and symbolized by the building of the Channel Tunnel (opened in 1994). Certainly many town and country planners found much to approve of in European regional and environmental initiatives and saw the Community as a partial antidote to the broad anti-planning ethos of 1980s' Britain.

The end of consensus
The political changes of the last two decades have therefore been profound. Most importantly the broad, if sometimes rather battered, all-party consensus over the mixed economy and the welfare state that had dominated British politics since the 1950s was finally broken. We have seen in the previous three chapters how the comprehensive approach to physical and land use planning had been an integral part of this broader consensus. Physical planning too was now remade as part of a wider political reorientation. Its story parallels the broader changes, beginning with increasing strains on the familiar policy nostrums during the later 1970s (Donnison and Soto, 1980).

THE CHANGING PHYSICAL PLANNING SYSTEM

The limits of Labour intervention 1974–9

'Positive planning'
Physical planning reflected exactly the same contradictory pressures that were apparent across the whole spectrum of Labour policies. By 1974 the left had been able to set a physical planning agenda for Labour that was much more radical than that of the 1964–70 governments. The property boom of the Heath years and its precipitate collapse amidst numerous scandals in 1973–4 had created public sympathy for a stronger dose of state control over the land market (Reid, 1982). The Conservatives themselves had moved in this direction in 1973, with proposals for a development gains tax, actually enacted by Labour in 1974. There was much talk of 'positive planning' (McKay and Cox, 1979), a theme which was developed in the White Paper *Land*, issued in the same year (DOE, SO and WO, 1974):

> Public ownership of development land puts control of our scarcest resource in the hands of the community, and thereby enables it to take an overall perspective. In addition, by having this land available at the value of its current use, rather than at a value based on speculation as to its possible development, the community will be able to provide, in places that need them, the public facilities it needs, but cannot now afford because of the inflated price it has to pay the private owner.
>
> (DOE, SO and WO, 1974)

This policy statement was largely authored by John Silkin, the left-wing Minister for Planning and Local Government at the DOE. It heralded the most radical attempted solution to the development land values question since Silkin's father, Lewis, had introduced the 1947 legislation (Ratcliffe, 1976; Ambrose and Colenutt, 1975). Indeed, it was in some key respects even more radical than 1947 and marked a return to the principles of the Uthwatt Report of 1942 (and the even older ideas of the pre-1914 land nationalizers and early planners such as Nettlefold). It was an approach which Environment Secretary Anthony Crosland, Silkin's superior, endorsed though the department's Civil Servants sought, unsuccessfully, to delay the proposals (Silkin, 1987, p. 109).

Community Land Act 1975 and Development Land Tax Act 1976

The 1975 Act gave local authorities greatly enhanced opportunities (and ultimately duties) to acquire development land by compulsion at (or initially near) the existing use value. The 1976 Act introduced a new tax on land value increases, initially set at 80 per cent (lower on the first £150,000), but with the intention that it would rise to 100 per cent as the Community Land Act became fully operational. The ultimate intention of these Acts was that development land would be publicly owned at the crucial point when the planning decision to develop was made. It might then be sold on, at development value, to private developers, but retaining the increase in land value that arose from the planning decision to allow development. Developers whose land remained outside public ownership would enjoy no financial advantage under the final scheme because of the 100 per cent development land tax which was ultimately intended. In the interim, however, local authorities were to build up their land acquisition capacity gradually as resources and expertise permitted. They were also supposed to improve their working relations with developers, so that the scheme would, it was hoped, ultimately take on something of the character of a partnership. There were many promised opportunities for exemptions and exceptions, especially of smaller and socially desirable developments.

The Community Land Scheme in operation

The Community Land Scheme, as the two Acts were termed, differed from the Land Commission by relying on local authorities as the land acquisition agencies (except in Wales where a principality-wide Land Authority was created). Local authority mistrust had been one of the main reasons for the failure of the Land Commission, as Labour itself readily acknowledged. Despite this, however, the new scheme fared no better than its predecessors, though for somewhat different reasons. The Conservatives, predictably, soon committed themselves to repealing it, but reaction from the property industry was surprisingly muted (Pilcher Committee, 1975). The new approach came at a time when the property market was seriously depressed after the Heath boom. There were, no doubt, hopes that it would act in the same way as the 1964 'Brown Ban', discouraging new commercial development and thereby protecting the profits of office schemes initiated before the downturn (Ambrose and Colenutt, 1975, p. 170). The promise of a more stabilized property market must also have seemed attractive to many commercial property developers who, rather like a group of wild party-goers, were nursing colossal hangovers from a boom that had turned sour.

Despite such unlikely allies, Labour's policies fell victim to the tighter restraints on public expenditure which were being introduced in 1975–6. Neither Crosland, who remained at Environment until 1976, nor his left-wing successor, Peter Shore, were able to give sufficient encouragement to local authorities to operate the Scheme at anything like the intended level (McKay and Cox, 1979). Both in effect subordinated their beliefs in positive planning to the new strait-jacket of monetarism. By 1979 little of significance had been achieved and the Community Land Act had become another failed solution, ripe for destruction. One of the most palpable signs of that failure was in the growing number of large derelict or vacant sites in the inner cities, and it was this which became an important new focus for planning concern under Peter Shore from 1976.

The origins of inner-city policies

The reorientation of urban planning towards the inner cities from this time was of such fundamental significance that it brought important intrinsic policy shifts within the planning system itself. We will address the detail of inner-city policies more fully in the next chapter, but they must also form part of the overall story.

The prelude had been a series of essentially social and community-based initiatives in the later 1960s, involving the Home Office and the Department of Education and focusing very much on the perceived inadequacies of the urban poor (Edwards and Batley, 1978). Such initiatives added a new emphasis to planning policies that were still dominated by clearance and redevelopment. However, a political reconceptualization of the inner city as a spatial coincidence of more fundamental social, economic and environmental problems began to occur in the 1970s (Lawless, 1986, 1989). We saw some of the early consequences of this in the last chapter, in a move towards gradual renewal, but it went much further after 1974. The redefinition was essentially driven by the specific local effects of wider economic problems within inner-city areas. As the location of many of the oldest industries and factories, inner cities were obviously most vulnerable to economic downturn. Closures and redundancies increased and unemployment in these traditional Labour strongholds rose well above the national average, worsening the already present problems of poverty and deprivation.

All this meant that inner-city policies were no longer just the prerogative of the social professions; they were becoming a problem for planners, requiring a more comprehensive approach. The DOE commissioned three important inner-area studies, of inner Liverpool, Birmingham and London, in 1972. And there was a growing volume of other work, notably from the Community Development Projects begun by the Home Office in 1969, stressing the inner city as a serious and deeply rooted problem. It was a conclusion which the DOE studies cautiously endorsed when they reported in 1977 (DOE, 1977a, 1977b, 1977c).

Inner Urban Areas Act 1978

Following an important 1977 White Paper (DOE, 1977d), a new policy framework emerged, the 1978 Act giving the necessary additional powers. The main intention was to empower inner-city authorities, largely Labour controlled, to encourage the economic development of their areas. In the longer term this approach was important because it showed a central faith in inner-city local government that was to be conspicuous by its absence in the years that followed. At the time, however, its chief significance was that it marked a clear break with the housing and roads emphases of earlier post-war planned inner-city change.

Three new categories of policy areas were created: partnership, programme and designated, to all of which additional funding and special new powers were granted. The partnership areas covered the severely deprived areas of the big cities and were managed by committees of representatives from central and local government, health and police authorities and voluntary agencies. Simpler arrangements were operated by local government in the other areas. The new powers available to all these areas allowed loans and grants for industrial building and the creation of Industrial Improvement Areas (DOE, 1986). The latter allowed for the renewal and improvement of run-down industrial areas in a way that had been pioneered under local powers earlier in the decade in Tyne and Wear and Rochdale.

Plan making and inner-city problems

Another feature of the 1978 Act was the power to expedite local plan preparation in inner-city areas, even in advance of approved structure plans. It was, admittedly, of minor significance, but it illustrated an attempt, at least, to integrate the new inner-city policies within mainstream plan-making activity. Paradoxically though, it also showed the strains that the new problems and policies were placing on the 1968 Act system. Inner-city disinvestment was bringing rapid changes that

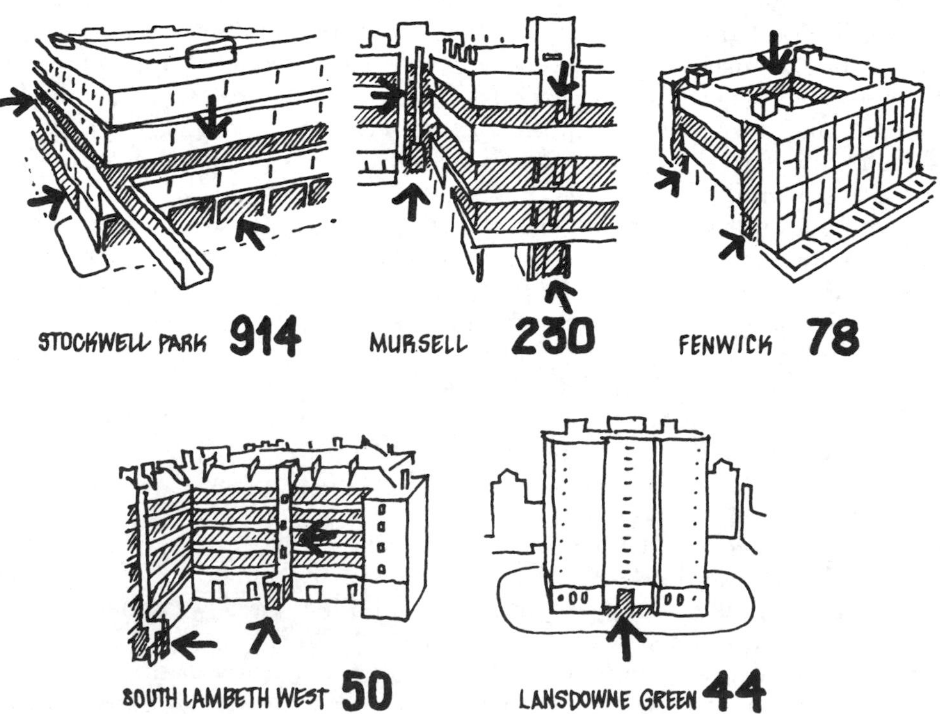

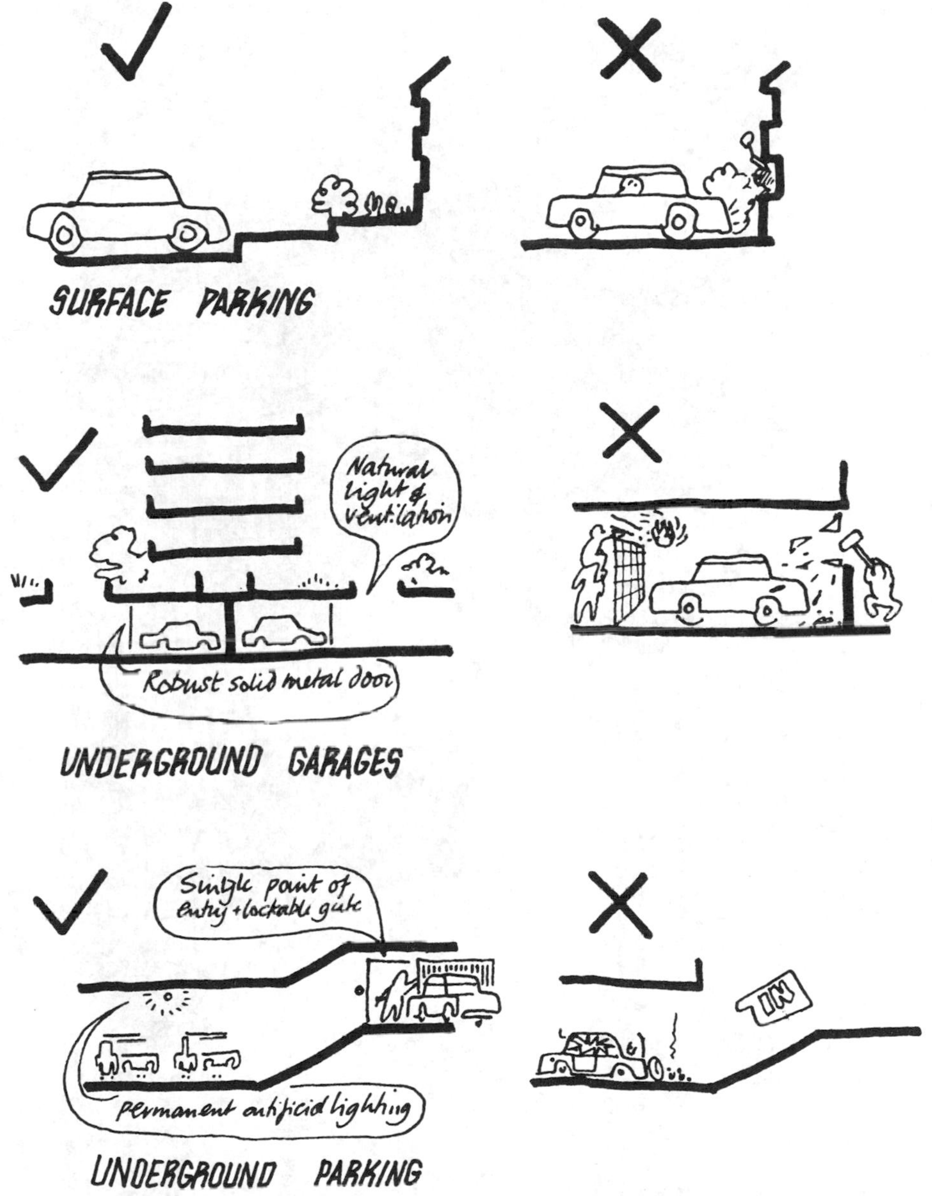

Figures 7.1 and 7.2 *One of the major discoveries of the 1970s was the 'inner-city problem' with its interlinked economic, social and environmental dimensions. These diagrams, from the 1977 DOE study,* Inner London, *show how previous planning efforts in the inner city were now being criticized. The level of crime on flatted estates was directly correlated with the number of dwellings accessible from a single entry point. The criticism of deck-access blocks (see Figure 6.8) was particularly damning. Planned parking arrangements on inner-city estates were similarly attacked as often aiding the vandal.*

TYNE AND WEAR ACT

Tyne and Wear Act, 1976

The Act, in part, enables the County and District Councils to tackle in a comprehensive way, the problems of older industrial areas and proceed beyond the mere identification and analysis of these problems towards implementing practical measures that will considerably enhance the employment opportunities in declared I.I.A.'s by providing a firm foundation for private investment.

Although the Act specifically mentions a number of works which may be carried out in I.I.A.'s, a wide range of works which will improve the amenities of an area or buildings within it can be undertaken either by the Council or in conjunction with individual industrialists. To maximise the potential of vacant derelict sites, for example, authorities may prepare them for development by a combination of reclamation works, provision of access roads and ensuring an adequate supply of services. The land is then available for either release to a private developer or for development by the local authority. The County Council has successfully provided a number of nursery and factory workshop units within I.I.A.'s and elsewhere. Similarly, redundant buildings may be refurbished and brought up to basic industrial standards for re-use or demolished to allow for the expansion of an existing industry, or redevelopment by a new industry.

In carrying out many of these works, the provisions of the Tyne and Wear Act 1976 are augmented by various other Acts. The main ones are:-

- **Local Authorities (Land) Act 1963 (for factory building).**
- **Inner Urban Areas Act 1978 (for grants and loans to industry).**
- **Local Employment Act 1972 (for grants towards the costs of the provision of access roads).**
- **National Parks and Access to the Countryside Act, 1949 (for site purchase in connection with reclamation schemes).**

***Figure** 7.3 The Industrial Improvement Area marked the beginnings of local economic planning in inner-city and older industrial areas. Tyne and Wear were one of the pioneers, using local powers, though the approach became widespread under the Inner Urban Areas Act 1978.*

could not wait for structure plans to be prepared. Many local authorities increasingly judged that they could not wait for local plans to be prepared either, or otherwise doubted the real value of statutory plan making in a period when the wider economic and public expenditure parameters were changing so dramatically (Healey, 1983; Healey *et al.*, 1989).

The structure plans

Despite these increasing doubts, the statutory structure planning process was in full swing during these years (Cross and Bristow, 1983; Bruton and Nicholson, 1987). By March 1979 fifty-one English structure plans had been submitted, including seventeen, mainly for the west midlands, before local government reorganization took effect in April 1974. Twenty-five of the plans submitted had received final approval by 1979. The first approvals were for Coventry (which also managed the first local plan), Solihull, Warwickshire and Worcestershire, in 1975. By 1979 twenty plans still had to be submitted, though most were in an advanced state. This rate of progress fell rather below the original hopes of PAG and the 1968 Act. In part this had institutional causes. As we noted in Chapter 5, local government reform had taken a rather different course to that presumed in the late 1960s. Instead of the unitary authorities assumed by the 1968 Act, a two-way split of planning between counties and districts tempted many counties to include excessive detail in their structure plans to pre-empt the districts' local plans and assert overall county dominance in plan making.

More seriously, structure planners in the main urban areas were also handicapped by the wider economic and expenditure uncertainties noted above. Forecasts had to be revised and the agenda for action had to be extended to address the new economic problems, even though financial restraint was encouraging greater caution. Tensions were therefore building up, especially in the metropolitan areas, between planning aspirations and expenditure realities. The presence of a Labour government in the 1970s still allowed local policy-makers to believe that resources would eventually be available when the economy improved. They were forced to relinquish such illusions in the following decade.

The emerging environmental agenda

In addition to the planning issues raised by a faltering economy and falling public expenditure, a third, less immediate, pressure for change was also beginning to manifest itself in these years (e.g. Nicholson, 1970). Environmentalist thinking and environmental control had a long history, though they became more prominent during the 1960s, reflecting both the waste-making affluent society and vaguer worries about possible limits to growth. What was new was the way that issues such as pollution, waste disposal, land and energy use were increasingly seen as parts of a single ecological whole, rather than as separate and specific topics. The establishment of the Royal Commission on Environmental Pollution in 1970, with a remit to report regularly on relevant issues, marked some official recognition, serving to inform and focus these mounting concerns (Cullingworth, 1976).

ADVERTISEMENT

Your Chance to Comment...

In preparing the Draft Structure Plan certain key choices had to be made.

They concerned for example:-

★ Which areas should have priority for investment? ★ Which facilities should have priority for investment?
★ Where should new development be located? ★ Which social groups should receive special attention?

This questionnaire is YOUR OPPORTUNITY to comment on the choices made in the draft plan. It is particularly important to know whether *you agree or disagree with the priorities and general approach* expressed under each topic below

Please fill in this questionnaire, go to one of the exhibitions* or to a public meeting* if you can.

1. **EMPLOYMENT** — *The growth of employment will be encouraged throughout the county but local authority resources will be concentrated on increasing the number and variety of jobs in areas with the worst unemployment and the worst prospects for the future.*

   Please comment

2. **ENVIRONMENT** — *Environmental improvements, such as land reclamation and increased urban space provision, will be concentrated in areas with the worst environmental problems.*

   Please comment

3. **TRANSPORT** — *Transport policies concentrate resources on maintaining cheap bus fares and improving services to provide an attractive public transport system.*

   Please comment

4. **HOUSING** — *Housing policies provide for small amounts of new housing throughout the county to cater for local needs and concentrations of new housing in a few attractive areas close to job opportunities and the four main centres.*

   Please comment.

5. **SHOPPING** — *Shopping policies allow some growth of the four main centres, encourage trade in the district and small town centres, and strongly discourage the development of large superstores outside existing centres.*

   Please comment

6. **RECREATION** — *Recreation policies give top priority to local authority schemes which promote the maximum use of existing facilities and provide more urban open space.*

   Please comment

7. **SURFACE MINERALS** — *Demand for mineral extraction will be catered for, but steps will be taken to keep environmental problems to a minimum and to restore land to a suitable after-use.*

   Please comment

If you wish to enlarge on any of the above subjects or, indeed, express opinions on any other policies please comment on separate sheets of paper. Send in the completed form to the County Planning Officer, FREEPOST (BY 158), South Yorkshire County Council, County Hall, Barnsley S70 2TN, by April 4, 1977. No stamp needed. *Please write your name and address here if you so wish*

Name Mr/Mrs/Miss ....

Address ....

• **SHEFFIELD**

Structure Plan Exhibitions

Sheffield Town Centre	
W. H. Smith, Fargate	Weds 9th March Tues 29th March

MOBILE CARAVAN EXHIBITIONS
(Opening most days 10 a.m.-5 p.m.)

STOCKSBRIDGE, Library	Thurs, 10th March
CHAPELTOWN, Chapeltown Working Men's Club	Fri. 11th March
DARNALL, Darnall Crossroads	Tues, 15th March Weds 16th March
MOSBOROUGH, The Royal Oak, High Street	Tues 22nd March
HILLSBOROUGH, The Old Blue Ball, Bradfield Road	Tues., 22nd March Weds, 23rd March
GREYSTONES, Challenge Supermarket, Ecclesall Road	Weds 23rd March, Fri. 25th March
GLEADLESS, Co-op, The Ridgeway	Mon, 28th March

PUBLIC MEETINGS
(Beginning 7.30 p.m.)

STOCKSBRIDGE, Stocksbridge Welfare Hall	Tues, 15th March
DORE Village Society Church Hall, Townhead Road	Tues., 22nd March, 8 p.m.
ECCLESFIELD, Ecclesfield Comprehensive School	Thurs, 24th March
MOSBOROUGH, Westfield School	Mon., 28th March
DARNALL, Greenlands Middle School	Tues., 29th March
FIRTH PARK, Firth Park School	Thurs., 31st March
SHEFFIELD, Channing Hall, Surrey Street	Fri., 1st April

• **ROTHERHAM**

Structure Plan Exhibitions

Rotherham Town Centre	
Rotherham Central Library -Foyer	Weds., 9th March Tues., 29th March

MOBILE CARAVAN EXHIBITIONS
(Opening most days 10 a.m.-5 p.m.)
will be centrally situated in the following towns:

Thurcroft, Spar Stores, Arbour Road	Thurs, 24th Feb.
MALTBY	Fri. 25th Feb. Sat. 26th Feb.
WICKERSLEY, Watsons Chemist, Bawtry Road	Weds 2nd March
SWINTON, Shopping Precinct, Station Street	Fri. 4th March
WATH, National Westminster Bank, Montgomery Square	Fri. 4th March
RAWMARSH, Parkgate Service Station	Sat 5th March and Mon 7th March
KIVETON PARK	Weds 16th March
DINNINGTON	Fri. 18th March
ASTON-CUM-AUGHTON, Wilsons Greengrocer, Aughton Lane	Mon. 21st March
BRINSWORTH, Allens Chemist	Mon. 21st March

PUBLIC MEETINGS
(Beginning 7.30 p.m.)

MALTBY, Maltby Grammar School	Fri. 4th March
WATH, Wath Public Assembly Hall	Weds, 9th March
DINNINGTON, Rother Valley College of Education	Weds, 23rd March
ROTHERHAM, Town Hall	Weds., 30th March

For details of Barnsley and Doncaster exhibitions and meetings, see local Press or ring Len Bartle, Barnsley 86141 ext 647.

Figure 7.4 Elaborate public participation exercises became an important feature of plan making in the 1970s, as here for the South Yorkshire Structure Plan in 1977. The issues highlighted illustrate how the urban economy was worsening even as the structure planning process was proceeding. This encouraged planners into more interventionist stances on the economy, social policies and public spending, tempered by increasing public concerns about the environment.

In 1973 the Conservatives introduced a Protection of the Environment Bill, but hopes that it would live up to its name, for example by requiring environmental impact assessments for major new developments, soon faded. The bill was enacted by Labour as the Control of Pollution Act 1974, its title reflecting its more limited conception. The truth was that no political parties were willing to see really decisive new intervention on these questions. Government environmental concern was still minor, even token, compared to the maintenance of economic growth.

Environmental concerns and roads

Far from lessening worries, the end of the long boom in 1973–4 seemed merely to heighten popular environmental awareness, particularly as the OPEC oil price increases reminded everyone of the perils of heavy reliance on the motor vehicle. Road-building plans, already under serious assaults from increasingly militant protestors, were thrown into increasing question (Tyme, 1978). Disruption of major public inquiries such as those for the Archway Road in London in 1976–7 became usual. The Leitch Committee (1977) Report on Trunk Road Assessment promised a more considered approach to new schemes. Public transport, rather unfashionable in the 1960s, now enjoyed a revival, something which most planners were glad to encourage. The new Tyneside Metro rapid-transit system became a showcase for the new approach. Significantly though, its wider emulation was prevented by public expenditure constraints that were pervading all aspects of planning action (Hill, 1986). Meanwhile Callaghan's pragmatic decision in 1976 to take transport planning away from the DOE, creating a new Department of Transport, seriously weakened the ideal of integrated strategic planning.

Planners and the new environmentalism

By the later 1970s planners were becoming increasingly affected by environmental concerns. Yet despite their traditional concerns for amenity and landscape protection, they actually played little part in actively developing the new and more fundamental environmental ideas. For all their late 1960s'/early 1970s' flirtation with quasi-scientific systems planning, they lacked the genuine scientific expertise to handle issues such as pollution and its ecological impacts. There were also more fundamental differences of ideology. Planners, and the policy system of which they were part, had always seen their role as much more than simply protecting the environment. They had to balance such concerns against needs for development and growth which were generally seen as essential for national social and economic well-being. Alone amongst the bodies of the planning movement, only the Town and Country Planning Association showed a real inclination to engage at the creative edge of the new environmental radicalism (Hardy, 1991b). It was newer pressure groups like Friends of the Earth, established in 1971, that set the pace and caught the imagination of the young with their single priority of saving the earth above all else (Porritt and Winner, 1988). In the eyes of these groups, planners often seemed to be little more than an institutionalized bureaucracy that legitimated ecologically wasteful development and road building.

Planners in the later 1970s

This might have been true in the mid-1960s, but by the later 1970s most of the old post-war planning certainties had either gone or were seriously in doubt. The faltering economy, public expenditure restraint, the end of clearance policies, the new economic focus on the inner city (with all its potential implications for the traditional decentralist thrust of urban planning), the increasing questioning of road-based planning and other changes threw many of the traditional planning responses into doubt (Donnison and Soto, 1980). There was widespread professional guilt at what had been done to cities in the name of slum clearance and redevelopment (Thomas and Healey, 1991). These wider changes combined with the continuing effects of adjusting to the new management structures of the post-reform local authorities to undermine the self-confidence and sense of manifest destiny that planners had seemed to possess in the 1960s.

Planning and critical theory

An important aspect of this uncertainty was the growth of critical planning theory in the 1970s. This added an intellectual dimension to the 'planner-bashing' that was already becoming more noticeable in newspaper journalism from the late 1960s. Several different intellectual traditions were represented, including anarchism (Ward, 1973, 1976) and free-market ideology, especially in the work of Americans such as Banfield (1970). In Britain, however, the main influences in the 1970s were the neo-Weberians, led by Pahl (1970), Dennis (1970, 1972), Davies (1972) and Saunders (1979). They based their interpretations on the work of the sociologist Max Weber, emphasizing the role of planners and other officials as 'gatekeepers' or 'urban managers', managing the consumption of urban space. Increasingly though, neo-Marxism became more fashionable (Cooke, 1983), with Harvey (1975, 1982), Cockburn (1977) and Castells (1977) as major influences. Their approach viewed planning as part of a state apparatus that essentially facilitated the reproduction of capitalism, while reflecting the strains of class conflict. It was an approach that acquired greater credibility as the structural economic changes of the later 1970s and 1980s became more apparent, reminding everyone how much planning was conditioned by the wider capitalist economy of the cities.

Corporate planning

The problem for practising planners was that critical theory, however intellectually important, had no immediately obvious application in their work, except to increase their self-doubt. This effect was greater because the supply of bright ideas that were applicable had rather dried up. Yet there was one important planning idea that temporarily captured mainstream professional attention in these years. This was corporate planning, a supposedly rational approach to intra-corporate organization, programming and budgeting (Friend and Jessop, 1969; Friend, Power and Yewlett, 1974). A product of American management theory, its coming had been encouraged by local government reform and specifically by the Bains Report (1972). It was not the exclusive property of town planners, but some thought that it would strengthen the profession's role in local government, a view which was encouraged as several town planners became chief executives of the new authorities, most notably Jim Amos at Birmingham and Ken Galley at Newcastle. The problem was that the approach added to the growing confusion of what planning actually was, without being an appropriate tool for the harsher climate of expenditure restraint that was just beginning.

In truth town planners were no better equipped to handle these unpleasant realities than were other professions. Rather they merely seemed to mirror the larger confusions. Not unlike Britain herself, they had lost an empire as the dreams of the 1960s faded but failed to find a real role in the troubled 1970s (Underwood, 1980). Their search was not made easier by the ascent of a new and strident vision of Britannia after 1979.

Thatcherite ideology and physical planning 1979–90

Planning and the enterprise culture

Consistent with her larger mission to give a new direction to post-war British politics, Mrs Thatcher's governments also imposed major changes on physical planning. These proved particularly damaging to the traditional comprehensive and strategic approach. They also weakened many detailed aspects of planning, though not as much as many critics have supposed. These changes essentially reflected an ideological reassertion of the legitimacy of market processes and individual freedoms in determining the nature of urban change. With a few exceptions, they did not draw on the mainstream traditions of British planning thought, though some North American right-wing planning thinkers such as Jane Jacobs (1964) were important inspirations. More specifically, though, it was the work of right-wing think-tanks such as the Institute of Economic Affairs and, even more, the Centre for Policy Studies (e.g. Denman, 1980) that developed the new policy agenda (Thornley, 1991).

From such standpoints, physical planners were obstructive bureaucrats, all too ready to stifle wealth-creating private enterprise by unnecessary curbs on development applications. As early as 1979, Michael Heseltine, Mrs Thatcher's first Environment Secretary, was complaining that 'thousands of jobs every night are locked away in the filing trays of planning departments' (Heseltine, 1979, p. 27). Though the critique was of dubious validity, Heseltine had articulated a deeply popular Thatcherite theme, blaming governmental regulation for market failure. It was a sentiment which recurred regularly in ministerial rhetoric and, as we will see, inspired specific changes in the planning system.

Planning and shrinking public expenditure

Planners, especially in the main (and largely Labour-voting) urban areas, were also perceived as being far too wedded to high public spending in their whole approach to plan making and implementation. Much was made of the unpopularity of 1960s' mass public-sector housing redevelopment of the inner cities: 'social vandalism, carried out with the best of intentions but the worst of results' (Thatcher, 1989, p. 128). The fact that the practice had by then changed was rarely mentioned. There were also profound central suspicions of the new 'municipal socialism' (Boddy and Fudge, 1985), especially the council-led initiatives to regenerate employment that were being launched in some metropolitan areas (Cochrane, 1986). The whole debate acquired greater piquancy as councils in several of the big cities moved decisively to the left during the early 1980s and mounted an unsuccessful challenge to central policies. Central government looked for ways to make such authorities toe the Thatcherite line or marginalize them. Many of the shifts in physical planning directly reflected these concerns.

Thatcherism and the public interest

Market principles were advocated as a way both to create new wealth and to give people what they actually wanted, rather than what planners or other professional bureaucrats thought they needed. The role of the planner as a guardian of a professionally determined public interest was weakened as the political advocacy of popular capitalism gave more scope to market processes to shape urban change (Thornley, 1991). The traditional planning mission to protect the wider public interest was increasingly supplanted by the enhancement of private interest. In

***Figure** 7.5 Although the practice had changed during the 1970s, the dire record of much post-war planning for the inner cities presented an easy target for Thatcherite criticisms of interventionist policies in the 1980s. Derelict blocks of flats were depressingly common, such as these awaiting their end in the London Borough of Newham in 1988. Sisters of the notorious Ronan Point (Figure 6.9), they had been strengthened following the disaster, but intractable structural, design and social problems remained.*

many respects, indeed, the public interest was almost redefined in terms of the functional requirements of private wealth creation.

Planning and 'nimbyism'

What mainly stopped this redefinition being complete was an ideological contradiction between the enterprise culture and traditional Conservatism. The fact was that many Conservative voters living in outer-city and urban-fringe areas wanted a planning system that would stop their areas being changed by large-scale developments. However much they might approve of the enterprise culture in other walks of life, they did not want it, in the form of opportunistic developers, shaping their immediate home environment. Even modest moves to relax planning controls in such areas by Heseltine and later Environment Secretaries, notably Patrick Jenkin and the ultra-Thatcherite Nicholas Ridley, quickly provoked opposition from their own side (Ridley, 1992). Significantly the American acronym 'nimby' – Not in My Back Yard – came into common British usage during the late 1980s, accurately describing this sentiment of selfish protectionism. It was the environmental embodiment of Galbraith's culture of contentment.

The extent of the market-oriented changes to the planning system in the 1980s should not disguise the continuing strengths of such pressures for negative planning (Healey *et al.*, 1989). If the developer-led approach was replacing the ideas of public-sector-led 'positive planning' in the urban cores, its writ certainly did not run in the Tory fastnesses of the outer cities and rural areas (Brindley, Rydin and Stoker, 1989). Even leading Conservative ministers such as Heseltine and Ridley were personally deeply imbued with this contradiction between enterprise and environment. In office they both favoured the former, though as private members they were at least as 'nimbyist' as their constituents. It was a dilemma that socialist author John Mortimer (1990) savoured, fictionally, in his satirical novel, *Titmuss Regained*.

Other pressures for intervention

Various other pressures also obliged the Thatcher governments to maintain a more interventionist approach than they would ideally have liked. Not the least of these was the increasing work of the EC in setting a new environmental agenda against which Britain had come to be seen as the 'Dirty Old Man of Europe' (Cullingworth, 1988, p. 213) by the mid-1980s (McCormick, 1993). Europe was also important in maintaining something of the traditional planning commitment to the older industrial regions, which suffered particularly from the effects of 1980s' economic change. Another international pressure, in the field of historic conservation, reflected the functional requirements of a growing reliance on foreign tourism in the now post-industrial service economy (Hewison, 1987). Historic townscapes, though valued by resident populations, especially in more affluent areas, were increasingly recognized as valuable dollar-, mark- and yen-earning resources. Meanwhile the increased emphasis on private transport in the 1980s, reversing the late 1970s' scepticism, necessitated a new round of roads planning, albeit with a growing interest in privately funded toll schemes.

Thatcherism and the fragmentation of planning

In planning as in other policy spheres, critics of Thatcherism waited for it to collapse under the weight of its own inner contradictions. Instead it was planning which began to fall apart. The strains which were already apparent by 1979 grew as planning action was obliged to become increasingly schizophrenic (Brindley, Rydin and Stoker, 1989). Faced with the contradiction between their own belief in a market-led promotional approach to planning the Labour-dominated urban cores and the vocal demands for a 'nimby'-appeasing approach in the Tory outer-city areas, together with the various other pressures to intervene, Mrs Thatcher's governments simply weakened the strategic strands of comprehensivity that had bound post-war planning together. Planning was allowed to break down into a series of disjointed and pragmatic initiatives, operating in very different ways in different areas and policy settings. This was entirely consistent with other strands of Thatcherism. If there was 'no such thing as society', then it made no sense to plan as if there were. We must now begin to examine in detail how this retreat from comprehensivity occurred.

Early Thatcherism and the changing planning system 1979–83

Local Government, Planning and Land Act 1980

The first decisive step was not long in coming. Here, in a single piece of legislation, were all the main elements of the Thatcherite approach to urban planning (Thornley, 1991). The repeal of the Community Land Act was a decisive rejection of public-sector-led planning (though the Land Authority for Wales survived, albeit in more market-sensitive guise). The extension to all areas of the power to adopt local plans in advance of structure plans (available in inner-city areas since 1978) and a decisive downward shift in development control powers from counties to districts was symptomatic of the retreat from wider strategies, giving comfort to 'nimbys' everywhere. The introduction of monetary charges on applicants for the processing of planning applications was part of the wider erosion of a state funded by taxation. It also openly challenged the generally held presumption of planning as a service run in the public interest in favour of one which explicitly serviced the private interests of developers. Much the most important innovations of the 1980 Act were, however, the Enterprise Zone (EZ) and the Urban Development Corporation (UDC).

Enterprise Zones

The former, which had been unveiled in the 1980 Budget, was heralded as the very essence of the new enterprise culture. Surprisingly though, the Chancellor had credited the planning academic Peter Hall, then a Labour supporter, as inventor of the idea (Butler, 1981; Hall, 1982; Massey, 1982). Although a geographer rather than a professional planner, Hall had been one of the most influential, prolific and persuasive writers on planning matters since the 1960s. An ardent advocate of strategic planning, it was perhaps his misfortune to enter the political spotlight with an idea that was widely seen as anti-planning. The Chancellor's reference was to Hall's occasional calls for experimental unregulated zones, latterly based on examples in Hong Kong and elsewhere in the Far East, where highly profitable, if rather low-wage, enterprises had flourished.

The 1980 Act exhibited a rather cautious implementation of this concept. An Enterprise Zone (EZ) was a run-down urban area where a much simplified planning regime was agreed by local planners and central government (DOE, 1987b). Thereafter there was none of the normal local planning discretion over details of site development. There were also important fiscal bonuses in the form of a ten-year exemption from local taxes and 100 per cent capital allowances against national taxation. Yet although the EZ was founded on the notion of freedom from governmental interference, many of the areas designated were actually owned by the local authorities. Moreover the DOE actually supervised their creation very closely.

Urban Development Corporations (UDCs)

In the case of the UDCs, the more authoritarian face of Thatcherism was plain (Imrie and Thomas, 1993b). The Corporations were to be development and planning agencies specifically intended to facilitate urban regeneration in a way that entirely by-passed local government. Based on the unelected development

***Figure 7.6** Thatcherite ideology emphasized the notion of planning as bureaucratic interference in wealth creation. Enterprise Zones were heralded, somewhat fancifully, as a decisive break with this tradition, a theme which was taken up in this 1981 photo-opportunity. Michael Heseltine symbolically cuts the 'last piece of red tape' to inaugurate Britain's first such zone.*

corporations created under the New Towns Act 1946, UDCs had even more draconian powers, especially in relation to planning. They were, moreover, exercised in the midst of existing cities, replacing much of the established frameworks of local democracy. From the government's point of view, the UDCs' main advantage was that they were able to do its bidding in a way that was uncompromised by local, almost invariably Labour, politics. In sharp contrast to the original New Town development corporations, the UDCs used their powers of land assembly and preparation to give unprecedented encouragement to the promotion of private development.

We will deal with the use of these two new planning instruments when considering the detail of inner-city policies in the next chapter. It is important to note, however, that they were initially used on a modest scale. Only eleven EZs were created in the first round and just two UDCs. They were essentially a demonstration of Thatcherite planning principles, rather more tentatively than its advocates were claiming. It was more the establishment of new styles of planning policies and practices, rather than creating an entirely new system.

Changes in development control practice

These trends were also apparent in other changes to the planning system in these early years (Thornley, 1991). And in 1980 important modifications were made to official guidelines on development control practice, discouraging detailed planning controls on design matters. The following year a new General Development Order was issued, shifting the definitions of what constituted development requiring planning permission. It brought additions to permitted development rights in relation to small house and industrial extensions. Yet although these changes were consistent with the general thrust of Thatcherism, they reflected a widely agreed shift. Thus although Crosland had baulked at the streamlining proposals of the Dobry Report (1975), Shore had in 1977 come close to accepting changes very similar to those now implemented by Heseltine.

Planning gain

Also in 1981 the DOE's newly established Property Advisory Group issued a highly critical report on the pursuit of planning gain. We have noted the origins of this practice in the 1950s in Chapter 6, but it began to acquire a new prominence in the later 1970s. Using Section 52 of the 1971 Act, growing numbers of authorities sought to mitigate the effects of declining public funds by making planning agreements with private developers. Typically these involved developer funding of infrastructure or amenities associated with the development, though occasionally straight cash payments were involved. Such instances prompted accusations that planning gain was simply 'selling planning permission', though advocates argued that it could be a way of securing 'betterment by stealth' (DOE, 1992b). It was against this background that the report was issued (Property Advisory Group, 1981). The Group sought to discourage local planning authorities putting pressure on developers by such bargaining. Yet the even greater strains on public spending in the 1980s meant that this was hardly realistic. Although cruder manifestations of planning gain were curtailed, the practice of negotiating for infrastructure and amenity provision has become commonplace and was officially endorsed in 1983.

The later structure plans

More immediately important was the mauling that Heseltine inflicted on the later structure plans, especially those for the metropolitan counties (Heseltine, 1979, 1987). The expunging of proposals to restrain development in buoyant localities to encourage faltering areas to revive was particularly notable, in effect asserting the primacy of market- rather than plan-led development (Jowell, 1983; Thornley, 1991). Many public expenditure commitments and vaguer social planning aspirations were also struck out, an early symptom of the wider battle that was being joined with the Labour strongholds in the big cities.

The emergence of heritage

Of longer-term importance was the retreat from comprehensivity signalled in the National Heritage Act 1983. Many matters related to conservation and historic buildings in England, including listings and area schemes, were now taken over from the DOE by a new, fairly autonomous body, the Historic Buildings and Monuments Commission for England, usually known as English Heritage (Cullingworth, 1988). In Scotland, however, the equivalent body remained a Directorate of the Scottish Development Department, though CADW, a direct parallel to English Heritage, was established in Wales.

We should not underestimate the significance of these changes. Historic conservation had been a core dimension of comprehensive planning since its inception. It had grown hugely in importance since the late 1960s. Now it was being hived off into the new policy area of heritage and pursued more single-mindedly. Those aspects of conservation traditionally most closely related to planning continued to grow apace. By 1985 there were about 5,500 conservation areas in England, growing to about 7,500 in 1993, when they included about 4 per cent of the total building stock. Yet the qualitative changes were also important; a new ideology of conservation was emerging and the new term 'heritage' was significant. It was widely taken as implying greater emphasis on marketing the historic environment rather than conservation for its own sake (Hewison, 1987). And it was a promise which English Heritage increasingly fulfilled.

Towards a new Thatcherite planning system 1983–7

The entrenchment of the Thatcherite approach

Following the 1983 election, Thatcherite thinking became firmly entrenched and its proposals for planning lost much of the tentative, experimental air of the early years. The first reforms began to make their mark. In addition there had, since 1981, been a multiplication of initiatives for the inner city, including the designation of more EZs in 1983–4. We will examine the detail of these changes in the next chapter, but note here how they helped further strengthen the credibility of the Thatcherite approach to urban planning. Planners were increasingly supposed to encourage rather than regulate private developers, still less to think in terms of old-fashioned public-sector development. By 1984–5 the scene was set for further, more ambitious, changes to the planning system. The first moves came as part of the final showdown with the metropolitan counties.

Unitary development plans

As a group, these metropolitan authorities were seen as wasteful and unnecessary by the Thatcher governments, a view that was encouraged by the fact that all had become Labour strongholds. Several, particularly the GLC under the leadership of 'Red Ken' Livingstone, had openly challenged central policies during the early 1980s (Livingstone, 1987) and had pioneered important radical planning initiatives, which we will discuss below. The 1983 Tory election manifesto promised to abolish the councils, implemented in the Local Government Act 1985 (DOE, 1983). Except for London, the loss was not in truth very deeply mourned by anyone. From the planning point of view, however, the abolition had the familar effect of weakening the strategic dimension of planning (Cullingworth, 1988). At a stroke it removed the structure planning authorities for England's main urban concentrations. A new type of unitary development plan for each metropolitan district, combining the characteristics of structure and local plans, was introduced, formally coming into operation during 1988–90. What remained of conurbation-wide strategic planning relied on inter-district consultations and central guidance. Meanwhile other changes had already been set in motion.

Lifting the Burden

The already familiar theme of planning as a burden on enterprise began to be voiced more prominently. Mrs Thatcher herself told the 1984 Conservative Party conference that it was 'one thing that can sometimes delay the coming into existence of new jobs' (Thatcher, 1989, p. 102). It was subsequently pushed by the ultra-Thatcherite Lord Young, Minister without Portfolio, and incorporated into the 1985 White Paper, *Lifting The Burden* (Minister without Portfolio *et al.*, 1985). Amongst many other proposals across the full range of government, this signalled several important ways in which the planning system was to be altered to make life easier for the developer and, by extension, the wealth- and job-creating entrepreneur. The most prominent proposals were the introduction of Simplified Planning Zones, an idea which had already been aired extensively in 1983–4, and the revision of the Use Classes and the General Development Orders that were fundamental to the operation of planning controls. It was also stated that approved plans would in future become just one of several 'material conditions' to be taken into account in determining planning applications. In other words, a plan could be ignored if the need to stimulate enterprise was more pressing. The potential this created for delays by encouraging general uncertainty and more appeals against planning decisions seems to have been rather overlooked.

Housing and Planning Act 1986

Many of the changes were introduced through DOE circulars, advising local authorities of the new emphases. The introduction of the Simplified Planning Zone (SPZ), however, required fresh legislation and the relevant provisions were included with a miscellany of other items in the 1986 Act. An SPZ was essentially a development and widening of the EZ idea, based on the same notion of a clearly defined scheme for land use which in effect gave permission for development that met the stipulated zoning (DOE, 1991). It had much in common with the zoning-based planning of other European countries (or much pre-1947 Act planning). Yet there were two important differences between EZs and SPZs. The latter could be initiated by the private sector, a token step nearer the enterprise culture, and more importantly they lacked the fiscal incentives of the EZs.

This absence of tax breaks together with rather cumbersome procedures discouraged widespread adoption of SPZs, particularly as the 1980s' property boom faded. By summer 1991 only two, in Derby and Corby, were operative and a further eleven were in various stages of preparation, mainly for industrial areas. Despite further moves to streamline procedures, the reality has hardly lived up to the promises of the mid-1980s. They have, though, encouraged local authorities to think about less cumbersome ways of speeding up planning procedures.

Revision of the Use Classes Order 1987

Rather more significant were the revisions of the Use Classes Order (UCO) and the General Development Order (GDO) (Home, 1989). The former specifies the different categories of land use used by planning. It is important because the tightness or liberality of the categories directly affects whether developers need to apply for planning permission for changes of use between, for example, different business uses. Relaxing the use classes effectively reduced the number of land use changes that required planning permission. The interest of the Thatcherites in such a relaxation is therefore immediately apparent, though its origins were economic as well as political. By the 1980s it was widely acknowledged that the existing UCO, introduced in 1948 and modified only slightly since then, was out of date. The late twentieth-century economy was beginning to order urban space in ways that no longer corresponded to the building types that had emerged in the late Victorian city. More specifically, some traditional land and building use distinctions, such as that between 'high-tech' manufacturing and offices, were becoming blurred (Goobey, 1992, p. 41).

Political and practical concerns came together in the mid-1980s and the DOE charged its Property Advisory Group with producing a new scheme. (It was incidentally very expressive of the changing times that planning should be re-shaped by a property group.) In the event the new UCO fell well short of what had been originally intended by the Property Advisory Group. Proposals to end the distinction between shops and other services (e.g. building society offices) and to relax the residential category to allow small businesses to operate from home met a good deal of widely based opposition, not least from Conservative areas. The ending of the distinction between light industrial and office uses was accepted, however, becoming the most important of the changes in the new UCO.

The future of development plans

Following hard on the abolition of the metropolitan counties and the 1985 White Paper came a 1986 DOE Green (i.e. consultation) Paper, *The Future of Develop-*

ment Plans (DOE and WO, 1986). In it the government rehearsed its desire to extend the principles of unitary planning to the non-metropolitan counties. Explaining these intentions, the Environment Secretary, Nicholas Ridley, criticized the existing two-tier system as 'too cumbersome' and 'overburdened with unnecessary detail' (1987, p. 41). Instead he proposed a system of mandatory district development plans, replacing the existing structure and local plans. There was certainly some validity in his criticisms of the existing system, though the frequently voiced disdain of Mrs Thatcher's ministers for development planning had hardly encouraged local efforts. It had not been until 1985 that the last of the 'first round' structure plans was finally approved. Progress on local plans was extremely slow. Ignoring subject plans, only about a fifth of England and Wales (by population) was covered by adopted local plans by mid-1988 (PPG12, 1988). Only 54 of the 333 non-metropolitan districts had local plans on deposit or adopted which covered the whole of their areas.

Yet although Ridley's mind seemed set, moves to reform the system proceeded only slowly. It was not until early 1989 that specific proposals for action appeared, in a White Paper also called *The Future of Development Plans* (DOE and WO, 1989). This stuck to what had been unveiled three years earlier, but the basic philosophy of the changes seemed to be shifting. The original Thatcherite project for a more deregulated, enterprise-oriented planning system was clearly being questioned. There was a growing (and rather belated) acknowledgement within government that a tighter planning system which provided more certainty would probably benefit developers more than a more deregulated system which increased uncertainty (Thornley, 1991). Some developers and property professional interests had actually been making this point for several years. They were now beginning to say it with an authority which could be heard above the Thatcherite rhetoric of total deregulation. It all pointed in a rather different direction to *Lifting the Burden*, to a more plan-led approach comparable to the zoning systems of other European countries, with rather less of the traditional planning discretion at the development control stage (e.g. BPF, 1986).

Yet the thrust of this important rethink was itself about to be challenged. If the Thatcherites had thought they could remake the planning system solely with the interests of developers in mind, another thought, derived from more traditional Tory values, was rapidly asserting itself. The whole Thatcherite project for planning its most serious challenge as the endemic 'nimbyism' of these years, noted earlier, began to be reflected in wider pressures about the environment. To understand this and set it properly in context, we must consider the ideas that were challenging the dominant thrust of policies for planning and development in the 1980s.

Green ideas and radical planning initiatives

The green movement

In the vanguard of these pressures was the new green movement, which had developed from the emergent environmentalism of the 1970s (Porritt and Winner, 1988). The movement was internationally based and had a genuine

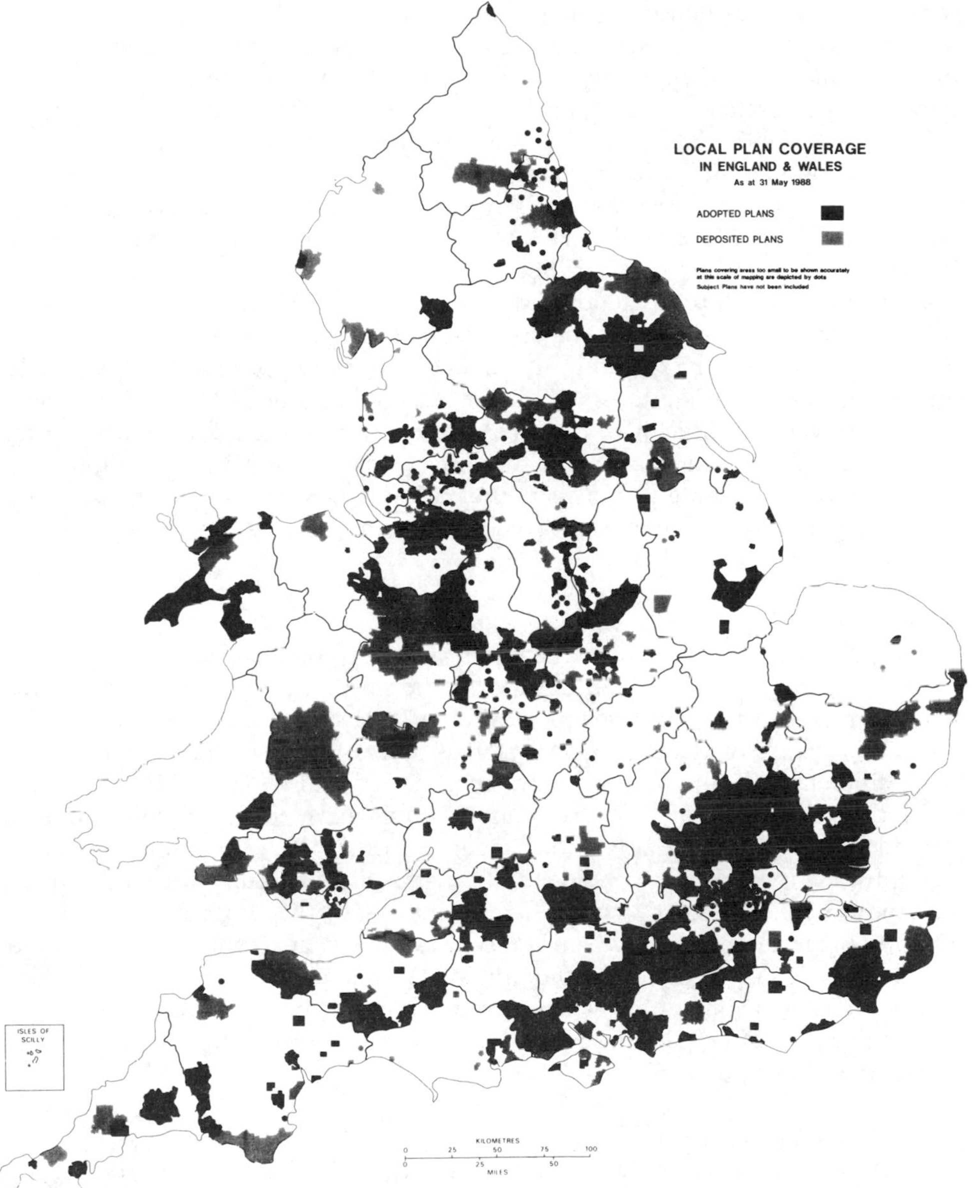

***Figure** 7.7 By 1988 Nicholas Ridley wished to abolish structure plans and replace them with district-wide local plans that would, it was hoped, speed the development process. Existing local plan coverage was, however, distinctly patchy.*

popular appeal in Britain far greater than anything mere planning had ever been able to muster, except perhaps during the Second World War. Membership of environmental groups in Britain doubled to about five million in 1990, with growth concentrated in the global and more explicitly green organizations (McCormick, 1993). This mass interest was manifest in a wide range of social, political, economic and cultural changes, from green consumerism and recycling to the formation of a British Green Party in 1985 to a growing ecological output in the mass media. The appeal of such themes to the young, already evident in the 1970s, increased greatly in the 1980s.

Conceptually the green agenda was more all-embracing than the environmentalism of the 1970s, encompassing a complete programme for world reform. Yet, although the movement scored some important victories, progress on the whole agenda was distinctly limited. With a few exceptions, the apostles of the new movement found it difficult to organize effectively and internal dissent was rife. There were serious tensions within the movement between left and right and, perhaps more damagingly, between the rationalist, scientific greens and the mystical, quasi-religious greens. As a whole, they were always too easy to mock to be taken seriously enough.

The green movement and planning ideas

Yet much of what the green movement was saying was of great interest to physical planners (Hardy, 1991b). As in the 1970s, the planning movement (which was tiny compared to the wider green movement) played only a small part in the conceptualization of the new green agenda. Ninety years earlier it might have been different. We might reasonably speculate that Ebenezer Howard, who suffered a fair degree of mockery himself, would have readily found common ground with a green programme to change the world by following environmental principles. As it was, the professionalization and institutionalization of the planning movement, particularly since 1945, meant that by the 1980s its independent tradition of radical thought and innovation was somewhat atrophied. As we noted in Chapters 5 and 6, planning thought had become a prisoner of the policy process. This was something which served it well from the 1940s to the early 1970s, but thereafter the triumph of anti-interventionist policies left planners confused and uncertain.

Radical planning initiatives

It would be wrong to leave the impression that all planners were devoid of alternative ideas, however. The scattered elements of a radical planning approach had begun to appear in the 1970s, concerned mainly with achieving a real public involvement, especially of less powerful social groups, in planning and urban change. In Macclesfield, for example, a young architect called Rod Hackney helped orchestrate a challenge to official planning, launching community architecture on the British scene (Knevitt, 1975). About the same time, in 1973, the TCPA initiated its planning aid service, helping to strengthen the representation of ordinary people in the planning process (Hardy, 1991b). There was a growth too in locally based anti-developer protests, most famously at Coin Street on

London's South Bank, where a combination of residents and younger planners and other professionals began in 1977 to challenge more powerful interests and establish a genuine community plan for the area (Tuckett, 1990). In 1979 the TCPA launched its proposal for a third garden city, asserting a need to return to the co-operative principles of Ebenezer Howard while embracing some of the new environmental concerns.

Radical planning under Thatcherism

Many of these initiatives began to receive wider attention in the 1980s. The concept of self-help that underlay some of them chimed in nicely with the new *laissez-faire* ideology of the Thatcher years. Community architecture (Wates and Knevitt, 1987) was particularly approved, not least by Prince Charles, perhaps because its main challenge was more to official planning than private development interests. Lightmoor in Telford New Town, a tangible, if minute, self-built outcome of the third garden city initiative, was admired on all sides (Hall, 1988). More controversial were those initiatives that sought to challenge the private development machine or to establish a genuinely popular basis for active local government. The new 'municipal socialism' of the big cities in the early and mid-1980s provided a conducive setting for such radicalism to flourish (Boddy and Fudge, 1985). Until its demise Ken Livingstone's GLC sponsored many such

***Figure 7.8** Lightmoor, in Telford New Town, was an attempt to recapture the community co-operation of Howard's original garden city vision and harness it to newer concerns for the environment and community architecture.*

initiatives (Mackintosh and Wainwright, 1987). A so-called People's Plan was prepared for the Docklands that challenged the legitimacy of the new UDC (Brownill, 1990a). GLC action secured the victory of the Coin Street Action Group in 1984 (Tuckett, 1990). Attention was focused on the specific planning concerns of women (Foulsham, 1990), black people (Grey and Amooquaye, 1990; Hossain, 1990), the disabled (Bennett, 1990) and other traditionally neglected groups. Other left-wing authorities such as Sheffield launched local economic policies based on worker co-operatives (Mawson and Miller, 1986). Yet others, for example Walsall, led the way in taking local service delivery out of the town hall and closer to the people (Burns, 1990).

To those who were affected by them, many of these initiatives seemed very important and relevant. But this alternative approach, usually associated with high local spending, was opposed by Conservative voters particularly in the suburbs and the outer cities. And because it failed to reach this contented majority, the flame of this radicalism never burned strongly enough to withstand the harsh wind of Thatcherism. The ideas and some of the initiatives lived on, but the policies had largely been extinguished by the late 1980s. This failure of radical planning rather contrasted with the spread of green ideas, which were both radical and capable of reaching out to the self-interests of the contented classes.

Sustainable development

The key green idea that began to permeate planning thought in the late 1980s was that of sustainability. The concept had grown from a synthesis of the ideas of the environmentalists and the development movement, concerned with improving the poorer countries of the world. Faced with the realities of destructive ecological regimes such as deforestation that quickly triggered serious problems, the appeal of a concept of ecologically sustainable development was readily understandable in such a setting. In 1987, however, sustainability began to be translated into a fully global approach, capable of being applied in developed countries such as Britain. The agent of this widening was the Report of the United Nations World Commission on Environment and Development, *Our Common Future*. Known as the Brundtland Report after its chairwoman, Gro Harlem Brundtland, the Prime Minister of Norway, the report gave the first clear definition of sustainable development: 'development which meets the needs of the present without compromising the ability of future generations to meet their own needs' (Brundtland, 1987, p. 8).

By the end of the 1980s it was being actively incorporated in planning thinking about the urban future, most notably in the European Commission's Green Paper on the Urban Environment, published in 1990 (CEC, 1990a, 1990b). This included the highly controversial prescription of a 'compact city', running counter to the traditional garden city-inspired decentralist thrust of British urban planning policies, which we will examine more fully in the next chapter. For the present we must now rejoin the main story and consider why it was that green ideas, which implied a much more interventionist approach than Thatcherism had ever been prepared to countenance, began to have a much more active influence on government proposals for planning and related policies in the later 1980s.

Greening Thatcherism and reprieving planning 1987–90

The 'turquoise tendency'

In 1985 the Centre for Policy Studies, the Thatcherite think-tank, had noted the lack of any obvious affinities between the green vanguard and their own supporters: 'For many Conservatives, the word 'green' conjures up the most disagreeable images . . . of eccentric, muesli-crunching peace fanatics in open-toed sandals . . . Germanic intellectuals of the far Left, the animal rights campaigners or, even worse, the benign socialist environmental planners with little respect for private property' (Porritt and Winner, 1988, pp. 81–2). Yet the same pamphlet perceptively recognized the potential appeal of green ideas to ordinary Tories:

> We are indeed a green nation, but not in the way the Left would like. We care about the details of our immediate surroundings, street corners and hedgerows, our parks and fens, our rivers and hillsides – small pockets of sustenance – which protect our sense of community, of history, and of beauty. We care too most of all for our own property – and widespread property ownership is the natural friend of environmental responsibility.
>
> (*ibid.*, p. 82)

Here then was the connection between narrow 'nimbyism' and wider environmental concerns, a potential link which began to come alive in the late 1980s. The main preconditions for this quite sudden transition were partly economic. By 1987 the economic problems of the early 1980s had receded and many Conservative supporters were enjoying the ephemeral rewards of the 'Thatcher economic miracle'. In such circumstances the continual need to stimulate enterprise appeared less of a priority and the question of how economic success was to be enjoyed began to assume greater prominence (Ridley, 1992). Given the heightened environmental awareness of these years, green issues provided an obvious focus for such concerns. It was, however, the unexpectedly strong showing of the Green Party, especially in Tory areas, in the 1989 European parliamentary elections which brought green issues decisively to the fore (McCormick, 1993). The electoral system denied the Greens any actual seats and the breakthrough was modest and very short-lived, but it sent out important signals. Mrs Thatcher moved quickly to appease the 'turquoise tendency' of green Tories, most immediately by replacing the abrasive Ridley as Environment Secretary with the more emollient and environmentally friendly Chris Patten.

Late Thatcherism and the planning system

The final months of Ridley's tenure had seen a few notable changes in planning policies (particularly in relation to new settlements, to be examined more fully in the next chapter). Generally though, central attacks on the planning system had slackened since the 1987 election. Earlier proposals continued to run their course, as noted above for the rather leisurely progress of the proposed revisions to the development plans system. The one remaining major pledge arising from *Lifting The Burden*, the revision of the General Development Order (GDO), was also implemented in 1988 (Thornley, 1991). It extended the permitted development rights introduced in 1981, mainly for extensions to business premises, though it also included some very un-Thatcherite tightening up in relation to building conservation. There were also other signs of a more interventionist approach developing for the inner cities, albeit still in a very fragmented and Thatcherite way, as we will see in the next chapter. Another important sign was the belated implementation of the 1985 European directive on environmental impact assessment for major projects in 1988 (DOE, 1989). More fundamental shifts were on the way, driven by this combination of domestic and European pressures.

This Common Inheritance

By the start of the 1990s it was clear that the language of the Brundtland Report had entered government policy-making. The 1990 environmental White Paper *This Common Inheritance* (actually a lavishly illustrated book, fortunately printed on recycled paper), marked a decisive shift in the presentation of central thinking about planning compared with *Lifting the Burden* five years earlier (DOE *et al.*, 1990). Its primary purpose was the promise of strengthened and integrated environmental policies and the Environmental Protection Act 1990 which soon followed gave some substance to that commitment (DOE *et al.*, 1991, 1992). Yet this was not a complete U-turn. The Act's new powers were important, though they continued to ensure that environmental planning was developing in a way that was partially detached from the main body of planning (and other) policies (McCormick, 1993). It was still therefore perpetuating the general trend of a fragmentation of planning into a series of disjointed initiatives.

But there was a definite sense within the White Paper that a strategic, comprehensive approach to planning as a whole was beginning to come back into fashion. In a marked policy reversal, structure planning was to be reprieved in streamlined form to supplement the new mandatory district development plans. Just a year earlier, it will be recalled, Ridley had wanted structure plans abolished altogether. There was also a commitment to develop a nationwide system of regional policy advice. Together with simultaneous shifts to revive strategic transport policies, it seemed to many commentators that the corner had been turned.

Town and Country Planning Acts 1990

In the same year the many changes that had affected the planning system since 1971 were consolidated. Although the new legislation was non-controversial, its general character highlights something of the changes of these years. The first point to make is that planning had survived and was, for most areas, recognizably the same system as that which had operated in the early 1970s. Thatcherism and the other pressures of these years had wrought important but not decisive changes. At the same time, though, the symptoms of planning's fragmentation were palpable even in the very form of the new legislation. Instead of one Act, there were now four. As well as the main statute, called simply the Town and Country Planning Act, there were separate planning Acts subtitled Listed Buildings and Conservation Areas (LBCA), Hazardous Substances (HS) and Consequential Provisions (CP).

The LBCA Act reflected the growing central separation of policy for heritage, as a marketable historic commodity, from the rest of planning (and leaving to local authorities the problems of reintegrating it). It continued the trend begun with the National Heritage Act 1983 (and was itself taken further when John Major created the Department of National Heritage in 1992, entirely divesting DOE of its historic conservation policy role and leaving it with only minor casework details). The HS Act arose primarily from provisions in the Housing and Planning Act 1986 that had never been brought into operation. Though partly a product of longer-term worries about health and safety risks in industrial planning, they also reflected the need for specific controls as the planning system was progressively deregulated for business uses, creating real risks that hazardous situations could arise. Finally the CP Act essentially reflected the tremendous complexity of the whole system!

The 1990 legislation was also important for what it left out. With the exception of specific planning instruments, the bulk of inner-city policy remained separate from the planning system. We have already noted the separate development of environmental planning powers. Even within the principal Act, important areas such as minerals were now virtually detached from the main body of planning legislation, following changes introduced in 1981. Taken together all this was ample testimony to a planning system that was less coherent than it had been two decades earlier. Overall therefore the 1990 Acts embodied and expressed both changes and continuity, weaknesses and strengths in the planning system at the end of the Thatcher era.

The planning system after Thatcher 1990–?

A new consensus?

The fall of Mrs Thatcher following a leadership challenge by her former ally Michael Heseltine, now apparently a staunch interventionist on planning matters, encouraged a belief that the trend towards strengthening planning would continue. There was talk of a new post-Thatcherite consensus that incorporated some of the most useful shifts of the Thatcher years into an increasingly more coherent planning system (see e.g. Thompson, 1990; Colenutt, 1993). Mrs Thatcher's successor, John Major, appeared to lend some initial credence to this, particularly since he appointed Heseltine as his first Environment Secretary.

Planning and Compensation Act 1991

The 1991 Act was the first (and, to date, the only) important sign of post-Thatcherite intentions for the planning system (Heap, 1991). It was very much on the lines promised by *This Common Inheritance*, so that it gave legal authority for the new streamlined structure plans and mandatory district-wide local plans. It also included many smaller measures that strengthened the position of planning and planning controls. Of most symbolic importance was the restoration of the primacy of the plan in the determination of planning applications. Such 'plan-led' planning was a clear retreat from the philosophy of *Lifting The Burden*, though motivated, as we have already implied, as much by a desire to reduce developer uncertainty (and planning appeals) as the assumption of a genuine environmental perspective.

It may be though that the Act's other main provisions, encouraging developer contributions to infrastructure and other aspects of the environment on a routine basis, will turn out to be the more profound change. This reflects an institutionalization of planning gain, renamed planning obligations, in effect completing the retreat from the principle of a general betterment tax. Final judgement on the effectiveness of this will have to wait until it has operated for some time. The early 1990s' property slump provides little real basis for judging its impact.

The future

Overall we should beware of immediately reading too much into what may well turn out to be a decidedly modest Act, in many ways an amalgam of principles drawn from the pre-1947 and 1968 Act systems. The fact that the 1991 system is plan led becomes a significant indicator of the power of planning only if the plans in question actually propose outcomes that differ from those of the market. Meanwhile continuing uncertainty remains about exactly how the new system will operate in practice if the proposals for local government reform gradually being unveiled in 1993 are implemented. These propose a varying mix of unitary and two-tier authorities, with joint arrangements for structure planning. They may prove as significant for the 1991 system as the 1972 reforms were for the 1968 system.

As to whether the 1991 Act will really be part of a wider conversion to a genuine environmental strategy, as the follow-up documents to *This Common Inheritance* would have us believe (DOE *et al.*, 1991, 1992), the jury must still be out. The environment and planning peaked as high-profile political issues in 1989–90, and the serious recession of the early 1990s restored economic worries to the fore (McCormick, 1993). Unlike the early 1980s, however, there was no longer any real political faith that further deregulation of the planning system could facilitate economic miracles. For all the promises of Thatcherite burden-lifting, the early 1990s' recession has been significantly worse than that ten years earlier. And, as usually happens when property booms burst, developers have become noticeably less resentful of planning restrictions. In particular they have been looking to government to inject some stability and confidence into the property markets, suggesting a potential role for more interventionist planning. On the whole though, despite a few hopeful signals, and a greater willingness to take advice from planners (including, most notably, Peter Hall), the Major governments had done little of substance to encourage or discourage the hopes of planners, environmentalists or developers by 1993.

OVERVIEW

During the 1960s and early 1970s it was increasingly necessary to remind observers that physical planning really was political, such was the strength and persistence of the bipartisan consensus on planning matters. Following the experiences of the late 1970s and 1980s, such reminders became redundant. Planning's politicization became totally obvious, though it was not the only source of change. An increasingly troubled economy, the associated crisis of Keynesian economic management, the emergence of monetarism, increasingly sharp political polarization, Thatcherism and the pressures of environmentalism have together wrought dramatic changes on the planning system.

Paradoxically, planners actually had wider duties in the early 1990s than twenty years earlier, but no one was quite sure any longer what it was that held all their manifold responsibilities together. It was green ideas that articulated the new radical path that began to offer new arguments for coherence. Essentially though, planners were waiting for someone to put planning back together again, with some forceful reassertion of the comprehensive and strategic dimension that had been so important in the heyday of the post-war consensus. There was, unfortunately, no new political champion of planning in sight at the time of writing in 1993. And the consequences of this fragmentation were all too apparent in the more detailed applications of planning policies, as we will see in the next chapter.

8

REMAKING PLANNING SINCE 1974: II. SPECIFIC POLICIES

INTRODUCTION

In Chapter 6 we examined the four major interrelated strategic planning policies that were applied with varying commitment over the post-war boom period. This chapter extends this examination from the mid-1970s, when the long post-war boom and political consensus that had sustained these policies were beginning to disintegrate. It is a story of both policy change and resilience that parallels that of the whole planning system over these same years. In some policy areas, such as the pursuit of regional balance, we find the story is one of wholesale retreat. Regional planning shrank in the 1970s and virtually disappeared in the 1980s and regional policy was dramatically diminished and partially reoriented. There are some similarities in the case of planned decentralization. The emergence of inner-city policies and the 1980s' assault on state spending brought a virtual end to the New Towns programme. Yet the continuing heavy development pressures in the outer-city areas brought an intriguing Thatcherite flirtation with the idea of private new settlements of various kinds.

Little very substantial came from these ideas, partly because of the immense strength of the third of the strategic policies, planned containment. Despite frequent alarms, the green belts remained the enduring rock of Britain's urban planning policies in these years. There is most to report, however, in the case of the urban cores where planned redevelopment had been the traditional policy objective. A considerable amount of central area private redevelopment continued, but much of the action now switched to the inner cities. The emphasis there shifted decisively from the public-sector housing-led approaches that had dominated in the 1960s and 1970s to one which relied increasingly on attracting private developers for a widening range of new activities. It is with these moves towards regeneration that we begin.

REGENERATING THE URBAN CORES

Implementing Labour's inner-city policies 1976–9

The new policy framework

Peter Shore's policy initiatives of 1977–8, noted in the last chapter, initially resulted in the creation of seven partnerships, for London's docklands, Hackney–

Islington, Lambeth, Newcastle–Gateshead, Manchester–Salford, Liverpool and Birmingham (Lawless, 1986, 1989). In Scotland, meanwhile, the Glasgow Eastern Area Renewal (GEAR) Project, an initiative broadly equivalent to the partnerships, had already been launched in 1976. A further fifteen programme areas and sixteen designated districts were also created in England, though significant alterations were subsequently made in 1986–7. Between them, programme and designated areas included most of the rest of the biggest cities with classic inner-city problem symptoms including Leeds, Leicester, Sheffield, Hull, Bradford and Nottingham. They also included many smaller older industrial towns and cities in both conurbation or freestanding settings, such as Sunderland, Oldham, Bolton, North and South Tyneside, Hartlepool and Burnley. By the early 1990s fifty-seven English authorities were benefiting under this Urban Programme, as it was collectively called. In Wales all local authorities were eligible to apply and the equivalent Scottish scheme was almost as wide ranging. Swingeing cutbacks were announced in 1992–3, however, and its future is unclear at the time of writing.

The new policies in operation

Even at the outset, the Urban Programme's most serious problem was that which bedevilled all Labour's urban and planning policies: inadequate funding. The very economic ills that were worsening inner-city conditions were also pushing the government towards public expenditure restraint (especially on capital projects), preventing a really bold government-funded response in such areas. The initial distribution of these limited funds was also questionable and the promised shift to an economic development focus was slow to appear. Local authorities at that time had little experience of economic promotion and development, so they found it difficult to generate credible projects. Work on Industrial Improvement Areas and other economic development measures eventually contributed to a widening of the relevant expertise among planners and other local officials, but there was an inevitable timelag (DOE, 1986). It was therefore social, community and recreation projects which dominated early expenditure.

The sources of inner-city action since 1979

A Thatcherite paradox?

Rather surprisingly, the Thatcher administration did not roll back the frontiers of state spending on inner-city programmes. Expenditure continued the upward trend established under Labour (Robson, 1987, 1988; Lawless, 1991; Middleton, 1991). Allowing for inflation, real spending in the 1983/4 financial year was more than twice the 1978/9 level. It dwindled very slightly until 1986/7, but then doubled again by 1990/1, when it exceeded £800 million. We should immediately add, though, that the Thatcherite commitment to reducing public spending brought a loss of other central government grants to inner-city authorities that has been many times greater than the specific assistance granted. Between 1979 and 1984 alone, the designated London boroughs gained roughly £300 million in inner-city money, but lost £1,530 million in rate support grant. Not that this was the whole story either, because some inner-city areas (particularly the

Urban Development Corporations) have certainly received considerable additional funding that has not appeared in inner-city programmes. Clearly then, a partial commitment to targeted public spending in the inner city was maintained and must be explained in view of Thatcherite ideology.

A worsening problem

One reason was simply that the problem got significantly worse after 1979. The acceleration of economic change brought very high unemployment to the inner cities (Robson, 1988). In 1986, the worst year of the decade, the jobless rate in the worst inner-city ward of Manchester, Hulme, stood at 48 per cent, over four times the UK rate. Most other inner ward rates in the city were roughly three or more times the national average. The same basic pattern was apparent in every large inner-city area, especially in the north. Even as recovery occurred nationally, inner-city unemployment remained obstinately high. In 1988 the worst wards in Nottingham and Middlesbrough recorded unemployment rates of 38 and 34 per cent, respectively, compared with a UK rate of 8.1 per cent (Audit Commission, 1989).

When the impact of this unemployment was added to already low economic activity rates (reflecting, for example, higher than average proportions of lone parents unable to work or, in some cases, old and chronically sick people) and the generally low wages of many of those in low-skill work, the impact on poverty can be imagined. In turn the erosion of the relative value of state benefits during the 1980s had an especially marked effect on such highly dependent populations. Cuts in housing grants to councils also further degraded the quality of the residential environment, to the point of dereliction in many areas. Economic change itself also added more dereliction in the form of abandoned factories, warehouses, docks, shipyards, etc.

A threat to social order

For all this, the Conservative Party, which drew its support largely from the outer-city and rural areas, remained woefully ignorant of the inner city. Many Conservatives were prone to attribute its problems merely to the inadequacies of individuals and the policies of Labour councils. The increasing marginalization under Mrs Thatcher of the 'one nation' Tories, who remained fully committed to a welfare state, made such viewpoints more influential than would have been possible under Churchill, Macmillan or Heath (Hall and Jacques, 1983; Deakin, 1987; Skidelsky, 1988). But if arguments for intervention that rested on the traditional post-war consensus were losing their potency, new ones based on real threats to urban social stability began to take their place. In 1980 came the first serious outbreak of inner-city rioting in the St Pauls district of Bristol. The following year much more serious disturbances occurred in several inner-city areas including the Brixton area of Lambeth, Moss Side in Manchester and Toxteth in Liverpool (Scarman Report, 1981). And there have been further outbreaks.

Often triggered by disputes over policing, the riots essentially highlighted the extent of social alienation in many inner-city areas. Unemployment and poverty, combined in many cases with the effects of racial discrimination, had together

Figure 8.1 *In the early 1980s the resentment of inner-city, mainly black, populations at the bleak economic prospects, run-down environments and, not least, heavy-handed policing of their areas finally boiled over into serious unrest. The photograph shows the Brixton riots of 1981.*

denied many young people the fruits of mainstream contented society. The riots were the inarticulate but unmistakeable voice of the substantial excluded minority. Other manifestations came in higher than average crime, including, by the early 1990s, the spectre of widespread drug trafficking and associated violence centred on inner-city areas such as Moss Side. All these images, with their overtones of the US cities, certainly provided powerful arguments for intervention.

Yet for all this the natural instincts of many Conservatives looked to a combination of tighter policing and individual self-help. These sentiments were certainly shared by some Cabinet Ministers. Speaking at the 1981 party conference, the ultra-Thatcherite Norman Tebbit summed up the prevailing political ethos of the 1980s when he recalled his father's unemployment in the 1930s: 'He didn't riot. He got on his bike and looked for work' (Tebbit, 1988, p. 187). To understand why inner-city policies moved beyond such Thatcherite nostrums, we need to give credit to the politician who carried the message, the Environment Secretary from 1979–83, Michael Heseltine.

The Heseltine factor

Like Mrs Thatcher, Heseltine was a charismatic and relentlessly ambitious populist (Critchley, 1987). Unlike her, though, he tempered his commitment to the new ascendancy of the market with a recognition of the real value of a pragmatic approach to state intervention (Heseltine, 1987). He was, in short, capable of reaching the parts other Thatcherites could not reach. As Shadow Environment Secretary at the time of the 1978 Act, he had actually criticized Shore for the inadequate funding of the programme (McKay and Cox, 1979). Not that he actually intended to increase spending at that time. What he sought was a redirection of public funding so that it would attract rather more private investment than Labour's proposals seemed set to achieve. Heseltine's initial understanding of the inner-city problem rested on the vast and very visible example of London's docklands. It was, in his perception, a problem to be addressed principally by a decisive approach to the promotion of development, so that dereliction could be banished and new activities established.

His fuller understanding of the social problems of the inner city came a little later in the immediate aftermath of the 1981 riots. In a move reminiscent of Lord Hailsham's temporary role in the north east in 1963 (though without the flat cap), Heseltine went to Liverpool as unofficial 'Minister for Merseyside', intent on seeking new remedies. It was, in his own words,

> one of those priceless formative experiences from which every politician takes strength; it tested many of my deepest political beliefs and instincts and intensified my convictions . . . it showed me how quickly we had to act and it greatly sharpened my appreciation of the scale and nature of the human and social crises with which the inner cities were struggling.
>
> (Heseltine, 1987, pp. 139–40)

Heseltine returned from Merseyside convinced that market processes were alone insufficient to address such problems. Significantly greater public spending, and much else, was essential. Putting all this to Cabinet in a paper provocatively entitled 'It Took a Riot', he won the support of the Prime Minister but not the Treasury. Only modest expansion of the programme was sanctioned, which then lost momentum after he left Environment in 1983.

'Those inner cities'

We can also detect a specifically political reason for the resumption of the upward movement of expenditure in the later 1980s, after several years of stagnation. The occasion was the immediate aftermath of the 1987 election, when a victorious Mrs Thatcher commented to her party workers (and a television-watching nation) that 'we have a big job to do in some of those inner cities' (cited in Robson, 1988, p. vii). Quite what she meant by those words was debatable (e.g. Robson, 1988; Burton and O'Toole, 1993). At one level there was crude party political concern. In stark contrast to most of the rest of Britain, the inner cities had decisively rejected Thatcher at the ballot box. This was irrelevant to the election outcome, but it acquired significance because Thatcherism was by then set on fundamental and enduring change to the whole political culture of Britain.

The inner cities stood as a direct challenge to complete dominance of the new ideology, their seemingly intractable problems a blemish on Thatcherite success. They had to be brought into the fold. Inner-city policy accordingly moved into higher gear during the third Thatcher administration (e.g. Cabinet Office, 1988; Thatcher, 1989, pp. 128–9). It is also significant that 1987–9 represented the highpoint of Thatcherite confidence about its economic success, perhaps encouraging a more charitable attitude to areas that were not sharing the good fortune (in much the same way as Special Area policies had grown as the economy recovered in the 1930s). Certainly it has been noticeable that spending on the inner city has languished as wider economic problems have again become prominent during the early 1990s under the Major administrations.

Inner-city policy since 1979

The restlessness of policy

There has been an extraordinary restlessness in inner-city policy-making since 1979, producing an ever changing array of initiatives (Lewis, 1992). Much of this reflected the reliance of policy development on the concerns of particular politicians at particular times. Shore had given the DOE an early co-ordinating role. This strengthened under Heseltine, but it then languished as his successors failed to match his enthusiasm for the inner city. Moreover as the content of inner city policies widened in the middle and late 1980s it was natural that the role of other central departments would strengthen. The land, planning and development initiatives favoured by Shore and Heseltine, such as IIAs, EZs and UDCs, were increasingly being supplemented by enterprise, business funding, training and other concerns by the middle and late 1980s. The roles of the Department of Trade and Industry, Employment, Education and the Home Office accordingly widened. By 1987 the Prime Minister herself had become personally involved through the Cabinet Office. It was often unclear which minister was actually in charge of policy and there were occasional inter-departmental disputes over this. By 1987 the DTI was in the lead role, but the end of 1990 saw the DOE, again under Heseltine, back in charge of inner-city policies.

Lifting the 'dead hand of socialism'

Despite this regular rearrangement of policies, several underlying themes were apparent. These translated the Thatcherite principles we identified in the previous chapter into the specific context of the inner city. Most prominent was the reduction in the role of local government, as the partnership and locally based inner-city programme set up by Shore became progressively less significant. Responsibility shifted directly to the centre or new agencies that were more amenable to central desires: 'the dead hand of Socialism should be lifted', as Heseltine (1987, p. 136) put it. Most prominent of the new agencies were the Urban Development Corporations (UDCs) noted in the last chapter. Others included the seven City Action Teams (created in 1985 and 1988) and the sixteen Task Forces (set up in 1986–7), both of which involved central officials shaping and co-ordinating policies at local level in ways that by-passed local government. The Urban Regeneration Agency, created under the Leasehold Reform, Housing and Urban Development Act 1993, seems set to become a kind of large, roving UDC.

The UDCs in operation

The UDCs themselves remained the showpieces of inner-city policies over these years. Just two were formed initially in 1981: for Merseyside (MDC) and London Docklands (LDDC) (Brownill, 1990b; Imrie and Thomas, 1993b). The latter covered an area of 2,150 hectares, with 40,000 population within its boundaries. MDC originally covered just 360 hectares (960 following enlargement in 1988) and an original population of 450. Five years on, these pioneer UDCs, especially the London Docklands Development Corporation (LDDC), had made a major impact. It was therefore natural to repeat the formula as the third Thatcher government sought to strengthen its inner-city policies.

New UDCs were designated for the Black Country, Tyne and Wear, Trafford Park, Cardiff Bay and Teesside in 1987, Sheffield, Leeds, Bristol and Central Manchester in 1988 and Laganside (Belfast) in 1989. Further UDCs, for Birmingham Heartlands and Plymouth Dockyard, were designated in 1992 and 1993. Though Teesside (4,858 hectares) and the Black Country UDC (2,598 hectares) covered larger areas than the LDDC, none of these later designations had budgets comparable with the first two. All had shorter life-spans (five to ten years compared to ten to fifteen for LDDC and MDC). Several of the later designations, from 1988, were dubbed 'mini-UDCs' because of their small areas and budgets. Central Manchester, for example, covered just 187 hectares and had a budget of just £44 million compared to MDC's £247 million and LDDC's £1,370 million.

Encouraging the private developer

A key part of the remit of the UDCs was the stimulation of profit-seeking private investment within their areas. Yet they had no monopoly on this objective, which has permeated all post-1979 inner-city initiatives to a greater or lesser extent (Healey *et al.*, 1992). It was apparent, for example, in a widespread desire to lure a more middle-class and entrepreneurial population back to live in inner-city areas. It was evident too in a host of specific policy measures (Heseltine, 1987). For example, it underpinned the recasting of derelict land grant (DLG) in 1980–1. New initiatives, such as the urban development grant (UDG), created in 1982, urban regeneration grant (URG), introduced in 1987, merged together as city grant in 1988, also embodied this same philosophy. It had, however, already been apparent for several years in the first major move in this direction, the Enterprise Zones.

The Enterprise Zones in operation

In 1981 the first round of Enterprise Zones (EZs) was announced (DOE, 1987b). They were not, however, a measure exclusively focused on the inner cities. Some, such as Salford/Trafford, Tyneside or London Docklands (the latter

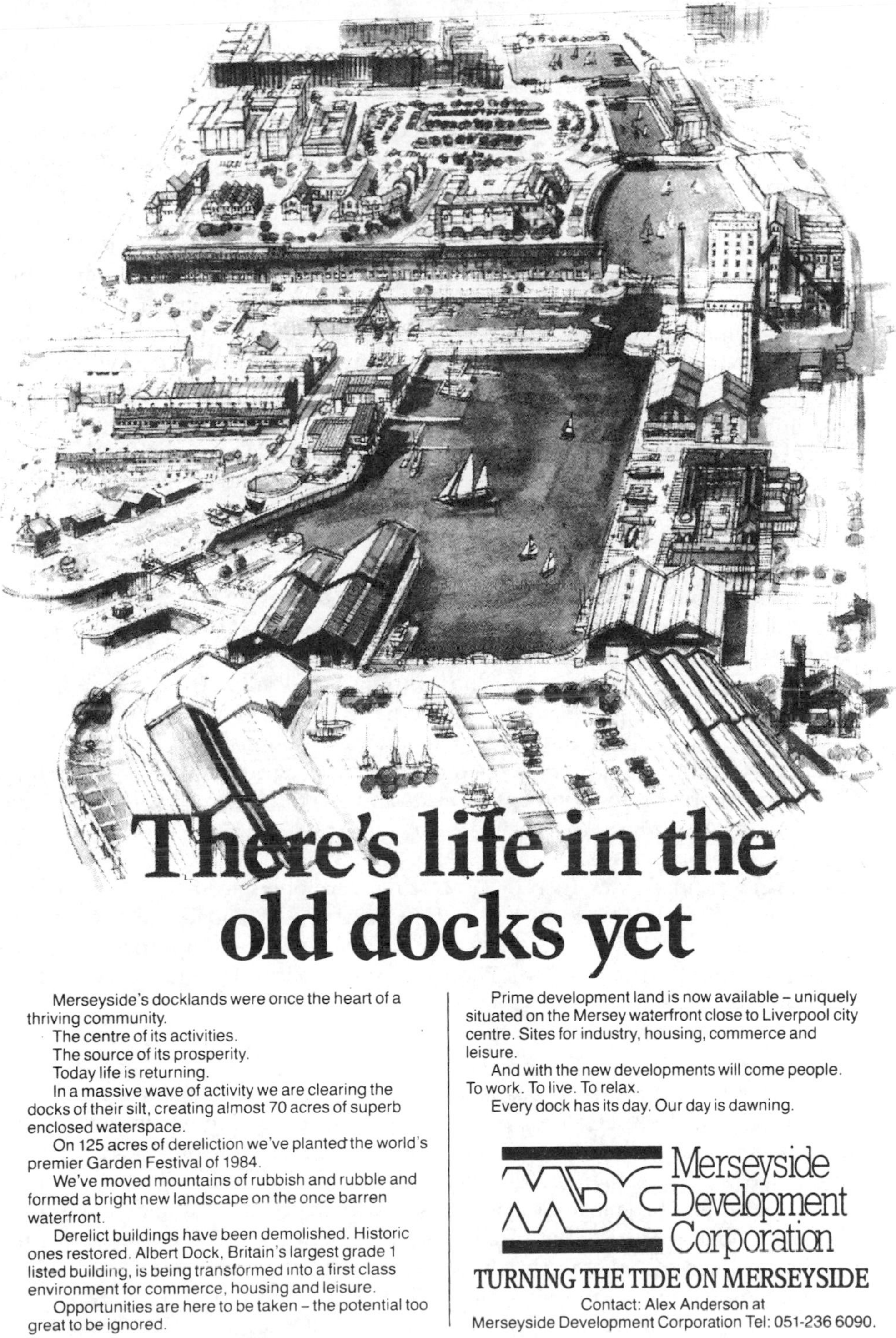

There's life in the old docks yet

Merseyside's docklands were once the heart of a thriving community.

The centre of its activities.

The source of its prosperity.

Today life is returning.

In a massive wave of activity we are clearing the docks of their silt, creating almost 70 acres of superb enclosed waterspace.

On 125 acres of dereliction we've planted the world's premier Garden Festival of 1984.

We've moved mountains of rubbish and rubble and formed a bright new landscape on the once barren waterfront.

Derelict buildings have been demolished. Historic ones restored. Albert Dock, Britain's largest grade 1 listed building, is being transformed into a first class environment for commerce, housing and leisure.

Opportunities are here to be taken – the potential too great to be ignored.

Prime development land is now available – uniquely situated on the Mersey waterfront close to Liverpool city centre. Sites for industry, housing, commerce and leisure.

And with the new developments will come people. To work. To live. To relax.

Every dock has its day. Our day is dawning.

TURNING THE TIDE ON MERSEYSIDE

Contact: Alex Anderson at
Merseyside Development Corporation Tel: 051-236 6090.

Figure 8.2 The Thatcher governments favoured a market-led approach to urban regeneration, epitomized by the promotional efforts of the Urban Development Corporations. This 1984 advertisement by Merseyside DC celebrates the Garden Festival then under way and anticipates the US-style restoration of the Albert Dock.

not finally designated until 1982) covered classic inner conurbation areas. In addition there were several run-down sites in smaller industrial centres such as Swansea or Dudley which had not featured in Peter Shore's initiatives. And a few were relatively good sites that could not be described as inner city, for example at Wakefield and Corby. This heterodox pattern was also apparent in the second round of EZs, designated from 1983. By 1985 there were twenty-five, but they fell from favour under Ridley and were used only rarely thereafter. The number diminished rapidly from the early 1990s as their ten-year life-spans expired. Although the zones have attracted a good deal of new development, they did not generally add to its overall amount. They either subsidised developments that were already decided upon, or that would have happened without EZs, or were simply diverted from nearby areas. Nevertheless they were an important early encouragement to private developers and a powerful device in place promotion.

Urban development grant and its successors

Another important new strand in the soliciting of private development in the inner cities was that represented by UDG, urban regeneration grant (URG) and city grant (CG). UDG grew out of Heseltine's post-riot initiatives as he searched for a way to persuade major financial institutions to put money into the inner city (Heseltine, 1987). They suggested something based on the American urban development action grant and used to apparently good effect in many older US cities (DOE, 1988). Unlike any previous British grants, these were available for a wide range of private development projects and were calculated on the basis that would allow the project to proceed. They addressed the problem of negative value, whereby the costs of undertaking a development exceeded the likely returns. Government was thereby protecting developers from losses. After the modest success for the original UDG, URG and CG extended the approach to larger schemes. By 1992, 267 city grant schemes had been approved since its inception in 1988, amounting to £238 million of public funds and £1,040 million of private investment.

Leverage planning

At the heart of these grants was a novel concept in British planning practice: leverage (Lewis, 1992; Healey *et al.*, 1992). It was something which increasingly permeated virtually all other aspects of urban regeneration during the 1980s. What it meant, essentially, was that development projects were conceived with a view to securing the maximum private investment for a minimum amount of public spending (a relationship often called the gearing ratio). Leverage (and the term itself, which served as both noun and verb) came from the USA, particularly the cities of Baltimore and Boston (Parkinson, Foley and Judd, 1988). The leverage approach there had rested on partnerships of private investors and municipal authorities, creating local growth coalitions. These had proved able to reverse urban decline and regenerate the cities as impressive post-industrial tourist and service centres. At the same time they had apparently been able to secure some important community benefits by bargaining and agreements.

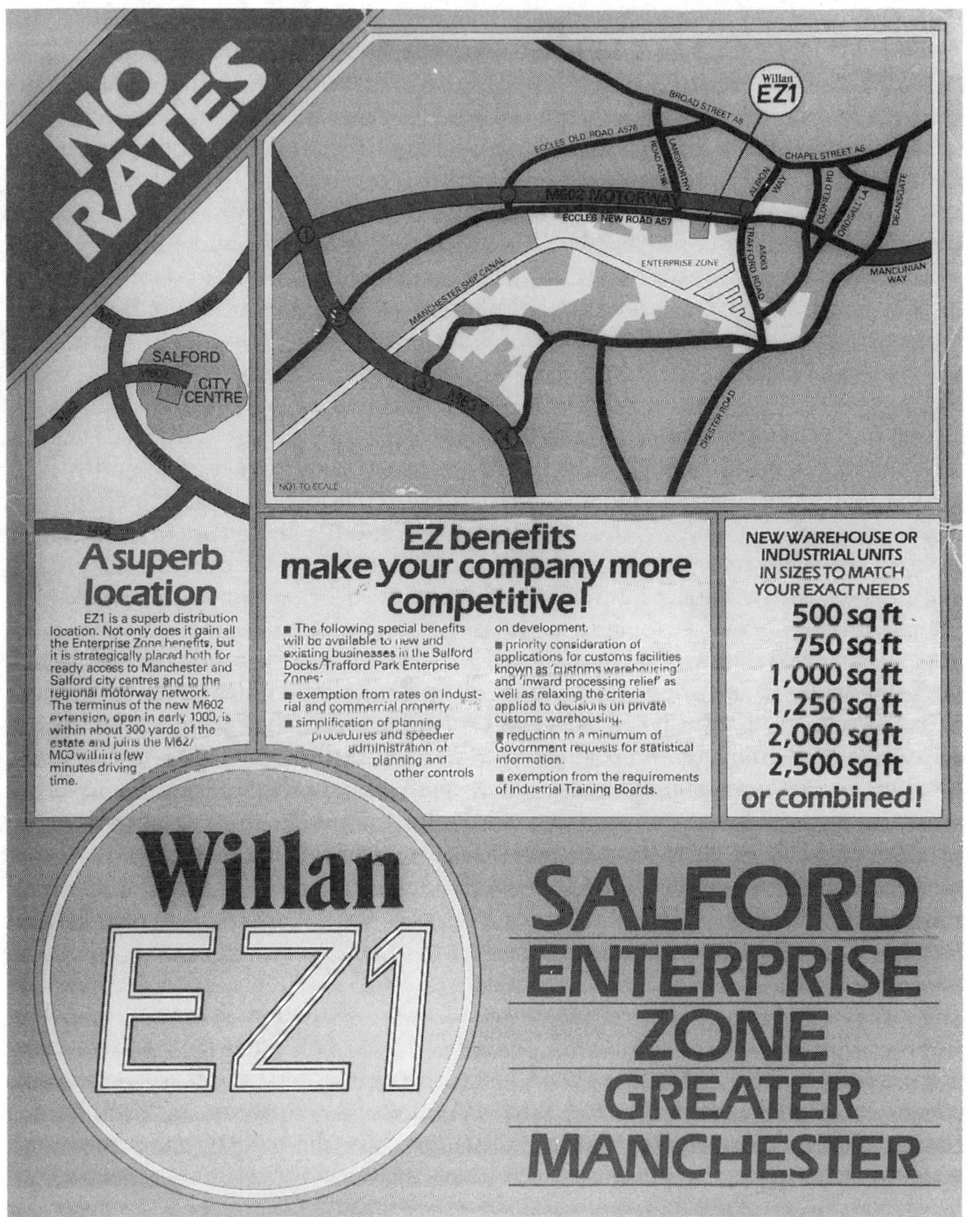

***Figure 8.3** Enterprise Zones became an important marketing device, applied with great enthusiasm by pragmatic Labour councils such as Salford. Note, however, how the absence of rates (i.e. local taxes) was the most powerful incentive. Streamlined planning procedures were of somewhat lower significance.*

Inner-city initiatives and promotional planning

Such a private development-oriented approach created a rather different style of planning to that which had traditionally been dominant. As we noted in the last chapter, inner-city policies contributed to the fragmentation of comprehensive planning into a series of more disjointed initiatives, requiring rather different styles of action. Most striking in the inner city and older industrial areas was the growth of local economic promotion and development activity, something which more planners became involved in during the 1980s (Mills and Young, 1986). Hitherto the preserve of regional policy, economic promotion and development now acquired a much stronger local dimension, operated by both the new central agencies, especially the UDCs, and local authorities.

Local authority regeneration policies

Local government used a varied mixture of powers, sometimes of dubious legality and subject to occasional challenge by the district auditor (Mawson and Miller, 1986). Some Labour metropolitan authorities such as the GLC and Sheffield saw it as a way of challenging the thrust of the Thatcher government's inner-city policies. In contrast to central emphasis on attracting private capital, mirrored in the policies of most local authorities of all political persuasions, they sought to extend municipal and co-operative enterprise. Yet this 'municipal socialist' challenge was always limited (Cochrane, 1983, 1986). By the late 1980s it had virtually been extinguished and Labour councils were generally pursuing the same broad urban regeneration objectives as central government (Audit Commission, 1989). By 1989 ministers were sufficiently confident that they could harness these local economic development initiatives to allow the practice (shorn of its more controversial elements) to be regularized under the Local Government and Housing Act. In practice the new measure made little difference to most inner areas which already had most of these powers under the 1978 Act, but it was an important symptom.

City Challenge

A shift of more practical importance was the City Challenge initiative, launched by the Major government in 1991 (Imrie and Thomas, 1993a; Colenutt, 1993; Burton and O'Toole, 1993). With considerable safeguards, it began to return more responsibility for regeneration to local authorities. They had to show that they had accepted the disciplines of the centre, demonstrating a genuine involvement on the part of the private sector, voluntary sector and local community. There was also competitive bidding against other areas for funds, encouraging local authority zeal in following the rules. But for the first time in many years, a new inner-city initiative vested a real measure of central trust in local authorities. Eleven authorities were involved in the first-round programmes which began in 1992, with central government committing £82.5 million a year for five years. There was a second round, involving more authorities, in 1992–3, but public expenditure cuts placed any further rounds in doubt.

The limited role of plans

The emphasis on promotion was very different to post-war planning's traditional mainstream roles (Lyddon, 1987). Instead of regulating private developers, they were encouraging them. Instead of actively shaping public-sector projects, they were negotiating the character of private development (Brindley, Rydin and Stoker, 1989). Yet the novelty of the planner–developer relationship was only one of degree. Since the late 1960s negotiation and bargaining within all parts of the planning process had become noticeably more common, evident for example in the growth of planning gain. In the inner cities of the 1980s it was perhaps at its ultimate, so much so that formal land use plans of conventional kind, reconciling a wide variety of public and private interests played very little part. Such plans as existed were largely promotional in character, development briefs designed to appeal to private investors.

A new partnership of planner and developer?

Nor was the wider approach entirely new. The buzzword of 1980s' and 1990s' inner-city policies was 'partnership' (Imrie and Thomas, 1993a; Lewis, 1992; Robson, 1988). This was not the same conception of partnership that Shore had unveiled in 1977, between central and local government. It was essentially between the public and private sectors and in that sense it had much in common with the favoured mechanism of planned central area redevelopment from the 1950s. It was, though, being extended into the much less favourable territory of the inner city, with the explicit intention (not always fulfilled) that the lion's share of the investment would be private capital (Healey *et al.*, 1992). Such locations were not traditionally profitable compared to the virtual certainties of the earlier city-centre partnership schemes. For several decades the private development industry had ventured into the inner cities only as contractors to local authorities, with public funds directly guaranteeing their construction profits. Moreover the new form of partnership envisaged the private developer carrying most of the risk, rather different to the typical city centre.

Not surprisingly, private developers did not immediately jump at the chance. For all the government's willingness to create a developer-friendly environment in the inner cities, a great deal of ministerial persuasion was necessary to change these attitudes in the early days. Much of this occurred away from the public gaze, but a notable and rather indicative public instance came when Heseltine got thirty top financial decision-makers together on a bus to show them the horrors of inner Liverpool in August 1981, soon after the riots (Heseltine, 1987). He won their sympathy and the secondment of some of their staff, but little else. No instant US-style growth coalitions emerged to provide the relatively spontaneous private sector solution the government wanted. For all the public investment, subsidies, tax breaks (and a virtual giving away of land) that we have identified, it also took a great deal of behind-the-scenes persuasion and arm-twisting to initiate private development even in areas, like London Docklands, which had most profit potential (Brownill, 1990b). Benefits to the wider community were accordingly slow to appear.

Late 1980s' success?

Thanks to such expedients and a rapidly accelerating property boom, an impressive momentum had been generated by the late 1980s that spread even into less promising inner-city areas (Healey and Nabarro, 1990; Goobey, 1992). Docklands attracted many established national and international developers, but this was unusual. Most inner-city areas were reliant on local and regional developers, though some of them certainly made their reputations on prestigious regeneration schemes. At the zenith of the property boom, it seemed that government really had forged a new kind of partnership with private developers that went beyond anything achieved during the 1950s or 1960s. Whereas the inner city had at that time been a no-go area for the private developer, the state was leveraging roughly £4 of private investment in the inner cities for every £1 of public investment by the late 1980s (Potter, 1990). LDDC managed, at best, a 12 to 1 ratio. Trafford Park and Sheffield UDCs managed similar rates in the late 1980s. The less-publicized MDC achieved only 1.2 to 1 in the same, very favourable period. Overall it had managed only £0.23 private investment for every £1 of public spending by the early 1990s.

Yet this was the exception. It seemed, for a moment at least, that the Thatcher 'economic miracle' might just be able to sustain a leveraged regeneration of the inner cities, successful in its own terms. There was even some promise that the trickle-down of social benefits might begin to flow more rapidly. That moment soon passed, however, as property boom turned to slump. The weaknesses of the policies soon became all too apparent. To understand why, we need to consider what was actually done.

Inner-city regeneration in practice since 1979

Land reclamation

Some 16,500 hectares of derelict land were reclaimed in England between 1979 and 1992, mainly in inner-city and older industrial areas (DOE, 1987a). This represented a marked increase on 1970s' reclamation rates, though in fact it barely kept pace with the national rate of new dereliction arising from economic change. In many local areas the amount of dereliction actually increased. Yet the local effects were often impressive. Most spectacular in the short term were the Garden Festivals held in Liverpool (1984), Stoke (1986), Glasgow (1988), Gateshead (1990) and Ebbw Vale (1992) (Middleton, 1991). The idea was German, developed as a device for regenerating post-war bomb sites by imaginative reclamation, attracting visitors and commercial sponsorship and stimulating permanent developments on profoundly unpromising sites. Their British innovator was Heseltine who initiated the first festival during his period as 'Minister for Merseyside'. In practice their planning was very rushed and they proved rather poor in securing long-term investment except where, as in Glasgow, they were part of a wider regeneration package. No further festivals were planned after Ebbw Vale. Yet they were generally popular events which certainly helped alter perceptions of former derelict sites.

New infrastructure

Many inner-city areas, especially docks, suffered from poor accessibility. The transport networks based on water and rail which had serviced their initial development were now barriers to their integration with the rest of the city. London Docklands particularly suffered from such problems (Brownill, 1990b). Fifty-five miles of new road had been built there by 1992, with further major links envisaged. The same area also saw the construction of the entirely new Docklands Light Railway (DLR) rapid-transit system. Improvements to the few parts of the existing London Underground system that served the area also occurred and an extension to the Jubilee Line was eagerly awaited in 1993. Finally a new urban airport, London City, was added to service the emergent business community of Docklands. No other inner areas faced quite the severity of these infrastructural problems, but several are intended to be served by new light rapid-transit 'supertram' routes.

Heritage, leisure and tourism

Improving accessibility encouraged leisure and tourism schemes of inner-city regeneration, especially in redundant dock areas. Again following the examples of Baltimore and Boston, the potential of waterfront-based development was quickly recognized, particularly when there were dock buildings of great historic and architectural quality (Middleton, 1991). The development at St Katherine's Dock adjoining the Tower of London and opened in 1973 was an early British example. Much more was done in the 1980s especially in London Docklands and, above all, the Albert Dock scheme at Liverpool, where Grade I listed Victorian dock buildings were imaginatively converted into a mainly leisure and cultural complex, opened in 1988. Another notable scheme with a significant leisure content was at Salford Quays (the former Manchester Docks), where a pragmatic Labour local authority, armed with Urban Programme and DLG funds, showed itself capable of outdoing even the favoured UDCs in promoting public–private development partnership (Robson, 1988; Law, 1992). Unlike most of the other schemes, the original rather newer dock buildings had no architectural merit. A few cranes, bridges, etc., were retained to provide some link with the past, but otherwise conservation played little part.

Post-industrial development

Alongside the leisure and tourism focus, one of the most striking features of 1980s' inner-city regeneration was the amount of commercial development, a marked break with previous precedents (Goobey, 1992; Healey *et al.*, 1992). The emphasis on both leisure and commercial services reflected increasing post-industrial pessimism about large amounts of traditional industrial employment ever returning to the urban cores. Some industrial development did occur, most notably in the movement of newspaper printing presses into London Docklands, but this was unusual. Manufacturing certainly remained important in some UDC areas, such as Trafford or Tyne and Wear, but there were few new manufacturing jobs. Many areas also had small warehousing developments, but they generated little employment. It was therefore in the new leisure

Figure 8.4 *Opened in their new guise in 1973, St Katherine Docks, London, were the first British example of US-style 'post-industrial' dock regeneration based on tourism, leisure and offices, together with some GLC housing. The location, adjoining the Tower of London, brought immediate success, though it was more difficult to apply the same formula to the huge dock areas further east.*

and tourist developments, shops and offices that the main potential for creating jobs lay.

Retailing

Shopping was a boom industry of the 1980s, a reflection of the national pursuit of private material contentment that was such a strong theme of the Thatcherite vision. This had important implications for urban change (Montgomery, 1991). Following the decentralization of food retailing which was already well established, retail multiples specializing in furniture, electrical goods and DIY were especially eager to secure non-central sites where they could have more space and cater more effectively for the car-using customer. This growing tendency offered major opportunities for the inner cities, particularly since firm planning controls tended to limit developments on the urban fringe. The first manifestation came as the new shopping phenomenon of the retail warehouse park. Bulkier household items were sold to car-based shoppers from warehouse-type buildings. Compared to traditional central area retailing, the cheapness of inner-city land was an important factor permitting larger, single-storey premises and ample car parking. Some of the EZs proved particularly popular for this kind of development, especially Swansea and Dudley.

Much more dramatic, however, was the creation of very large out-of-town shopping developments on the North American pattern (DOE, 1992a). These were purpose-built centres based on the same kind of shopping mix which traditionally dominated central areas, though with the added benefit of vast amounts of free car parking. In Britain the planning system had traditionally discouraged such developments, to protect the established centres. The first such scheme, an 0.79 million gross square feet centre at Brent Cross in north-west London (opened in 1976) was allowed only because it would not damage existing centres.

Regional shopping centres and urban regeneration

Much more favourable conditions for out-of-town developments arose in urban regeneration areas in the 1980s (Middleton, 1991; Goobey, 1992). The weakening of planning controls, especially in the EZs, made it more difficult for local planning authorities with established shopping centres to object. Moreover this deregulation came at just the time when easy consumer credit was intensifying the demand for new shopping space. The first of the new wave was the Gateshead MetroCentre (1.36 million gross square feet) in the Tyneside EZ, created by local developer John Hall, and opened in 1986. Other large schemes (over 1 million square feet each) were in the Dudley EZ at Merry Hill and at Meadowhall in the Sheffield UDC area, both opened in 1990. Like the MetroCentre they were also brainchilds of local entrepreneurs (the Richardson brothers and Eddie Healey, respectively). Many similar schemes were proposed on inner-city or derelict sites but planning objections and the recession signalled the demise of most of these. The value of these new assets was also placed in doubt as the Environmental Protection Act 1990 required that sites thought to be contaminated be placed on a register. In a falling property market, this was sufficient to thwart the sale of Merry Hill, pushing the company which owned it into bankruptcy in 1992.

Office development

Something of the same fragility of the inner-city property market was apparent in office development (Healey *et al.*, 1992). The 1980s saw a huge growth in the financial sector as exchange controls were removed (Diamond, 1991). The 1986 'Big Bang' deregulated London financial dealings and introduced electronic trading. In turn this growth triggered a huge office development boom, especially in London, following the abolition of office development permits in 1979. Docklands and other inner-city areas were marketed as new office centres, with rents well below those of city centres. Nearly 9 million square feet of office space, much of it high quality, was started in the LDDC area between 1988 and 1990 alone. It was, however, in the nature of how inner-city regeneration occurred that little strategic thought had been given to the relationship of these new developments to existing central business districts. The presumption was that the developer knew best; planning in any wider sense was not deemed necessary.

Canary Wharf

Nowhere illustrated this more graphically than Canary Wharf in Docklands (*Building*, 1991; Goobey, 1992). The proposal was conceived in the mid-1980s by American G. Ware ('Gee Whizz') Travelstead, but taken over in 1987 by Canadian developers Olympia and York (O & Y), headed by the Reichmann

Figure 8.5 *The property-led urban regeneration of the 1980s weakened traditional planning concerns to protect established retailing areas (evident in Figure 7.4). Several American-style out-of-town shopping centres appeared, such as here at Meadowhall, opened in 1991 in the Lower Don Valley, Sheffield's former steel belt.*

brothers. Benefiting from location in both the LDDC and the Docklands EZ, the development proposals as a whole involved the creation of 10.5 million lettable square feet of office space. It was nothing less than the establishment of an office centre to rival the City of London itself. Fresh from huge successes on a comparable development at Battery Park in New York, the Reichmanns were able to do what Travelstead never could: establish the financial credibility of the scheme. The first two phases, including the fifty-storey centrepiece tower, took shape in 1987–90, during the heady post-'Big Bang' years of the Thatcher 'economic miracle'.

Unfortunately the boom had evaporated by the time the buildings were finished in 1991. The City of London had not idly accepted its new rival and had sanctioned major new office developments in the meantime (Diamond, 1991). There were some 2 million square feet of vacant office space in the City and the rent advantage of the off-centre location had diminished. Quite apart from the reduced demand for new office space, potential tenants were still worried about poor transport links. The capacity of the DLR was inadequate and the Jubilee Line extension was still awaited (dependent in fact on contributions from O & Y, drawn from rental income on the first phases of Canary Wharf). Not even the legendary Reichmanns were able to overcome this combination of problems and the development, still largely empty, went into receivership in 1992. What had been heralded as a monument to the market-led regeneration of the 1980s became a cautionary tale of the 1990s.

Housing

Another important contrast with earlier inner-city redevelopment came in the field of housing. The differences were essentially twofold. First, there was a downgrading of the overall emphasis on housing consequent on the shift to economic regeneration (Potter, 1990). By 1992 total housing completed or started in all UDCs together numbered only about 25,000. Secondly, there was little concern to provide social housing. New council housebuilding dwindled to practically nothing and private housing developers were dominant, in some cases taking over older council housing for refurbishment and disposal. Residential development was viewed in the same way as other developments, via the demands of the market. Once initial reluctance was overcome, often with free or very cheap land, a new inner-city housing market was created, based on attracting back the middle classes. UDCs, especially the LDDC, and local councils such as Salford were heavily criticized for subsidizing 'yuppie' housing and neglecting the social dimension of inner-city policies. Despite the promotional rhetoric and apparent success in the late 1980s, this new housing market also proved very fragile, especially in London (Brownill, 1990b). By the early 1990s local councils were offering support to developers by leasing unsold dwellings in Docklands to meet their housing needs.

***Figure 8.6** Canary Wharf was another monument to the weakening of traditional planning controls. Developed as a huge new office centre in the London docklands, its post-modern design looked back to the classically inspired boosterist aesthetic of American 'City Beautiful' planning and provided London with a striking new landmark tower. Yet by 1993 it was still largely empty, its developers virtually bankrupt.*

The balance sheet

Overall therefore inner-city property-based regeneration was one of the major aspects of 1980s-style planning. The new concepts of promotion and leverage were used, with subsidies, to attract private developers into what had historically become very unpromising territory for them. It was difficult not to be impressed by the scale and panache of regeneration, especially in the UDC areas. Yet it is very doubtful whether all this really addressed the social issues of unemployment and multiple disadvantage that lay at the heart of inner-city problems (House of Commons Select Committee on Employment, 1989; Burton and O'Toole, 1993). By 1992 something like 100,000 extra jobs had reportedly been created in the UDC areas since their designation, an impressive total though puny by comparison with total unemployment. Moreover many of these were transferred from elsewhere (77 per cent of new jobs in the LDDC area 1981–90, for example) (Potter, 1990). And it was generally only lower-paid employment, such as cleaning, security work, etc., that was available to existing inner-city populations who lacked the skills to do the higher-paid jobs of the post-industrial economy. The story was much the same in housing; in effect the poor could only benefit indirectly, after the well off had been assisted.

But the most damning criticism of inner-city regeneration, 1980s style, was that it failed on its own terms because of a chronic failure to think strategically. LDDC in particular showed the effects of a failure to plan comprehensively that followed from the belief in letting market forces rule. Developers found they could get subsidies for what they wanted, to enhance or guarantee the profitability of their specific developments, but not for what they needed, a proper transport and social infrastructure. They also learned a more general lesson: a planning system which let them individually do as they wanted reduced overall certainty and increased risks in what was already one of the most risky business undertakings. By the early 1990s inner-city skies were dark with the wings of Thatcherite chickens coming home to roost.

Changing city centres

Inner-city regeneration and the city centres

The push for inner-city regeneration was often undertaken with little consideration of its wider impacts. Yet adjoining areas, particularly the city centres, were clearly affected, sometimes directly so. Although there has been nothing comparable in scale to inner-city changes, it is important to consider briefly the central area changes of this period. As we saw in Chapter 6, the great emphasis during the 1950s to the 1970s had been given to redevelopment and intensification of established central office and retailing functions. This had taken the form of large new purpose-built developments, often undertaken with a degree of public–private partnership. Yet now the inner-city areas were threatening to scatter some of these traditional functions. The example of American cities with their declining city centres – the 'hole in the doughnut' – did not go unnoticed (Davies and Champion, 1982; Law *et al.*, 1988; Healey *et al.*, 1989).

There was another effect. It was perhaps inevitable, given the prevailing property emphasis of much inner-city urban regeneration, that some of these initiatives would themselves blur into the traditionally more promising territory of the city centres (Law, 1992). Birmingham's International Convention Centre (1991) was a good example of central area fringe development, in this case strongly promoted by the local authority. The mini-UDCs established for Central Manchester and Leeds (Roberts and Whitney, 1993) also promoted development in the equivalent areas of their cities, building on earlier local initiatives.

Planning for traditional commercial functions

To address the underlying threat posed by decentralization, city-centre planners were forced to widen their horizons. There was much greater emphasis on the historical and cultural associations of the central areas. Conservation and reuse became the typical pattern for new shopping and commercial developments in the big cities. Thus, in contrast to its earlier purpose-built Arndale Centre, it was the restoration of the redundant Royal Exchange and its conversion to shopping and theatre use, the restoration of the Victorian arcades and the creation of GMEX from the defunct Central Station that set the pattern for 1980s' change in central Manchester (Healey *et al.*, 1989). Leeds also made much more of its Victorian buildings, refurbishing and extending its arcade complex, converting the Corn Exchange and retaining facades in new developments. The pattern was similar in cities such as Glasgow and Bristol (Jones and Patrick, 1992; Punter, 1991). In contrast, however, the City of London saw another round of entirely new major developments, notably at Broadgate, prompted by the removal of the ODP and 'Big Bang' (Diamond, 1991; Goobey, 1992).

Promoting new central area activities

There was also a growing emphasis on heritage to encourage leisure and tourist use of the central areas (Middleton, 1991). Manchester's Castlefields urban heritage park was a notable pioneer, but the trend was general. The renewal, largely through conservation, of dock areas that were close to the centres of cities as varied as Liverpool, Hull and Gloucester became an important element in the reinvigoration of the centres themselves, very much on the Baltimore and Boston pattern. We have already noted Birmingham's efforts to raise its profile as a convention (and cultural) venue. Another widespread trend was that to encourage a modest reversal of the long-term decline in central area residential population with the creation of new housing, often in converted warehouses or similar, as in Glasgow's 'Merchant City' (Jones and Patrick, 1992).

Public transport and city centres

Finally we can note the moves to improve public transport access in central areas, particularly the welter of 'supertram' schemes of the type that became operative in Manchester in 1992. Although their impacts were wider, major concerns were the improvement of central area accessibility and environmental quality. Despite much Thatcherite trumpeting both had been noticeably worsened by the undermining of the 1970s' cheap fares policies of South Yorkshire, Lothian, the GLC and other authorities and the progressive deregulation of bus services under the Transport Act 1985 (Hill, 1986; Pickup *et al.*, 1991). Yet central areas could never compete with out-of-town developments for car-based access. With so many traditional remedies ruled out, in public transport as in many other aspects of policy, the continued health of the central areas was depending on ever more imaginative approaches (e.g. Bianchini *et al.*, 1988).

ACCEPTING REGIONAL IMBALANCE?

The erosion of regional assistance

The north–south divide

As we have seen, the shift to urban regeneration policies was accompanied by cuts in more traditional types of local spending. Established approaches to exerting influence on spatial economic development also suffered. Regional planning virtually disappeared and regional aid was sharply curtailed from 1977, especially during the 1980s, when it was halved in real terms (Townroe and Martin, 1992). This was despite worsening unemployment everywhere, but especially in the traditional depressed areas. In 1986, the worst year of the decade, Scotland recorded 13.3 per cent unemployment, Wales 13.5 per cent, the north west 13.7 per cent, the north 15.3 per cent and Northern Ireland 17.0 per cent. Yet even a former prosperous area such as the west midlands was now at 12.9 per cent, significantly above the UK average of 11.1 per cent. Contrasted with the perceived impact of the 'economic miracle' on the south, there was much public discussion of the reality of a 'north–south divide', expressed in wealth, well-being, politics, culture and even diet (Lewis and Townsend, 1989; Balchin, 1990). But post-1979 governments, consistent with their general philosophy, put little faith in regional policies.

Regional policies and planning 1974–9

The impact of this was all the more profound because Labour, at least until 1977, had managed to maintain its traditional strong commitment to regional policies (which largely aided the very regions where it was strongest, of course). Initially Labour enhanced the already elaborate regional aid structure that it inherited from the Conservatives in 1974 (McKay and Cox, 1979). The downturn began in 1977, a product of wider public expenditure restraint and a growing recognition of the new spatial policy paradigm of the inner city. The latter cut across some of the traditional thrusts of regional policies, which had tended to encourage outer-city developments in the assisted regions. But this new thrust was still quite minor by 1979, limited mainly to a partial reversal of the Location of Offices Bureau's traditional concern to promote dispersal from the cores of big cities. But the changes were sufficient to prompt the pessimistic but prescient comment from an academic commentator that 'a consensus of nearly fifty years in the making is probably collapsing' (McCallum, 1979, p. 38).

The Scottish and Welsh Development Agencies

Yet a very significant reinforcement of regional policies had come with the creation of a partial framework for more interventionist economic planning at regional level. We saw in the last chapter how the National Enterprise Board (NEB), created under the Industry Act 1975, was conceived as the centrepiece of a national economic strategy. NEB was also intended to have a specifically regional development function within England, though this withered along with the wider strategy (Holland, 1976). More enduring were the Scottish and Welsh Development Agencies (SDA and WDA), also created under the 1975 Act (Wannop, 1984; Balchin, 1990). These were conceived as, and have remained, executive economic development agencies for the two most distinctive regions within Britain.

The claims for special treatment had been heavily reinforced since the late 1960s by the growth of active separatist sentiments in both the Celtic nations of Britain (Parsons, 1986). This reached a peak in 1976–9 as the Labour government attempted to meet nationalist sentiment halfway with proposals for devolution of power to Edinburgh and Cardiff (Glasson, 1978). It was a solution that satisfied few and failed to gain sufficient support in the 1979 referendum, with the result that it was never implemented. But the devolution question served to remind politicians of the special regional status of Britain's subordinated nations. The SDA and WDA are the enduring legacy of this awareness, allowing more unified approaches to economic development that have survived even the Thatcher era. As bodies that played key roles in spatial economic development and urban regeneration in the 1980s, they promised and, in SDA's case, delivered a rather more integrated and strategic quality to planning than was apparent in England (Donnison and Middleton, 1987; Boyle, 1988; Hague, 1990). The WDA has by contrast been rather more opportunistic in its work, though not so much as to negate all the attractions of the development agency model (Lewis, 1992). There have been many, as yet unheeded, calls for similar bodies for the English regions.

The end of negative regional controls 1979–81

Most of the key changes to regional policy after 1979 were quick to appear (Balchin, 1990). As noted earlier, the office development permit, introduced in 1965 to encourage regional dispersal of offices, was immediately jettisoned. The Location of Offices Bureau continued to play its changed promotional role until it too was abolished in 1981. The industrial development certificate (IDC), one of the linchpins of post-war regional policy, was also suspended in the same year and finally abolished under the Housing and Planning Act 1986. The certificate had allowed national control over the location of factory building, intended to discourage new factories in the buoyant areas. Now growing unemployment throughout the country, notably in parts of the generally prosperous south, had made it increasingly difficult to countenance IDC refusals. Between 1975 and 1981 there had been only 28 refusals out of some 7,000 applications. It had also become very easy to evade the controls by using existing vacant premises, which were increasingly easy to find as economic change intensified.

New policies for regional aid

More significant than the loss of the negative controls was the reduction in regional assistance. The Thatcherites had little economic faith in regional aid (Parsons, 1986). Consistent with their overall philosophy, they saw market forces as the long-term solution to regional problems and saw no reason why they should indefinitely subsidize industry to move to areas of cheap labour. Policy, in this view, fulfilled an essentially short-term, residual role, with social rather than economic objectives. They preferred to see areas marketing themselves for new investment rather than being feather-bedded by generous regional grants. The reality was a good deal more pragmatic than the philosophy, especially when a nationally important investment such as the Japanese car company Nissan was being pursued in 1981–5, but the ideals were certainly reflected in actual policy changes.

A changed pattern of regional aid

In 1982 and 1984 important changes were made in the structure of regional assistance (DTI, 1984). The 1982 changes dramatically reduced the extent of the assisted areas. These now included only 27.5 per cent of the working population compared to 44 per cent in 1978. Yet the established assisted area categories of Special Development, Development and Intermediate Areas were retained, for the moment at least. There was, however, some acknowledgement of the changed economic circumstances of former prosperous areas in the midlands as the steel closure area of Corby received Intermediate status. The changes of 1984 went much further, abolishing the Special Development Areas to leave just two categories, though the total assisted area was increased to cover 35 per cent of the working population. This apparent paradox reflected the granting of Intermediate Area status to the economically depressed and politically sensitive west midlands. The significance of this small increase in extent was, however, outweighed by the reduction in the level of automatic grant and much greater emphasis on selective assistance.

This new pattern operated until 1993, when even more striking spatial changes were introduced. The present pattern accounts for some 34 per cent of the working population. Consistent with the traditional conception of regional policies, assisted status has been newly granted to colliery closure areas. Yet granting it to parts of East Anglia and the once mighty south east, including inner east London and several coastal areas is an important departure. It reflects the particularly harsh impact of the 1990s' recession on formerly prosperous regions and is the clearest evidence yet that a new political geography of the regional problem is being fashioned. There is no evidence, however, that regional aid will regain its former important position under this new pattern.

The European Regional Development Fund

In contrast to the shrinking UK commitment to regional policies, the European Community's Regional Development Fund (ERDF) has grown markedly since it was formed in 1975 (Mackay, 1992). Its objectives are the redress of regional imbalances, assisting areas of serious industrial decline and structural change in agriculture and combating unemployment. In view of the extent of change in its economy during the 1980s, the UK did rather well out of the ERDF. Between 1984 and 1987, for example, it received nearly 606 million ECU (roughly £420 million), compared to Germany's 102 million. Only the larger countries of southern Europe, Italy and Spain, with their extensive residual rural poverty, did better. In theory ERDF assistance was based on the principle of 'additionality'. This involved substituting European in place of national aid for approved schemes, with the intention that national aid would go towards additional projects. The principle was, however, rather difficult to enforce and became something of a charade. Despite increasing protests from the EC, UK governments in the 1980s and 1990s used European monies simply as a device for cutting national spending on regional aid.

The disappearance of regional planning

Planning the south east

Yet at least regional assistance survived. Regional planning shrank during the 1970s, as growth and public-spending projections withered, and was then all but abolished in the 1980s (Hall, 1989, 1992). The plans for the most populous region, the south east, focus of so much planning attention in the 1960s (MHLG, 1964; SEEPC, 1967), were subject to successive downward revisions in the later 1970s (SEJPT, 1970, 1976). On each occasion the documents became more slender. By 1978 it was being dealt with in just thirty-seven pages (DOE, 1978). Yet this was nothing compared to the 1980s, when a series of three ministerial letters in 1980, 1984 and 1986 were supposed to cover it all. Collected together as a Planning Policy Guidance Note they amounted to just nine A4 pages (PPG9, 1988). It had, of course, very little to do with established ideas of regional planning as a reconciliation of varied sectional interests. This was regional direction from the centre. The only non-central planning input was heard through SERPLAN, a non-statutory agency maintained by local authorities to give the region a voice. It was all a far cry from true regional planning.

The other regions

Although the south east had important peculiarities, the pattern was broadly similar in the other English regions. The Regional Economic Planning Councils, forged in the white heat of the 1960s' technological revolution, had proved a flawed product, always denied real importance by established planning and governmental interests. Yet they had done some useful work in formulating thinking about regional development, especially in the 1960s. To the incoming Thatcher government they were, however, expendable and were abolished in 1979 (Cullingworth, 1988). Along with the abolition of the metropolitan counties, the weakening of the planning role of the non-metropolitan counties and the erosion of structure plans, it amounted to the abandonment of all pretence of regional planning. Only Scotland and, to a lesser extent, Wales were able to maintain some degree of a coherent regional approach, a lasting institutional legacy of the 1970s' devolution debates. Some English regions (such as the west midlands) fared better than others, reflecting locally orchestrated collaboration rather than any real central encouragement. By 1990 there were some vague central promises of greater attention to strategic regional planning guidance, notably in *This Common Inheritance* (DOE *et al.*, 1990) and in PPG15 (1990), which addressed the relationship between regional guidance and statutory plan revisions. But by 1993, at least, there was little to show for it.

The need for regional planning

The retreat from strategic regional planning was all the more remarkable in view of the dramatic urban and regional economic changes of these years (Hall, 1989). As we have seen, the primary response of the Thatcher governments was to encourage a patchwork of economic promotion and regeneration, with no overall strategic planning framework to give coherence (Brindley, Rydin and Stoker, 1989). Localized place marketing took over from regional planning. The spectacle of Docklands competing as an office centre with the City arose because strategic planning policies had been removed, as became all too apparent in the property collapse of the 1990s. Meanwhile outer-city areas, especially in the south, were facing formidable growth pressures that showed no signs of abating. There were major infrastructure developments, including the completion of the M25 London orbital motorway in 1986, the continued growth of the region's international airports and the construction of the Channel Tunnel. Had this been the 1960s, we may be certain that regional planning studies would have been commissioned to devise strategies to address all these diverse regional needs and pressures. Yet, as we have seen, the spirit of Thatcherism proved allergic to such devices. In the absence of real regional planning strategies, the accommodation of outer-city growth pressures became one of the most vexed planning questions of the 1980s.

Decentralization after the New Towns

The end of the New Town programme

New Towns and inner cities 1976–9

The main 1960s' planning response had been to create New Towns, or major expansions of existing towns. These solutions had fallen from grace by the 1980s. The first shift had come in 1976, when Peter Shore's new concern for the inner city brought a questioning of the role of New Towns (Aldridge, 1979). Shore clearly acknowledged that inner-city decline could not be blamed on the New Towns, because the major part of metropolitan decentralization went elsewhere. But he had serious doubts as to whether the New Towns could be any long-term part of a solution to the newly identified problem. Reflecting this new mood, the already designated (but barely started) Glasgow New Town of Stonehouse was abandoned and its expertise diverted into the new GEAR inner-city project. Further south, the proposed new city of Maplin in Essex ('Heathograd' as

Crosland dubbed it) had been quietly dropped, along with the airport that would have provided its *raison d'être* (Hall, 1980).

New Towns and Thatcherism

Apart from these shifts, Shore contented himself with trimming the ultimate targets of the New Towns. The new Thatcher administration, while fully sharing the new inner-city focus, also saw the New Towns as profitable, publicly owned plums ripe for privatization (Hebbert, 1992). They accelerated the winding up of the remaining development corporations. By early 1994 only those of the five Scottish New Towns remain, with the last (Irvine) scheduled to go in 1999. Traditionally the assets of the development corporations had gone to the Commission for the New Towns. Yet this too was to be wound up, under the New Towns and Urban Development Corporations Act 1985. Valuable commercial assets were progressively sold off and private developers took on increasing importance in the completion of the unfinished New Towns.

The changing New Towns

Meanwhile existing tenants began to become owner-occupiers, a process which predated, though was greatly encouraged by, Mrs Thatcher's 'right-to-buy' legislation of 1980. In the south particularly the public housing emphasis which had characterized New Towns from the outset was finally shed. This change highlighted the broader process of upward social mobility amongst New Town residents which had begun when they (or their parents) sought to better themselves by moving from London. Given the early perceptions of the New Towns as Labour-voting enclaves in Conservative shire counties, it is not without ironic significance that it was the New Town (or former New Town, as it was now termed) of Basildon that was widely held to epitomize Thatcherite values by the late 1980s.

After the New Towns

The pressures for outer-city development

The fact that there were no longer huge public programmes of planned redevelopment in the inner cities creating a need for overspill to New and Expanded Towns did not greatly diminish demand for additional land for private housebuilding in the outer-city areas, especially in the south. Population migration from the bigger cities continued, but the most important pressure now arose from the population and economic buoyancy of the outer cities themselves (Herington, 1984). Accordingly rates of new household formation in the rural counties surrounding the main metropolitan areas were appreciably higher than elsewhere. Job growth was also higher as new 'high-tech' manufacturing and service industries developed in these same areas (Phillips, 1993). Symptomatic of this emergent pattern were Cambridge Science Park in the 1970s (Carter and Watts, 1984) and the Aztec West Business Park, north west of Bristol in the 1980s. A more urban, but still outer city, variant was Stockley Park, abutting the green belt near Heathrow Airport (SPCL, c.1992).

Accommodating growth

The problem, especially severe for a Conservative government whose main support came from these very areas, was how to accommodate all this growth. The 1970 Strategic Plan for the South East, perhaps the last round of the 1960s' commitment to active regional planning, had identified several important areas. These were being built on through the late 1970s and 1980s, but it was clear that more were needed to meet continuing demand. Without major growth areas being identified, the solution inevitably involved piecemeal development around existing settlements, putting strain on existing infrastructures and services.

It was a solution which spread the misery and the political damage. County councils and even more the districts, which of course became much more powerful players in the planning game during the 1980s, frequently railed against growth projections and sought to allocate less development land than was necessary. The housebuilders' organizations were quick to press the opposite view with equal vigour (Rydin, 1986). Not surprisingly, therefore, Mrs Thatcher's Environment Secretaries were sometimes attracted to planning solutions which concentrated development, where they felt it was appropriate. Large developments were also much cheaper for the housebuilders and therefore offered the possibility of private provision of infrastructure, services and amenities as planning gain. Inevitably though the growing reliance on such development methods strengthened the overall influence of the private developer on the planning process, compared

Figure 8.7 *Stockley Park, near Heathrow Airport, was one of the new generation of 1980s' business parks, bridging the traditional divide between industry and office activities. It was developed on a former derelict site abutting the inner edge of the metropolitan green belt and offered a superb environment, designed to attract prestige companies.*

with the traditional reliance on public-sector provision of roads, schools, amenities, etc.

Large private developments

A perceptible shift towards private new towns began to occur during the later 1970s, for example at South Woodham Ferrers in Essex, developed from 1975 for a population of 17,500 (Neale, 1984; Hebbert, 1992). Yet the hand of planning was still relatively strong here. The land was originally owned by the county council and the provision of roads, public sites, etc., was undertaken conventionally, by the usual public agencies. The first large example of a more completely developer-oriented approach soon followed in the Lower Earley area, adjoining Reading, in the growth corridor along the M4 motorway (Healey *et al.*, 1982; Hall, 1989). At one level we can see Lower Earley as a product of early 1970s' regional planning, yet the manner of its detailed planning and implementation made it the harbinger of a new approach. Over 6,000 dwellings were built between 1977 and the early 1990s by a consortium of private builders, on land entirely owned by them. The planning gain negotiated amounted to 8 per cent of the selling price of the houses, used to finance roads and leisure facilities, together with land gifts for open space and school sites.

***Figure 8.8** Lower Earley emerged from 1970s' regional planning proposals for a growth area around Reading, yet as the largest single private-sector housing development in the UK, it presaged a new, more developer-led approach to new settlement building, laying much of the basis for the ill-starred Consortium Developments initiative. The photograph shows part of the south-eastern corner of the development.*

'Heseltown'

Development had already begun when, in 1980, Michael Heseltine forced Berkshire to find land for 8,000 extra houses in its structure plan (Short, Fleming and Witt, 1986; Short, Witt and Fleming, 1987). The fiercely contested target area for most of these dwellings, north of Bracknell, was quickly dubbed 'Heseltown', though local resistance prevented it from becoming a reality. Instead, the county attempted to meet some of the new housing requirement, at least in the short term, by accelerating development at Lower Earley. And when critics looked for a tangible example of what Thatcherite principles might mean for the outer city, for a real 'Heseltown', it was invariably Lower Earley they chose, despite its planned origins.

Consortium developments

The next step was completely developer-initiated private new towns, scaled-down models of American settlements such as Reston, Virginia or Columbia, Maryland (Ward, 1992b). It was in 1983, at the high point of Thatcherism, that the main private housebuilders formed Consortium Developments Ltd (CDL), with the express intention of creating new, relatively self-contained, country towns of at least 5,000 dwellings, with employment and social amenities (Lock, 1989; Northfield, 1989; Hebbert, 1992). There was no regional planning direction in the location of the intended settlements. They arose, as RTPI President Francis Tibbalds put it in a 1988 Channel 4 television programme, from 'rather opportunistic developers waving chequebooks at farmers' (Dispatches, 'The Battle for Stone Bassett').

The intention was a variant of the formula Ebenezer Howard had followed: buying land rather cheaply, nearer agricultural than usual development land values, then benefiting by both the economies of scale in development and the rise in land values that accrued from development itself. Unlike Howard, however, the bulk of the betterment would be retained by the developers rather than being passed to the community, though infrastructure and public facilities were to be developer provided, as planning gain. The problem, from CDL's point of view, lay in the traditionally obstructive power of the planning system. CDL made the mistake of entirely believing Thatcherite rhetoric about deregulating the planning system. And it took them some time to realize their mistake. Over the next few years CDL initiated specific proposals to build four new towns: in Essex ('Tillingham Hall'), Hampshire ('Foxley Wood'), Oxfordshire ('Stone Bassett') and Cambridgeshire ('Westmere').

The fate of the new country towns

The first, which was in the London green belt, was rejected by Nicholas Ridley (Herington, 1989). He made it clear though, even in his refusal, that he strongly sympathized with the concept of new, privately developed settlements, a view which was elaborated in several DOE statements during his tenure. With supreme irony, the third of these proposals, Stone Bassett, was in the constituency of Michael Heseltine, then out of office. Predictably neither he nor his constituents wanted to see another 'Heseltown' in their backyards. He and Ridley fought publicly over the issue in 1988, personalizing the essential planning dilemma of 1980s' Conservatism, between the enterprise culture and 'nimbyism'. For a time, in 1989, it seemed that enterprise had won the day (as it did in John Mortimer's 1990 satirical fictionalization of the battle for Stone Bassett, renamed 'Fallowfield').

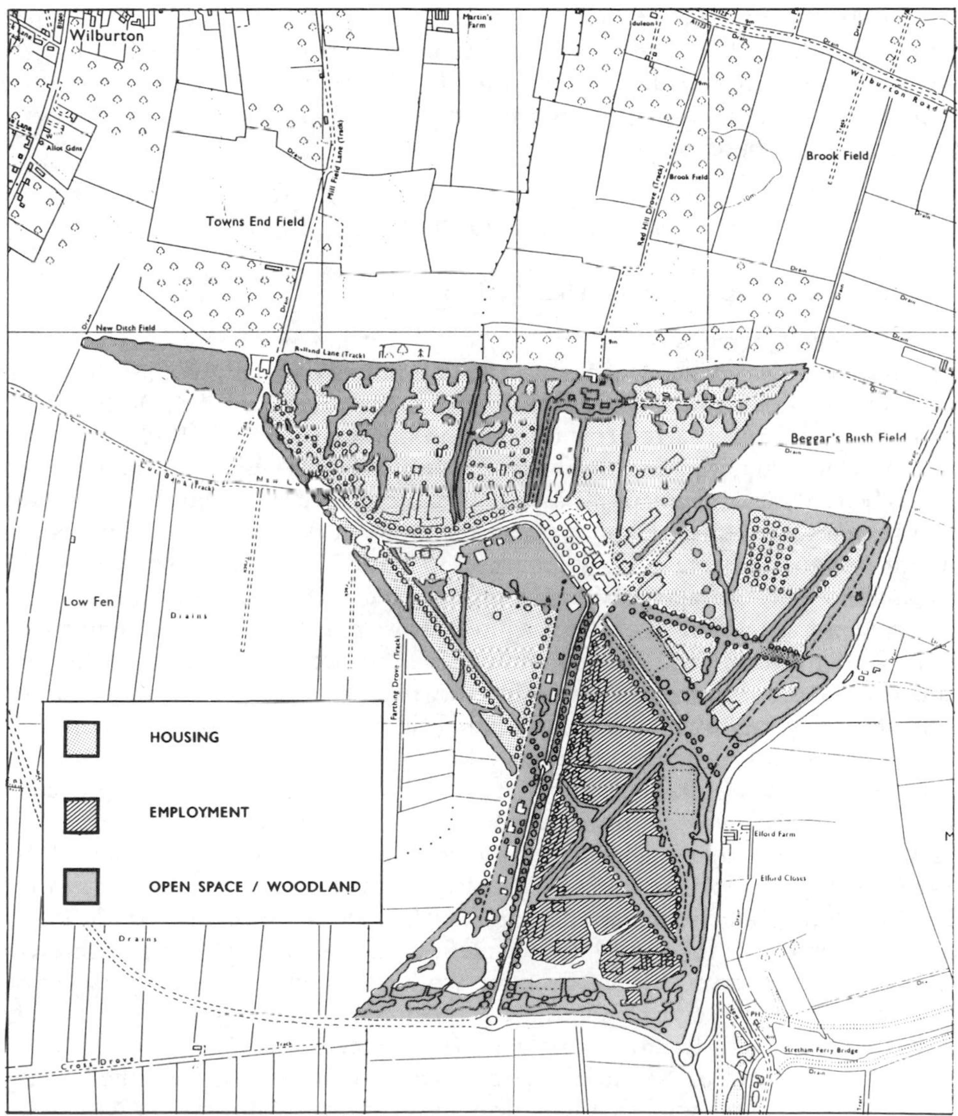

Figure 8.9 *'Westmere', near Cambridge, was the last of Consortium Developments' attempts to promote a new country town. Despite sympathy from Cambridgeshire planners, the proposal was rejected following an inquiry in 1991.*

Yet the real breakthrough for the developers seemed to come on the second proposal, Foxley Wood. The area was, in Ridley's own words, 'hardly a jewel of the English countryside' (Ridley, 1992, p. 116) and he provisionally approved CDL's proposal, in July 1989. Approval was short-lived, however, as the pendulum swung sharply in favour of green issues and the Conservative 'turquoise tendency'. Within a few months his successor, Chris Patten, had overturned the decision. Stone Bassett and Westmere also suffered rejection, despite support from the local county council in the latter case. Overall the new country towns were a 1980s' experiment that failed, leaving only a few, much smaller, new villages and fewer freestanding suburbs to be implemented (Amos, 1991). They were a sign of what might have been if Thatcherite commitment to enterprise and deregulation really had ousted traditional Conservative faith in regulative planning in the outer cities.

Planned Containment

Green belts: the exception to Thatcherism?

The expansion of green belts

The most obvious evidence that this tradition remained relatively unscathed was the increased dominance of green-belt policies (Elson, 1986, 1993; DOE, 1993). From being merely a sacred cow of planning in the 1950s and 1960s, the green belt became a supreme deity in the 1970s and 1980s. The amount of approved green belt increased dramatically as structure plans were approved. In 1974 there had been 14,468 square kilometres subject to green-belt policies in England, of which only 6,928 were actually approved. By 1987 this had grown to 18,192 square kilometres of approved green belt (PPG2, 1988). Most of this increase reflected the extension of existing approved or outline green belts, embracing more settlements in outer metropolitan areas. Distinctively new green belts were added around the north-east Lancashire towns, Lancaster, the Fylde Coast and Burton-on-Trent, however. Taken together, approved green belts amounted to some 14 per cent of the area of England.

In Scotland, a 1,000 square kilometre Glasgow greenbelt (*sic*) was approved in the 1980 Strathclyde Structure Plan. Much larger and more continuous than that previously approved, it finally fulfilled the promise of the 1946 Clyde Valley Plan. Yet the Scottish conception of the green belt remained slightly different to the English, its symbolic status not quite so exalted. Dundee actually abandoned its formal green belt in the late 1980s, seeing fit to manage the urban fringe through countryside protection policies. Even for Scotland this was a singular perspective, however. Virtually everywhere in the UK green belts or quasi-green belts were in the ascendant. In Wales, for example, Clwyd's 'green chains' were an important new initiative. And in Northern Ireland, Belfast's 'stopline', established in the mid-1960s, was upgraded to a fully fledged green belt (Murray, 1992).

Attempts to erode the green belt 1983–4

Yet this overall trend of expansion concealed a serious attempt by the second Thatcher government to loosen the grip of the green belt. It was part of the wider

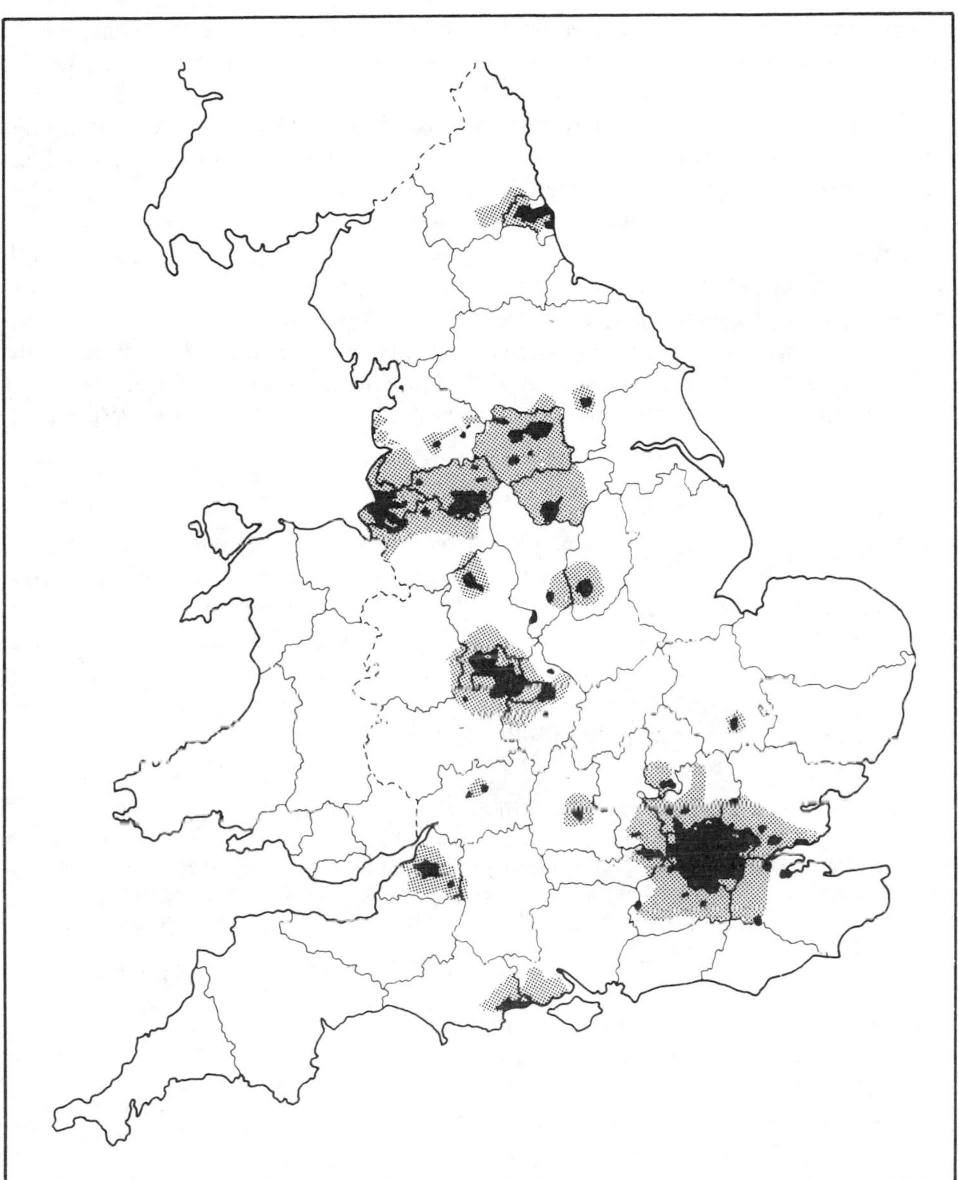

Figure 8.10 *Following Patrick Jenkin's ill-judged moves to relax green-belt policies in 1983–4, they became progressively more entrenched during the rest of the decade, surviving all deregulatory impulses. The challenge for the 1990s is to make them into something that is more than just a 'stopper', with a more positive role in strategic planning and the promotion of sustainable development.*

Thatcherite project to give greater freedom to the private developer. The agent of this was the then Environment Secretary, Patrick Jenkin (Elson, 1986). In 1983 he issued draft circulars on the green belt and housing land. These were far from being an overt frontal assault on green belts. An innocent reader, unused to the coded nature of the public discussions of such politically sensitive issues, might easily have assumed they were offering unswerving support for the green belt.

In fact the circulars were advocating a loosening of green-belt boundaries. As Jenkin himself said, 'they must not be pulled so tightly that there is no room left for development in the future' (Jenkin, 1984, p. 17). Yet to urban-fringe Tory MPs and conservation lobbies such as the Council for the Protection of Rural England, this was the thin end of a very fat wedge. Certainly the creation of CDL in 1983 (noted above) suggested that the major private housebuilders were preparing to move into the urban fringe in a big way. A flurry of Commons motions followed and Jenkin was forced to retreat, with lasting damage to his political career. The final circulars in 1984 duly sustained the sense of certainty and permanence which had been such a key feature of green belts since their inception.

The significance of the retreat

The truth was that many politicians (and planners, for that matter) would have dearly loved to have a more flexible interpretation of the policies, with less sense of it merely as a 'stopper' of development. Yet no subsequent minister was foolish enough to repeat Jenkin's mistake of trying to do something about it. Even Ridley, often accused as an enemy of the green belt, largely respected it in practice. The political realities were stated with great clarity by a junior Environment Minister of the mid-1980s, William Waldegrave:

> You would need a brave and confident minister to try and change the policy now, to point out that there is actually a lot of derelict land in the Green Belts, so why do we not try and create something attractive instead of, for instance, these terrible old gravel pits. The trouble is, everybody believes this would be seized upon as a wonderfully convenient opening by the developers. It's like the First World War: any movement might produce a huge collapse. So it has become almost impossible to change it.
>
> (Porritt and Winner, 1988, pp. 83–4)

In a wider sense, 1983–4 saw the first real sign of the limits to Thatcherite deregulation of planning, at least when it applied to the outer cities.

New justifications for containment

Green belts and the green movement

Although it was scarcely needed, another line of support for the traditional conception of the green belt as a 'stopper' came from the strengthening green movement of the 1980s. As we noted in the last chapter, green sympathies flourished on the interface between narrow 'nimby' sentiments and growing awareness of wider global issues. They allowed an aura of spurious radicalism to surround the contented society of the outer cities, defending not just their own backyards but the altogether grander notion of 'the environment' from the incursions of the builder. To the more committed greens the green belt gave a glimpse of planning at its best, stopping the sprawling city.

Green belts and compact cities

Their views received powerful conceptual reinforcement in the early 1990s as the European Commission (CEC, 1990a, 1990b), Friends of the Earth and the Policy Studies Institute (Elkin, McLaren and Hillman, 1991) pressed the case for compact cities (Dantzig and Saaty, 1973). There were harsh criticisms of the traditional decentralist thrust of British planning and its rigid separation of land uses. Instead a case for higher-density urban living was articulated, allowing a better access to urban services, closer proximity to work and less commuting. Naturally the prevention of urban sprawl was to be an integral part of this approach, underpinning the radical changes in the planning of urban core areas. This new approach was not accepted wholeheartedly, not least because it appeared to point to an unacceptable quality of urban living to those who remembered British planning's disastrous dalliance with high-density social housing in the 1950s, 1960s and early 1970s (e.g. Breheny, 1992a; Breheny and Rookwood, 1993). Yet to many concerned with regenerating the urban cores, the emphasis on urban values struck a sympathetic note. The multi-use regeneration of the inner cities seemed to encapsulate part of the argument. More importantly for present purposes, some modest signs of such compact-city thinking had already begun to permeate green-belt policies during the 1980s.

Green belts and urban regeneration

The most important formal addition to green-belt policies in the 1980s was to assist the regeneration of the inner cities. Effectively this became part of policy from 1980, though the 1984 green-belt circular gave it formal status (Elson, 1986). It was further reiterated in PPG2, *Green Belts*, published in 1988. The logic was that harsh restriction of land availability in the urban fringe would force developers to the 'brown' land of the inner city. This was a politically neat solution, since it united the inner- and outer-city authorities in defence of the green belt. It also gave a veneer of coherence that bound together the promotional planning of the inner city and the negative planning of the outer city. Yet the link was valid only to the extent that inner-city problems could be solved within the existing urban cores.

The lack of strategic planning for containment

In practice only a very small part of total housing demand can be met in the inner cities at anything like acceptable densities. More importantly, many new sources of employment demand locational characteristics found only in the urban fringe. Major international airports, perhaps the most potent employment generators of the late twentieth century, provide a general example of this (not contradicted, incidentally, by the very small London City airport in the LDDC area). More specific instances were Birmingham's National Exhibition Centre and the Nissan car plant at Washington, Tyne and Wear. Both were 'flagship' investments of immense symbolic and actual importance for local and regional economies. Both involved major relaxations of green-belt policies, sought by urban local authorities for developments of major economic importance.

While maintaining a formal commitment to tight green-belt policies, urban authorities facing major regeneration tasks have tried to be pragmatic in practice. We can readily concede that complete abandonment of the green belt would certainly have made it more difficult to regenerate that inner city, simply because it would have allowed extensive urban-fringe development. Yet, as we have seen, this has never been a realistic option. The more typical question has not been whether a tight green belt might deflect development to the inner city, as intended, but rather push it beyond the green belt or choke it off completely. Taken together, these points highlight the simplistic nature of the new policy emphasis, looking at issues on too localized a basis. A proper and coherent policy response would have been to strengthen the strategic and regional planning dimensions of urban containment, a line which post-1979 governments have conspicuously failed to follow. A new PPG note on green belts will be issued soon, however, which might well propose a more positive strategic use of green belts in this and other respects. Elson and his associates have certainly recommended that DOE move in this direction in their recent research study for the Department (DOE, 1993).

CONCLUSIONS 1974–94

Taken overall, the evolution of the main concerns of urban planning policies over the last two decades reinforces the broad conclusions of the last chapter. There were important continuities, especially in urban containment policy. But the impress of economic and political change was palpable in the other three policy areas. This was particularly evident in the inner cities, which both bore the brunt of economic change and became a vehicle for the most radical Thatcherite planning experiments. The demise of traditional approaches to regional policies and planned decentralization also owed something to economic changes, though much was certainly politically motivated. It is difficult to imagine that policies in these two areas would have followed quite the path they have under a Labour or coalition government after 1979. Overall the consistent political antipathy to coherent strategic planning remains the strongest impression of these years. And, although the general decline of economic optimism and certainty played a part in the retreat from strategy, it would certainly have been possible for governments to have pursued a more comprehensive and strategic approach. The weakening and fragmentation of planning was a matter of political choice at least as much as of predetermination by wider economic forces.

The effects of this broader political choice were felt further down the line. In contrast to the importance of officials and professionals during the 1960s, when planning actions had rested much more on agreed general principles, policies in the 1980s became far more reliant on politicians than at any other time in the post-war period. Moreover the detailed shaping of planning policies also became much more obviously political in a wider sense. In the urban fringe and outer cities, many planning battles became undisguised tussles between developers and 'nimby'/environmental lobbies and were often resolved on grounds of political pragmatism rather than broader planning principles. In the urban cores (and what remained of the assisted regions), developers and investors found themselves freer than ever before to determine the character of urban and regional change.

Planners meanwhile found themselves forced to become active promoters of development (in the inner cities and the assisted regions) or rather ineffective umpires (in the urban fringe and outer cities). Either way they were bit-players in a confused and disjointed game that was not of their choosing. Gone were the relative certainties of the hitherto dominant comprehensive and coherent approach to planning policies for which they had been trained. What they did not appreciate was that the planning of the post-war heyday was merely a historically specific approach. It had been invented by their forebears in the early planning movement who had eventually convinced governments, assailed by successive experiences of depression, war and a long post-war boom, that a modified version of their ideas was worthy of adoption as policy. For good or bad reasons it then came to be seen as markedly less relevant in the last quarter of the century.

Rather than relying solely on political changes (or changes of heart) to reinvigorate planning policies, the planning movement of the 1990s ought perhaps to develop its own renewed and more coherent approach to planning ideas, to establish a new vision that is more relevant to the late twentieth and twenty-first centuries. This is not easy because professional planners who dominate the movement are primarily equipped intellectually to operate the planning system as they find it. Moreover, it is a suggestion that assumes planning has brought worthwhile benefits that are worth trying to recapture. It is to this question of impacts that we must turn before we can develop this suggestion further.

9

PLANNING IMPACTS SINCE 1945 AND THE FUTURE

INTRODUCTION

This final chapter focuses mainly on the cumulative impacts of planning policies since the Second World War and then offers some last thoughts on the future. There has been a great increase in officially commissioned studies of the impacts of individual planning policies since the 1970s, a welcome by-product of Thatcherite scepticism about all planned intervention. But any kind of overall assessment, comparable to the Barlow Commission in the late 1930s, has been lacking. We have had to rely on unofficial, and inevitably incomplete, assessments. The nearest thing to an exception to this generalization was Political and Economic Planning's weighty study *The Containment of Urban England* (Hall *et al.*, 1973).

Yet there has been nothing to match it over the last two decades. It is, quite simply, very difficult indeed to assess overall planning impacts, because they involve an interaction between the intentions and instruments of policies and the other powerful and diverse economic, social and political forces which shape urban change (e.g. Cooke, 1989; Fielding and Halford, 1990; Whitehand, 1992). Since planners have, in the long term, tried to plan with, rather than against, these broad trends of change, it not easy to determine what part of urban change is attributable to planning policies, or what would have happened without planning. But the issue is such an important one that we must try, using both the evidence of previous chapters and such impact studies as exist. We focus initially in planning's effects on the spatial arrangement of cities and regions.

SPATIAL IMPACTS OF PLANNING

The city centre

Protecting commercial functions

Planning's main impacts here were the protection and enhancement of established commercial functions. As we have seen, especially in Chapter 6, planners played important roles in facilitating the renewal and extension of the commercial fabric of the city centres, creating new shops and offices. Already declining central area land uses such as manufacturing and residential were further diminished. The principal planning mechanism of this change involved the use of planning powers to create large redevelopment sites from a multiplicity of ownerships. Such land assembly would certainly have been very difficult for individual private developers to achieve. Without such planned intervention, it seems likely that private redevelopment efforts would have been either rather piecemeal or limited exclusively to parts of city centres where large land holdings were typical. Inevitably off-centre and suburban commercial development would have been much more attractive, as is visible in most North American cities. Yet in Britain the positive role in site assembly was accompanied by strong negative controls. Until recently at least, planning policies have discouraged commercial development outside the established centres and especially on the urban fringe.

Relative decline and decentralization

This is not the whole story, however. Despite planned redevelopment, central areas in big cities have not matched national trends of growth in retailing and office-based activities. Such statistics as there are indicate a marked relative decline, and some hint at absolute decline in cities such as Liverpool and, to a lesser extent, Manchester, though not yet on the scale of many cities in the USA (Law *et al.*, 1988). By contrast the centres of smaller towns and cities with buoyant local economies have performed better. Yet everywhere food and bulky item non-food retailers have shown marked decentralizing tendencies away from central areas, especially since the 1970s (e.g. Dawson, 1991). Office employment has also shown somewhat weaker shifts in the same direction. These trends are also to some extent planning impacts. Thus policies to encourage lower population densities in the inner areas of the big cities, which we will consider more fully below, have reduced the number of people for whom the central area is the most convenient place to shop. Paradoxically though, the more recent attempts at inner-city regeneration in the 1980s and 1990s have also encouraged more commercial development outside central areas.

Unplanned processes of change

Yet we must also recognize the crucial role of more general factors in this decentralization of traditional central area functions. Even without planning there would still, for reasons we will outline below, have been marked declines in inner-city populations, with knock-on effects for central areas. And the growing post-war reliance on the car has had a huge negative effect on the relative attractiveness of central areas for both shopping and office development. Central commercial pre-eminence had always reflected its optimum accessibility within the city, based on public transport. Now mounting traffic congestion and parking difficulties in central areas were shifting the optimum locations for the car-based shopper and office worker to cheaper non-central land, near major road junctions. Limited public transport improvements offered a partial antidote in some cities. But these have not so far been on a scale sufficient to counter Britain's becoming a car-based society, less dependent on traditional central areas. We are still a long way off the

US-style 'hole in the doughnut', but recent policy shifts promoting leisure, tourism, speciality retailing and the restoration of some residential population in central areas, noted in Chapter 8, hint at a sense of vulnerability that was not apparent in the first two decades after 1945.

The inner city

Replacing the slums

Planned redevelopment also had a profound effect on the inner cities during the two decades from the mid-1950s. Many of the oldest and most decrepit dwellings, the street corner shops, many public houses and other smaller places of business were removed. They were replaced by entirely new urban landscapes with new roads, more open space, new areas of municipal housing, often provided as multi-storey flats, new schools and new service centres. This was only part of the impact of redevelopment, however. Despite a strong policy push for high-density housing, population densities in areas of redeveloped housing were usually between one-third and two-thirds lower than before clearance. This loss was highest in the biggest cities (e.g. Smith and Farmer, 1985; Coates and Silburn, 1980). There was, accordingly, a very marked population decline directly associated with redevelopment. In the same way the removal of shops and backyard industries also reduced business activity in these areas, though to a lesser extent than population. Bigger plants were usually untouched by the clearance process and an estimate of a 25 per cent inner-city job loss attributable to redevelopment in Manchester (Law *et al.*, 1988, p. 127) was probably fairly typical of the bigger cities.

Housing, planning and population decline

Yet again, redevelopment was only part of the story. Other factors have contributed to a huge decline in inner-city populations over the post-war period. Between 1961 and 1981 alone, populations in the urban cores (that is inner and central areas, mainly the former) of Glasgow, Manchester, Birmingham and London declined by 46.1, 43.4, 39.9 and 28.4 per cent, respectively (Law *et al.*, 1988, pp. 89–91). This was far more than could be explained by redevelopment, which affected only the most decrepit parts of the inner city. In most English cities, these declines were the net result of complex inward and outward movements, as Caribbean and south Asian immigrant communities established themselves in areas which were being left in even larger numbers by indigenous white populations (Redfern, 1982). By contrast, cities in the less prosperous regions, such as Newcastle, Liverpool or Glasgow, received fewer immigrants so that net inner-city migrational decline was often more precipitate.

The growth of owner-occupation was particularly important in encouraging outward movement since inner-city areas were dominated by rented, increasingly public rented, housing (Herington, 1984). The problem was compounded by an unwillingness of building societies to advance loans to buy older housing in inner-city areas (Boddy, 1980, pp. 68–71, 176). Although such 'red-lining' practices have now largely ceased, they undoubtedly had an important effect on inner-city decline over several post-war decades. Often it was only Commonwealth immigrant communities who bought housing in these areas, reflecting their exclusion from the main housing markets and use of fringe housing finance. Meanwhile for white inner-city populations who were outside clearance areas but eligible for better housing on general needs criteria, the availability of publicly rented housing on suburban estates or, to a lesser extent, in New Towns was also often an important factor encouraging migration (Herington, 1984; Ermisch and Maclennan, 1986). Thus, although greater in the public-sector housing system than elsewhere, the role of the planner was still only an indirect one in all these general housing and migration processes which were producing inner-city population decline.

Economic change and planning

Planning's role in the decline of the inner-city economy was also limited. In addition to the effects of redevelopment, noted above, planners certainly eradicated many other non-conforming industries based in areas zoned for housing, particularly during periods of aggressive council housebuilding programmes (Buck *et al.*, 1986, pp. 56–7). Planned industrial moves to New or Expanded Towns also accounted for an estimated 14 per cent of employment decline in London as a whole between 1960 and 1978 (Fothergill, Kitson and Monk, 1982).

None of this was sufficient to account for the scale of inner-city job losses. In the six main conurbations of London, Greater Manchester, West Midlands, Merseyside, Tyneside and Clydeside, manufacturing employment in the inner areas (defined more widely here as including centres, inner cities proper and some inner suburban areas) declined by 8 per cent in 1951–61, 26.1 per cent in 1961–71 and 36.8 per cent in 1971–81 (Begg, Moore and Rhodes, 1986). The main sources of decline were broadly structural, so that inner-city areas contained older industries and less efficient plants, more vulnerable to closure or job loss. But perhaps planners could have played a more encouraging part in replacing manufacturing jobs in the inner city. Compared to all the effort that went into planned redevelopment for housing, little was done for manufacturing in the inner cities.

Yet against a national trend of dwindling manufacturing employment, it remains very doubtful whether enough new manufacturing jobs could ever have been attracted into the inner cities on a scale large enough to dent the statistics of their decline (Fothergill and Gudgin, 1982). This certainly was the experience of the conurbations in the assisted regions, as we will see below. Inner-city areas had few really large sites, certainly none that could be used without extensive site preparation prior to reuse. Inner cities were, moreover, characterized by workforces with low or redundant skills, inappropriate to the requirements of new 'high-tech' manufacturing (e.g. Buck *et al.*, 1986). We can conclude that planners, by pursuing policies of metropolitan decongestion, did have some supporting role in the decline of the inner-city employment base during the first three post-war decades. Yet as more recent policy initiatives have shown, it was not easy to swim against the growing tide of de-industrialization.

The impacts of inner-city regeneration

In conjunction with the ending of crude 'red-lining', rehabilitation policies showed signs of being able to slow the 'white flight' to the suburbs during the 1970s. Small-scale middle-class 'gentrification' of some more attractive inner-city enclaves in prosperous cities was recognizable (e.g. Ferris, 1972; DOE, 1977a). (This did not always stem population decline, however, since the more affluent usually occupied their homes less intensively than poorer households.) On top of this the more dramatic initiatives of the late 1970s and 1980s seem to have had some impacts. There was much evidence during the 1980s that the rates of decline were slowing down (e.g. Robson, 1988). The 1991 Census, though it has not yet been subjected to small area analysis, also points in the same direction (OPCS, 1992). In inner London (central area and inner city), for example, a population decline trend that had been 13.2 per cent in 1961–71 and 17.6 per cent in 1971–81 had slowed to only 5.9 per cent in 1981–91. Moreover, Tower Hamlets, the inner London borough most affected by the LDDC, actually showed a 7.3 per cent increase in 1981–91, reversing a decline trend of very long standing. Yet this was the only inner borough showing growth, suggesting that regeneration policies will need to be on a very much larger scale if they are to reverse rather than simply slow the overall rate of population decline.

On the economic side the evidence is still very limited. Certainly the various localized initiatives have succeeded in creating some new jobs in the areas where they operate, as we noted in the last chapter. Many though have clearly been relocations from other parts of the urban core, so the impacts on overall employment trends must necessarily be very limited. Undoubtedly though, inner-city planners have been at pains to protect existing businesses and nurture the growth of new small firms. There has been much improvement of infrastructure and accessibility of inner-city industrial areas. Yet it is clear that this counts for little compared to the general health of the national economy. Despite all the restructuring and policy initiatives of the 1970s and 1980s, the early 1990s' recession has still put large numbers of existing inner-city jobs at risk and seriously damaged prospects for the creation of new jobs.

The suburbs

Infilling and intensification

The remainder of the continuously built-up area of cities has been less dramatically affected by planning. In the biggest cities, the essential character of such suburban areas was largely set before 1945 (Hall *et al.*, 1973). Planning, particularly in cities where tight containment policies have operated, has encouraged the infilling and intensification of land uses in the suburbs (Whitehand, 1992). Thus the remaining large areas of open land suitable for building within the pre-1945 urban area were typically developed in the first half of the post-war period, usually for public-sector housing. In many cities there were also significant extensions to the suburban area, usually initiated in the first post-war decade. The creation of green belts for many provincial cities and the reinvigoration of the London green belt in the later 1950s was an important factor curtailing this expansion. Meanwhile those towns and cities without strict containment policies, or where growth was actively encouraged, have continued their peripheral expansion, albeit at higher densities and in a more tightly controlled way than occurred before 1939 (e.g. Herington, 1989).

Planning and the suburban environment
Such controls meant that post-1945 peripheral suburbs showed more of the impress of planning than their pre-1939 equivalents. In public housing areas particularly, neighbourhood principles were usually adopted, producing more careful grouping of dwellings around shared facilities or amenities. Planned employment areas were created in some estates, drawing on the inter-war satellite town idea, though rarely producing anything resembling employment self-containment. On the LCC's post-war 'out-county' estates, for example, the ratio of resident workers to local jobs was about 4 to 1 (TPI, 1956). All suburbs showed much greater attention to the planned provision of shopping areas and communal services, though adequate facilities did not always materialize, especially on the more isolated peripheral council estates. Gradually too more attention was given to the accommodation of motor vehicles, with greater pedestrian–vehicle separation. Inevitably, though, public-sector suburbs remained very dependent on public transport.

Demographic and economic changes
Generally speaking, the more freestanding cities, where planning constaints on peripheral growth have tended to be slacker, have experienced growth in their contiguous suburban areas for longer. By the 1970s, however, virtually all the biggest cities including most conurbation cities were showing marked declines in suburban populations and in economic well-being, if not yet on anything like the scale of the inner cities (Redfern, 1982; OPCS, 1992). And it was usually the most planned parts of the suburbs, the public-sector estates, inhabited by the least affluent suburbanites, where these problems were greatest (e.g. Smith and Ford, 1990).

Suburban problems
In extreme cases, such as Glasgow or Liverpool, problems are at least as severe as in the inner cities proper (e.g. Meegan, 1989). Some such areas have been defined as inner cities for policy purposes. Many of the problems are familiar: unemployment arising from a changed economy and planned living environments that never quite lived up to the promises, not least in poor community facilities. Yet such problems were compounded by the greater isolation of suburban locations. Nor were these issues confined to cities with serious economic problems. Thus the rather more prosperous cities of Oxford, Bristol and Cardiff experienced serious social unrest on peripheral council estates during 1992.

The outer city

The physical character of the outer city
Beyond the continuously built-up area of the big cities a new outer city has developed, functionally related to the urban cores (though less obviously than was

the suburban ring) yet physically separate and lacking in coherence (Herington, 1984). For this reason it has proved difficult to conceptualize. It is, in the 1990s, a landscape of villages and small and medium-sized towns within a setting of green fields (and other space-extensive land uses including airports, garden centres and gravel pits). Most settlements are the results of organic growth, and their historic character is typically protected, to some extent, by conservation area policies. Yet most of the towns are also swollen with more recent accretions of private housing estates, industrial parks and rustic superstores.

Functional patterns in the outer city

This combination of new and old urbanization within a rural setting has seen the highest rates of population growth of any parts of the urban system since 1945 (OPCS, 1992). Yet the physical concentration of building and emphasis on the prevention of urban sprawl has rarely produced the urban functional coherence or self-containment that planners have traditionally sought to create. Patterns of working, shopping and use of services are not as localized as the physical structure of settlements in the outer city superficially suggests. Workers typically seek employment across wide areas that include conurbation core areas, rather than just in the small town or village where they live. Shopping normally involves lengthy journeys to a large superstore or comparison shopping between several centres. The corollary of this is that the motor car is an all-pervasive influence, underpinning journeys to work, shop or school to a far greater extent than in concentrated city areas. New motorways and other road improvements, relatively easy to insert in the comparatively open landscape of the outer city, have provided the infrastructure for such movements and, in turn, further stimulated outer-city developments.

Planned decentralization?

This is all very different to what planners intended in the early post-war years or the 1960s. At those times it was assumed that planned decentralization of the concentrated cities would play a key role in outer-city development. In the 1990s we find that some of the larger towns in the outer-city areas, typically around the biggest cities, have indeed been deliberately planned as New or Expanded Towns. But these are very much exceptions, accommodating less than 5 per cent of the total UK population and accounting for small proportions of out-migrants from the concentrated metropolitan areas, far less than was originally intended.

The Greater London Plan of 1944 (Abercrombie, 1945) had envisaged that almost three-quarters of migration from the capital would be to new satellites or planned country town expansions. By the 1960s the expectations of planned decentralization were being lowered, so that the South East Study (MHLG, 1964) envisaged it accounting for only a third of outer-city growth. Yet such revisions were still well in excess of the reality: only about 15 per cent of migrants from London went to such settlements up to the 1970s, a proportion which has fallen dramatically since then (Buck *et al.*, 1986, p. 59). Nor was the pattern very different elsewhere. The 1946 Clyde Valley Plan (Abercrombie and Matthew, 1949) envisaged 60 per cent of Glasgow's out-migrants, virtually all those remaining in the region, going to New Towns. Even during the 1960s, however, when planned decentralization was at its peak, the actual proportion was barely 30 per cent (Smith and Farmer, 1985, p. 63). Significantly no other city with formal decentralization policies was able to better Glasgow's record. Overall demographic evidence suggests that planned decentralization policies of all kinds had only a rather limited impact on outer-city development.

Yet to concentrate too much on population data probably understates planning's role. New Towns and many Expanded Towns were relatively large growth centres in the outer city and they have done more than just house people. Despite a relatively small direct role in the relocation of existing jobs from the big cities, they have performed relatively well in attracting new employment (e.g. Buck and Gordon, 1986) and, because of their size and planned development, they also have good shopping and community facilities that have served more than just their own residents (Donnison and Soto, 1980). Though local employment and facilities were provided in the name of self-containment, these characteristics have also meant that the New and Expanded Towns have become foci for outer-city growth in many areas. On balance it seems that without them outer-city development would have been even less co-ordinated, with fewer urban facilities and even heavier dependence on extensive car use than has actually been the case.

Planned containment

There is little doubt about the long-term impact of containment policies (and more general planning restraints on development of rural land) in the outer city. As we have seen, restraint policies, particularly green belts, have been pursued with a sustained rigour that, though not absolute, has not been equalled in any other major aspect of planning (DOE, 1993). The long-term effects are clear. We must recall here that the average annual transfer of land from agricultural to urban use in England and Wales, largely on the urban fringe, had been running at 25,100 hectares per annum in the 1930s (Best, 1981). There was then a post-war decline from 17,500 hectares per annum in 1945–50 to about 15,000 per annum in 1950–65 and a slight rise to 16,800 in 1965–70 (a period which included the all-time housebuilding peak). This was followed by a marked decline to 9,300 in 1975–80, with further falls to about 6,500 hectares per annum in 1985–9. Over time this diminishing agricultural land loss has moved progressively further away from the big cities, reflecting the impact of green belts and avoiding inter-war-style continuous suburbanization (e.g. Gregory, 1970; Longley *et al.*, 1992).

The dimensions of the general trend have been contested, however, particularly when proposals to weaken restraint policies were mooted in the 1980s. Official figures were challenged as reliable indicators of the loss of rural land. A recent study for the Council for the Protection of Rural England (CPRE) alleges an annual loss of 12,000 hectares of rural land in England during the 1980s, compared to official figures of only 5,000 hectares (Sinclair, 1992). Yet this alleged disparity is of a far greater order than that for earlier post-war decades. Even the CPRE does not quarrel with the long-term trends of *relative* loss over time. All parties are agreed that, compared to pre-1945, post-war planning and policies for agricultural support have greatly reduced rural land loss.

Outer-city growth management?
Such policies have had much wider effects, most of which arose in conjunction with forces over which planners had little or no effective control. The already noted weaknesses in policy commitments to planned decentralization were crucially important here. More pervasive, however, were the weaknesses of later and less ambitious attempts at outer-city growth management through structure plans such as those for Oxfordshire or Hertfordshire (Elson, 1985, 1993). The broad aims were to produce a planned spatial coincidence of (largely private) housing, jobs and services in towns, deflecting development pressures from the fringes of London, larger cities and villages. Such strategies drew on the same conceptual tradition as the New and Expanded Towns, but involved little by way of formally planned overspill from larger centres. Yet these apparently neat solutions have also proved elusive. The location of employment has been impossible to control effectively in a period of wholesale economic change, major changes to the outer-city transport, mainly road, infrastructure, and weakening planning controls over business activity. Traditional patterns of shopping are also beginning to change in ways that seem to contradict planning intentions.

Planned or unplanned?
All this begs a similar question to that posed in Chapter 3 about inter-war suburbanization: how far should planners carry any real responsibility for the creation of what is frequently, beneath the outward attention to physical design details, a confusing functional mess, seemingly remote from any stated planning intentions? In other words, is outer-city growth essentially unplanned? The answer, even more now than then, is that planning has actually created several of the important preconditions of the present pattern of outer-city change. Private housing developers have followed, grudgingly in some cases, some of the growth management cues provided by planners, while also opportunistically seeking (and finding) areas beyond or in gaps and weak points within green belts. Reflecting the general restrictions on land availability, residential densities have typically been somewhat higher than those in pre-green belt, traditional suburbs (e.g. Winter, Coombes and Farthing, 1993).

Yet the impacts of planning have been somewhat weaker in the location of shopping and employment, particularly outside the New and Expanded Towns. This has prevented the creation of communities with a functional balance of activities, giving a more piecemeal quality to outer-city urbanization than planners have consciously sought. In turn it was this lack of spatial congruence of homes, jobs and shops that necessitated more daily travel, both to the big cities, but increasingly to other parts of the dispersed outer city. But, of course, it was containment policies, maintaining the open character of much of the immediate urban fringe, that facilitated the expansion of the road infrastructure which carried most of these daily movements. This in turn did much to allow the bewildering changes in outer-city living, employment and shopping patterns (e.g. Headicar and Bixby, 1992).

Outer-city development clearly owes much to the dominance of a private development process, with few opportunities since the 1970s for central or local planning agencies to give positive directions. Also of immense importance has been the growing availability of private cars. But neither of these allows us to acquit planning of responsibility. Certainly the planned mechanisms for decentralization and growth management have proved wanting in either quantity or quality, but strong planned containment policies have created significant, if not always fully understood, impacts. One of the major research tasks of the next few years will be to seek better understanding of the interrelationship of home, work, transport, shopping (and leisure) in the outer city and planning's actual and potential role in shaping this interaction.

The regional and national dimensions

Regional policies and employment

Thus far we have treated the city as a standard national entity. We have, though, shown how growth trends were more pronounced in the most buoyant and prosperous parts of the UK and the decline trends in the more depressed and poorer areas. Since part of post-war planning's mission was to address such regional unevenness, it is also important to consider these spatial impacts. Although there are many counter-claims (e.g. Martin, 1988), the most careful analysis of the subject (Moore, Rhodes and Tyler, 1986) has shown that in the years 1960–81 604,000 manufacturing jobs were directly created in the assisted areas by regional policies. About one-quarter of these jobs did not survive to 1981 (roughly the same loss rate as elsewhere), leaving a net job creation of 450,000, at a cost to the Exchequer of roughly £40,000 per job (1982 prices). The standard regional multiplier of 1.4 indicates a further 180,000 service jobs as a result of this, making a grand total of 630,000 in 1981.

By any standards this was a hugely impressive achievement, sustained even in the worsening economic climate after 1976. It was a powerful, but largely ignored, case for more active regional policies in the 1980s. Less successful, however, were office dispersal policies. Only 25 per cent of the 160,000 private office jobs leaving central London between 1963 and 1979 actually left the south east; only 9 per cent actually reached the assisted regions (Balchin, 1990, pp. 66–7). This was partly counteracted by government office dispersals, which moved about 55,000 jobs from London in the twelve years from 1963, many to assisted areas.

A regional solution?

Yet all this effort had been far from sufficient to 'solve' the regional problem, as conventionally defined, by bringing unemployment levels in the assisted areas to the same level as the national average. To achieve this, regional policies would have needed to have been roughly two to three times more effective than they actually were in the 1960s and three times more effective than they were in the 1970s. Clearly they would have needed to be very much more than this in the 1980s. Moreover the regional problem was more complex than disparities in unemployment. After Massey (1979, 1984), it was more commonly seen as disparities in economic power. In such a view there was a southern core region dominated by business headquarters, research and development, and key financial services, and the rest, characterized by more dependent 'branch plant' economies reliant on decisions made in Britain's (or other countries') core regions. In this view, regional policies, which tended to encourage branch plants of multinational companies as a replacement for older more regionally based industries, were part of the problem, not the solution (e.g. Hudson, 1989). And although office dispersal policies had some (very modest) successes in encouraging office development in some assisted regions, these too were occupied by the office equivalent of the branch plant, undertaking routine functions such as issuing driving licences (in Swansea) or various state social security benefits (in the north east).

Regional policy and urban change

A key question is also how regional policies affected cities. Clydeside, Merseyside and Tyneside were the main conurbations in the regions that received most assistance, yet they accounted for only one-third of the manufacturing jobs created by regional aid in 1961–81, less than would have been expected on the basis of their existing share of manufacturing jobs (Moore, Rhodes and Tyler, 1986). This was particularly evident for new firms moving into the assisted areas, underlining our earlier point about the unattractiveness of the inner city for new manufacturing. Nor is this point contradicted by the rather more limited effects of office dispersal policies. Some central or core areas of cities in the assisted areas or other less prosperous regions have certainly benefited by new office developments. Yet it is also clear that many government offices moving from London have favoured suburban or outer-city areas in their new provincial locations.

More generally, regional policies also played a modest role in the run-down of employment in the inner cities of the prosperous regions. Dennis (1980) has estimated that 9 per cent of London's job loss 1960–74 was attributable to shifts encouraged by regional policies. The main spatial impact of regional policies on urban areas has, then, been to reinforce urban decentralization processes, encouraging the development of suburban and outer-city areas in the less prosperous regions at the expense of core, especially inner-city, areas everywhere.

Unresolved questions

It is uncertain how far regional policies prevented jobs being created by refusing permission to build at preferred locations if these fell outside the assisted regions. This argument has long been voiced in the west midlands, asserting that regional policies prevented a post-war diversification of the economy of the kind that had characterized the half century before 1939 (Sutcliffe, 1986). This then made Birmingham and its neighbours, which enjoyed great prosperity based on the motor and metal industries from the 1940s to the 1960s more vulnerable to economic changes from the 1970s onwards. This may be so, but to sustain the general argument would imply fairly wholesale changes in government policies, including other aspects of planning. In particular we need to ask where the people to sustain the extra growth would have come from and how the physical growth would have been accommodated.

It was this kind of rather open-ended hypothesizing that led Hall *et al.* to suggest, somewhat tentatively, that post-war urban and regional planning has broadly sought to promote social stability, if at the expense of overall national economic growth (Hall *et al.*, 1973, Vol. II, pp. 375–7). That was in 1973. Two decades and three very serious economic downturns later, the question has become even more difficult to answer. It serves to remind us, however, that spatial impacts are not the whole story.

Economic and Social Impacts

Planning, land and capital

Planning and development land prices

Planning's most important net effects on the national economy arose directly from its restriction of the supply of development land. Between the wars, before an effective planning system operated, land costs typically accounted for about 5 per cent of the price of new private suburban housing in provincial towns and cities. The proportion was often greater around Greater London, especially at the height of the 1930s' boom, when it was sometimes in double figures (Jackson, 1991). Yet everywhere this land component began to rise from the mid-1950s as large-scale speculative housebuilding reappeared and development plans began to impose restrictions on the supply of development land. By 1960 land costs as a proportion of new housing prices had not apparently moved much beyond inter-war levels in the north, though in the west midlands and south east there had already been a marked increase to 10–12 per cent (Hall *et al.*, 1973, Vol. II, p. 214). By 1970 the proportion had dramatically increased to 20 per cent or over in many parts of provincial England and 30 per cent in the London fringe. By the late 1980s' boom, land costs were accounting for well over 40 per cent of new house prices in the south east (DOE, 1992c). Even Yorkshire and Humberside, the English region with least restrictions on land supply, had reached 20 per cent.

There is little doubt that this escalation, paralleled for all types of development land, was largely due to planning restrictions. Although there was a general rise in agricultural land prices, it was far less than the rate of increase for development land, suggesting that the right to develop, namely planning permission, was the crucial consideration. This then became a factor in the general escalation in the costs of housing and indeed all other kinds of building, with knock-on effects for the wider economy. One recent, rather alarmist, estimate has suggested that higher land costs alone accounted for a 4 per cent national reduction in real incomes in 1983 with a cumulative loss of perhaps 10 per cent of national income (Evans, 1988). Against this, other commentators (e.g. Bramley, 1993) have pointed out that it would be difficult to produce very marked changes in house prices by suddenly easing planning restrictions. This emphasis on the comparative insignificance of marginal changes does not, however, contradict the basic point which is a reflection of the cumulative and ingrained impact of policies pursued over several decades.

Developer certainty and planning

Yet by restricting permissions (and by encouraging public–private partnerships in development in some circumstances), planning has also had a role in providing a measure of certainty within the land market that has reduced the risks of unprofitable investment. It is almost impossible to estimate precisely the overall importance of this. Since 1959 the principle has been that where planning does not create certainty, in situations of planning blight, then compensation up to the value of the established use of the land may be payable. Such exceptional circumstances are not an adequate surrogate for the more general uncertainties of a free land market, though they do suggest that certainty is worth a great deal to developers.

More anecdotally, we might point to the Canary Wharf fiasco, discussed in Chapter 8, as a cautionary tale of what happens to developers liberated from the otiose certainties of the planning system: other developers have the same idea, increasing the chances of speculative overprovision and undermining the basic calculations of the original rental advantage of the new location. Developers, of course, face many more uncertainties than those decreased (or increased) by planning actions, but we should be in no doubt that the certainties traditionally provided by planning are significant.

Other economic impacts

There have been many other assertions about the overall national economic impacts of planning, especially in the political climate of the 1980s, as we saw in Chapter 7. As might be inferred from earlier discussion, it is very difficult to make an assessment of the overall impact of planning policies on manufacturing and service industries. As we have already seen, planners have certainly played a part in extinguishing jobs in some locations and creating new ones elsewhere. At times they have undoubtedly prevented some manufacturers and office developers building and operating in what were seen, rightly or wrongly, as optimal locations, where greatest profits could be extracted. Such studies as there are inevitably focus on what planning has prevented, giving little credit for what it has allowed or created.

It is, however, difficult to believe Thatcherite claims that planning has played any significant part in worsening the economic problems of Britain as a whole. (As we noted in the last chapter, ministers no longer seem to believe it either.) Indeed, the most complete exercises in town planning, the New and Expanded Towns, were amongst the finest settings for capital accumulation anywhere in the urban system of post-war Britain. Good, well-serviced employment sites were created in close proximity to good-quality, low-cost housing with good community facilities. Not surprisingly, the labour market characteristics and general ambience of such towns were particularly attractive to the larger manufacturers. Workforces were flexibly skilled and less wedded to restrictive practices than in established industrial locations in the older cities. They looked forward to bettering themselves in the new post-war affluence and placed less emphasis on old allegiances of class.

Had such new planned environments formed a more significant part of post-war urban change, it is then arguable that the national economy would have been commensurately stronger. Against this though the spatial problems of the inner city might well have been relatively worse as a result. And of course the opportunity costs of building many more new settlements would have been very great. It is this need to consider the impacts of alternative courses of action that make it difficult to push the more general economic arguments very much further. Certainly the state of current work does not allow us to come much nearer to any definitive answers. Although we can point to important direct impacts on the land market, the broader question of effects on the national economy must remain unresolved.

Economic gainers

From the above discussion we can conclude that developers and property owners have derived tangible net benefits from post-war planning, chiefly through the greater certainties provided by the system and the possibilities of large increases in land value arising solely from the grant of planning permission. It is important to note that such benefits have been characteristic of the whole post-war period and may even to some extent have been slightly undermined by the limited deregulation of planning since 1979. Equally important is the point that land value policies have generally meant that gainers have not usually borne the costs of these benefits themselves. We have also suggested, though with less evidence, that other forms of capital have also experienced net benefits.

Social impacts

Different tenure groups

(i) The owner-occupiers: It is somewhat easier to grasp the essentials of the social balance sheet. A convenient way to begin is to focus on housing tenure groups, qualified where necessary by income and other criteria (Hall *et al.*, 1973, Vol II, pp. 405–8). The main beneficiaries of post-war planning have undoubtedly been affluent owner-occupiers. Whether they have chosen to live in an outer city or green-belt village or small town, expensive suburban areas or inner-city enclaves, they have benefited particularly from the protection of their residential environment conferred by planning. Compared to a more free-market system, they have more security against unwelcome developments. Many of them have thus been able to enjoy enhanced social exclusivity of their residential environment in the outer city. Moreover they are less troubled by the higher land costs associated with planning since they find least difficulty in becoming owner-occupiers and are able to pay more to buy larger houses. They are also more able to bear any extra costs of travelling that may arise from their residential choice. Finally they have derived particular benefit from rising house values, fuelled partly by high land costs.

By contrast, the less affluent owner-occupiers have done less well. Higher land costs essentially increase the price of crossing the threshold into ownership, implying higher borrowing and greater financial vulnerability, even though once established they too benefit from house price inflation. This will probably be less impressive than for the more affluent, however, because their lower incomes mean that they are less able to afford to live in areas on which planning has conferred particular protection or amenity value. To a much greater extent than the better off, they are forced to weigh general environmental attractiveness against other desirable qualities such as dwelling size or convenience for work, social amenities, etc., in the search for affordability (Pahl, 1970). Higher land and housing prices mean that more of them have to live in smaller dwellings on smaller plots than their equivalents in the 1930s. For many, the expense of travelling is also likely to be considerable and commuting distances (and times) longer because of the impact of containment policies on outer-city development.

(ii) Public-sector renters: Their benefits from planning vary considerably. At one extreme the residents of the New and Expanded Towns have secured far better living and working environments than would have otherwise been possible. In most cases they have good-quality housing with better internal and external space standards than the lower end of the private sector. Employment and social amenities are convenient; environmental qualities are usually protected. Until recently, however, these tenants have not been able to benefit by rising house values. Moreover recent purchasers, under right-to-buy legislation, have found it difficult to escape the legacy of the public sector so that, despite space and design standards that are often superior, market prices have not so far matched those commanded by the cheapest privately built houses. The general point remains that their benefits from planning have been expressed in the use value of their living environment, rather than its capital value. Many of these benefits are shared by tenants on low-density suburban estates, especially in smaller towns. In bigger cities though they are likely to face rather longer journeys to work and services than in the New or Expanded Towns.

At the other extreme are those public-sector tenants who have paid the greatest price for restrictive land policies by being rehoused in generally unpopular high-density estates, usually in inner cities, though occasionally in suburban areas (Coleman, 1985). Although inner-city estates were usually convenient for employment and social amenities when they were built, this is generally less true today because of economic change, declining public expenditures and continued population decline in these areas, undermining service viability. Moreover the dwellings were all too often built by methods that were inadequately tested and with poor quality controls. Fewer tenants have sought to exercise their rights to buy in such settings (Sewel, Twine and Williams, 1984). Overall they are planned environments that have given little by way of use or capital values to their inhabitants.

(iii) Private tenants: Planning has done least for private renters and has directly contributed, with mainstream housing policies, to their further marginalization in the housing market. By promoting first redevelopment and then rehabilitation, planning has contributed directly to the decline of the tenure in favour of public-sector renting and owner-occupation, respectively. The housing situation of many former private renters has clearly benefited as a result, if sometimes at immense costs to community cohesion. Moreover, those unable to become owners or less eligible for what is today increasingly scarce public-sector housing, have suffered declining choice and value for money. Beneath all others are the rapidly growing homeless, denied any living space. For them too, an over-restrictive planning has now become part of the problem not, as it once promised to be, part of the solution.

Other social criteria

Tenure and income have, then, been important primary determinants of the extent of social benefit deriving from planning. But this assumes that every member of a household in the various categories benefits equally. Recently, mainly under the influence of equal opportunities concerns, more attention has been given to differential impacts of planning within households or for social categories more specific than tenure or income. By considering criteria such as gender, age, disability or ethnic origins, we are able to conceive a picture of social impacts altogether more subtle and varied (Montgomery and Thornley, 1990). A general point is that the interests of such social groups have rarely been directly represented within the planning process and have therefore been interpreted in rather idealized or insensitive ways, or simply ignored, at least for most of the post-war period.

All, moreover, tend to have lower personal mobility and greater likelihood of being poorer than basic indicators of their social class position might suggest. Both these factors have made them less likely to benefit from planning. Yet it is clear their own distinctive needs or requirements have interacted with planning intentions in specific ways. Without reciting chapter and verse, we can note that planning has had some detrimental impacts for all these social categories. Thanks to much recent research, we can demonstrate this particularly well for what is numerically the most important category: women (*Built Environment*, 1984 and 1990; Little, Peake and Richardson, 1988; Roberts, 1991; Greed, 1993).

Planning and the domestic ideal

Perhaps because it has been a predominantly male dominated activity, planning has generally responded only slowly to the tremendous and permanent changes in women's lives since 1945. Despite the huge wartime importance of women in paid employment, generally perceived as temporary, the initial post-war planning stereotypes reflected the ideal of women only as wives and mothers. Accordingly, where planning was actively shaping new places, in New Towns and new estates, the focus was on perfecting a domestic environment centred on the home, but embracing schools and other local social facilities (Attfield, 1989). Land use zoning, however, substantially separated residential neighbourhoods from the realm of paid work, seen more as a male domain.

Yet within the confines of the domestic ideal, such early post-war planning did acknowledge many important needs of the domestic woman (Bowlby, 1984, 1988). Thus local shops were provided in residential areas and major shopping continued to be located in the parts of cities most accessible by public transport. This did not especially disadvantage the transport-poor and often child-accompanied woman. Women as childcarers, however, suffered particularly from the progressive degradation of external private space in the home environment produced by restrictions on land supply from the 1950s. Smaller gardens and the acute difficulties of child supervision in high-density flatted estates intensified the problems of mothers.

Working women and planning

Meanwhile the whole planning approach established in the 1940s also created more fundamental problems for women as they increasingly sought paid employment in addition to their domestic labour during the following decades. The planning-sponsored spatial separation of the realms of home and work now posed serious difficulties for working wives, particularly since formal arrangements for childcare were largely non-existent (Roberts, 1991). Planners lacked any formal authority to insist on such provision (though we can note their relative success in securing (unsupervised) play areas on public and private housing estates). They might, however, have been able to help change the climate of thinking on this question, not least in recognizing that childcare needed to be more than just a female duty.

All these problems were compounded by the simultaneous growth in motor car use and changes in retailing, which began to undermine the commercial basis of small local shopping provision during the 1960s (Bowlby, 1988). Planners increasingly acquiesced in the shift towards less localized shopping provision, which by the 1980s embraced out-of-town centres and retail parks. Such shifts particularly disadvantaged women who generally had much less access to cars than men.

Women's safety and planning

Finally we should note how the increasing emphasis on planning for the motor vehicle from the late 1950s brought changes that have made cities seem less safe for women. This issue arose partly because women's increased economic role has both required them to use cities more independently of husbands or families and created a growing expectation of their rights to be able to do so. In one respect, mothers at least benefited by central area traffic-free shopping precincts since they required less close supervision of children (Roberts, 1981). Yet the growing separation of pedestrian and traffic circulation systems in city centres and Radburn-inspired residential areas often brought a growing reliance on underpasses, footbridges and landscaped footpaths that were often ill lit and remote from roads or populated areas. Frequently unpleasant for any pedestrian, they were particularly disliked by women as rendering them more vulnerable to attack. Yet women were more likely to have to use them unless they had personal access to a car or were able or willing to limit their use of the city.

THE FUTURE

A balance sheet

This glimpse of the differential social effects of planning completes our review of its post-war impacts. We can therefore now return to the question posed at the end of the last chapter: has the comprehensive system of physical and land use planning established since 1945, but lately become more fragmented and disjointed, had tangible enough benefits to warrant attempts at restoration? Or should we think in terms of some more complete dissolution and reformation? The answers are somewhat equivocal. Planning has been, at best, an important minor influence on the overall course of urban change. With few exceptions, it has tended to benefit most those economic interests and social groups which would probably have done well even in a process of urban change without planning.

In short, we cannot claim that planning has been an unqualified success. Like the curate's egg, it has only been good in parts. In some of the instances where it has been good, like the early New Towns, it has clearly been very, very good. Yet in some of the instances where it has been bad, such as much planned redevelopment of housing, it has been truly horrid. Much of it, though, has been between those extremes. Most planners for most of the time have been concerned with adjusting a pattern of urban change that was essentially shaped by powerful economic, social and political processes that were largely beyond their control. This minor role of planning has been further diminished in recent years. And in contrast to the comprehensive approach that was characteristic of planning's heyday, the whole activity has become more fragmented as its political saliency has diminished.

Conceptual benefits of fragmentation?

Yet, if we discount political marginalization, it is possible to represent fragmentation in a more positive light, promoting more conceptual diversity and variety. It was, for example, the breakdown of the monolithic comprehensivity of the 1960s which both stimulated and gave conceptual space to the radical and more socially responsive planning initiatives of the 1970s and 1980s, such as planning aid, community planning, etc. (Montgomery and Thornley, 1990). Such initiatives are relatively insignificant weighed against all planning actions. Yet they promise the actual social outcomes traditionally sought but never fully realized by planners, empowering disadvantaged groups and improving their living conditions.

The role of consensus in planning history

The problem is that such conceptual initiatives, however exciting, have come at just the time when public funds to assist the less advantaged have been diminishing. Moreover, support for the new approaches has not managed to move beyond the margins, to reach the consciences or the self-interests of mainstream contented society and dominant economic interests. This has severely limited the potential of the new planning ideas. There is an important historical lesson in the pre-1914 planning movement which quickly gained limited, but very real, support for its ideas from powerful economic interests. It was this backing, though it toned down some of the original radicalism, that began to take town planning from the margins into the social and political mainstream. This process was completed as a much wider political consensus was created during the Second World War out of the experiences of the inter-war years and war itself.

If we accept the validity of this historical parallel, it follows that a strengthened planning system that embodies the new radical planning ideas of today would depend on the acquiescence or even the enthusiasm of mainstream society. As those who sought to challenge the dominant political mood of the 1980s found, it is very difficult to assemble an effective and reliable radical coalition based on the disparate interests of marginalized groups. Traditional elements of social and political cohesion that might have provided a framework to bring together these interests, such as class consciousness, have been eroded over the post-war years. We cannot realistically expect a strengthened planning to arise solely from the margins, however much they might conceptually inspire its practice and policies, or from traditional class politics.

Green ideas and a new planning consensus?

It is therefore important to establish a new coherence capable of transcending the narrow self-interest of contented society and the powerful economic interests that together shape the political agenda, without pre-empting or precluding those new ideas which have grown on the social and political margins of planning in the 1970s and 1980s. The only body of ideas which looks at all capable of forming this basis of a new, widely based consensus seems to be that which has grown out of the green movement. The most important reason for such a belief is that green ideas have achieved widespread social adherence in a very short period (McCormick, 1993). Much of this adherence is distinctly modest, even superficial, but it still far outshines any genuine popular approval that planning has ever achieved, except perhaps in 1940–5.

Obstacles to the new consensus

Yet although the wide social reach of green ideas is important, it does not automatically mean they can form the basis for a new and viable planning consensus. The tensions between the long-term abstractions of saving the planet and the more immediate and tangible concerns of planning are real enough. Many planners, including those who have done most to pioneer new concepts in the 1980s, are rightly suspicious of green ideas that sometimes appear to be too ready to sacrifice jobs or houses for those who need them in favour of some ill-defined long-term environmental good. In this sense many green ideas seem to imply an intensification of the most socially regressive features of post-war planning, pointing to further restrictions on the supply of development land with implications for housing costs and living space.

Sustainability: a basis for reconciliation?

The concept of sustainability potentially shifts the balance of green thought by acknowledging the requirement to meet present needs for development, without compromising the ability to meet future needs. Such a concept is far more acceptable to planners, yet its vagueness needs to be made more precise if it is to become the key element of a new intellectual cohesion (Blowers, 1993a, 1993b). In particular planners will want to assert that the needs of the present involve more than just 'narrow' environmentalism, concerned with natural resources, pollution and biodiversity. They involve decent houses, rewarding jobs, fulfilling leisure and recreation activities, safe and pleasant cities and a freedom from the burden of unnecessary, routine travelling (Breheny and Rookwood, 1993). They involve too a sense that people can participate directly in shaping their own places and taking responsibility for their own environment. All this will probably also involve a need to return to the idea at the heart of town planning: land reform (Hall, Hebbert and Lusser, 1993). In short, they involve asserting much that is important from planning's traditional conceptual repertoire, together with the newer radical planning ideas, yet integrating these into the new global environmental concerns.

The anti-vision

Without such a conceptual renewal, planning thought will be condemned merely to follow rather than give a lead to policies. The likelihood will be that mainstream planning ideas will reflect a post-Thatcherite policy consensus that is not remotely radical in any sense. This anti-vision will take its main cues from the shifts of the 1980s, though it will be more coherent, more practical, less rhetorical and (let us be honest) more boring than at the high point of Thatcherism. But essentially it will involve doing little more than quite openly turning the handle of a market led process of urban change, neither disturbing the self-interests of contented society nor offering much benefit to the weak and disadvantaged. It will acknowledge green issues only where these reinforce (or at least do not challenge) its central imperatives.

Visions and policies for the twenty-first century

If planners aspire to do more than this, as they should do if they pay any heed to their own reformist traditions, they cannot do it without a new vision to challenge present policy limitations. The call is for a new comprehensivity and a new consensus, a reformulated vision of planning that incorporates sustainability and social responsiveness as new and central themes. Visions alone are not sufficient to produce fundamental policy changes, but they are a necessary precondition. Thus the clarity of the original comprehensive planning vision, honed and rehearsed between the wars, was a key factor in its 1940s' political success.

We may reasonably doubt whether wider conditions quite as conducive to radical reform as those of the 1940s can possibly arise in the near future, but there are some modestly hopeful signs. Anti-planning sentiments are no longer declaimed with the same enthusiasm as in the 1980s. The imminent prospect of a new century will certainly encourage greater than normal social willingness to embrace thoughts of the collective future. The recognition of a need for strategic action comparable perhaps to that which occurred in the early 1960s is not entirely inconceivable. In such uncertain but not completely unhopeful circumstances, the best thing that the planning movement can do is ensure that it has a coherent and relevant vision that reaches out to society as a whole, addressing its wants, needs and insecurities. There is much still to be done.

Bibliography and Sources

A Note on Sources and Further Reading

This bibliography consists solely of the works cited in the text, which were included largely to advise the reader on further reading. A great many other more original (and more inaccessible) sources have been consulted in the preparation of this study. In the interests of brevity (and reflecting this book's general role as a text rather than a scholarly monograph) I have not included these types of source except where they are directly quoted or where they are particularly necessary for the development of the argument. They include the RTPI archives from 1914 to 1973, various other national records and numerous more local sources from various towns and cities that I have worked through for research or teaching purposes over more than two decades. Other exclusions from the list are local statutory plans and, with a few exceptions, other documents produced by non-central planning agencies, which I have made considerable use of in some sections. National and local press articles, television and radio programmes and personal interviews with many planning practitioners, mainly in connection with teaching, have also played their part, especially in relation to the recent past, but are not specifically acknowledged here. Nor is reference made to the standard statistical sources *Regional Trends* and *Social Trends*, which are occasionally used in the later chapters.

If any readers require further details about sources used in this book, they are welcome to contact the author at the School of Planning, Oxford Brookes University, Oxford OX3 0BP.

Bibliography

Aalen, F. H. A. (1992) English origins, in S. V. Ward (ed.) op. cit.

Abercrombie, P. (1945) *The Greater London Plan 1944*, HMSO, London.

Abercrombie, P. and Kelly, S. A. (1932) *Cumbrian Regional Planning Scheme*, Hodder & Stoughton/Liverpool University Press, London/Liverpool.

Abercrombie, P. and Matthew, R. H. (1949) *The Clyde Valley Regional Plan 1946*, HMSO, Edinburgh.

Adams, T. (1929) The origin of the term 'town planning' in England, *Journal of the Town Planning Institute*, Vol. XV, no. 11, pp. 310–11.

Addison, P. (1975) *The Road to 1945: British Politics and the Second World War*, Cape, London.

ALDOUS, T. (1972) *Battle for the Environment*, Fontana/Collins, Glasgow.

ALDRIDGE, H. R. (1915) *The Case for Town Planning: A Practical Manual for the Use of Councillors, Officers and Others Engaged in the Preparation of Town Planning Schemes*, National Housing and Town Planning Council, London.

ALDRIDGE, M. (1979) *British New Towns: A Programme without a Policy*, Routledge & Kegan Paul, London.

AMBROSE, P. (1986) *Whatever Happened to Planning?* Methuen, London.

AMBROSE, P. and COLENUTT, B. (1975) *The Property Machine*, Penguin, Harmondsworth.

AMOS, C. (1991) Flexibility and variety: the key to new settlement policy, *Town and Country Planning*, Vol. 60, no. 2, pp. 52–6.

ANSTIS, G. (1985) *Redditch: Success in the Heart of England – The History of Redditch New Town 1964–1985*, Publications for Companies, Stevenage.

APT (Arndale Property Trust) (c.1965) *Arndale in Partnership*, APT, Bradford.

ARDILL, J. (1974) *The New Citizen's Guide to Town and Country Planning*, TCPA/ Charles Knight, London.

ASHWORTH, W. (1954) *The Genesis of Modern British Town Planning: A Study in Economic and Social History of the Nineteenth and Twentieth Centuries*, Routledge & Kegan Paul, London.

ATTFIELD, J. (1989) Inside pram town: a case study of Harlow house interiors, 1951–61, in J. Attfield and P. Kirkham (eds.) *A View from the Interior: Feminism, Women and Design*, Women's Press, London.

Audit Commission (1989) *Urban Regeneration and Economic Development: The Local Government Dimension*, HMSO, London.

BACKWELL, J. and DICKENS, P. (1978) *Town Planning, Mass Loyalty and the Restructuring of Capital: The Origins of the 1947 Planning Legislation Revisited*. Urban and Regional Studies Working Paper no. 11, University of Sussex, Brighton.

Bains Report (1972) *The New Local Authorities – Management and Structure*, HMSO, London.

BALCHIN, J. (1980) *First New Town: An Autobiography of the Stevenage Development Corporation 1946–1980*, Stevenage Development Corporation.

BALCHIN, P. N. (1990) *Regional Policy in Britain: The North–South Divide*, Paul Chapman, London.

BALL, M., GRAY, F. and MCDOWELL, L. (1989) *The Transformation of Britain: Contemporary Social and Economic Change*, Fontana, London.

BANFIELD, E. C. (1970) *The Unheavenly City: The Nature and Future of Our Urban Crisis*, Little, Brown, Boston.

Barlow Commission (Royal Commission on the Distribution of the Industrial Population) (1940) *Report* (Cmd 6153), HMSO, London.

BARTY-KING, H. (1985) *Expanding Northampton*, Secker & Warburg, London.

BEATTIE, S. (1980) *A Revolution in London Housing: LCC Architects and their Work 1893–1914*, Greater London Council/Architectural Press, London.

BEAUFOY, S. (1952) Presidential address, *Journal of the Town Planning Institute*, Vol. XXXIX, no. 1, pp. 2–7.

BEEVERS, R. (1988) *The Garden City Utopia: A Critical Biography of Ebenezer Howard*, Macmillan, Basingstoke.

BEGG I., MOORE, B. and RHODES, J. (1986) Economic and social change in urban Britain and the inner cities, in V. Hausner (ed.) op. cit., Vol. I, pp. 10–49.

Bellamy, E. (1888) *Looking Backward*, Ticknor, Boston.

Bendixson, T. (1988) *The Peterborough Effect: Reshaping a City*, Peterborough Development Corporation.

Bendixson, T. and Platt, J. (1992) *Milton Keynes: Image and Reality*, Granta, Cambridge.

Bennett, T. (1990) Planning and people with disabilities, in J. Montgomery and A. Thornley (eds.) op. cit.

Benton, S. (1987) Death of the citizen, *New Statesman*, Vol. 114, no. 2956, pp. 21–2.

Best, R. (1981) *Land Use and Living Space*, Methuen, London.

Beveridge Report (1942) *Report of the Interdepartmental Committee on Social Insurance and Allied Services* (Cmd 6404), HMSO, London.

Bianchini, F., Fisher, M., Montgomery, J. and Worpole, K. (1988) *City Centres, City Cultures*, Centre for Local Economic Strategies, Manchester.

Blackaby, F. T. (ed.) (1978) *British Economic Policy 1960–74*, Cambridge University Press.

Block G. (1954) *The Spread of Towns*, Conservative Political Centre, London.

Blowers, A. (1993a) The time for change, in A. Blowers (ed.) op. cit.

Blowers, A. (ed.) (1993b) *Planning for a Sustainable Environment: A Report by the Town and Country Planning Association*, Earthscan, London.

Boddy, M. (1980) *The Building Societies*, Macmillan, London.

Boddy, M. and Fudge, C. (eds.) (1985) *Local Socialism*, Macmillan, Basingstoke.

Bogdanor, V. (1970) The Labour Party in opposition, 1951–1964, in V. Bogdanor and R. Skidelsky (eds.) op. cit.

Bogdanor, V. and Skidelsky, R. (eds.) (1970) *The Age of Affluence 1951–1964*, Macmillan, London.

Booth, A. (1978) An administrative experiment in unemployment policy in the thirties, *Public Administration*, Vol. 56, pp. 139–57.

Booth, A. (1982) The second world war and the origins of modern regional policy, *Economy and Society*, Vol. 11, no. 1, pp. 1–21.

Bor, W. (1974) The Town and Country Planning Act, 1968, *The Planner*, Vol. 60, no. 5, pp. 696–702.

Bowden, P. (1978) The origins of Newton Aycliffe, in M. Bulmer (ed.) op. cit.

Bowlby, S. (1984) Planning for women to shop in post-war Britain, *Environment and Planning D: Society and Space*, Vol. 2, pp. 179–99.

Bowlby, S. (1988) From corner shop to hypermarket: women and food retailing, in J. Little, L. Peake and P. Richardson (eds.) op. cit.

Bowley, M. (1945) *Housing and the State, 1919–44*, Allen & Unwin, London.

Boyle, R. (1988) Private sector urban regeneration: the Scottish experience, in M. Parkinson, B. Foley and D. Judd (eds.) op. cit.

BPF (British Property Federation) (1986) *The Planning System: A Fresh Approach*, BPF, London.

Bramley, G. (1993) The impact of land use planning and tax subsidies on the supply and price of housing in Britain, *Urban Studies*, Vol. 30, no. 1, pp. 5–30.

Breheny, M. J. (1992a) Contradictions of the compact city, in M. J. Breheny (ed.) op. cit.

Breheny, M. J. (ed.) (1992b) *Sustainable Development and Urban Form*, Pion, London.

Breheny, M. J. and Batey, P. W. J. (1982) The history of planning methodology: a preliminary sketch, *Built Environment*, Vol. 7, no. 2, pp. 109–20.

Breheny, M. and Rookwood, R. (1993) Planning the sustainable city region, in A. Blowers (ed.) op. cit.

Briggs, A. (ed.) (1962) *William Morris: Selected Writings and Designs*, Penguin, Harmondsworth.

Briggs, A. (1968) *Victorian Cities*, Penguin, Harmondsworth.

Brindley, T., Rydin, Y. and Stoker, G. (1989) *Remaking Planning: The Politics of Urban Change in the Thatcher Years*, Unwin Hyman, London.

Brittan, S. (1971) *Steering the Economy: The Role of the Treasury*, Penguin, Harmondsworth.

Brown, G. (1972) *In My Way: The Political Memoirs of Lord George-Brown*, Penguin, Harmondsworth.

Brownill, S. (1990a) The people's plan for the Royal Docks: some contradictions in popular planning, in J. Montgomery and A. Thornley (eds.) op. cit.

Brownill, S. (1990b) *Developing London's Docklands: Another Great Planning Disaster*, Paul Chapman, London.

Brundtland, G. H. (1987) *Our Common Future: Report of the World Commission on Environment and Development*, Oxford University Press.

Bruton, M. (1981) Colin Buchanan 1907 – , in G. E. Cherry (ed.) op. cit.

Bruton, M. and Nicholson, D. (1987) *Local Planning in Practice*, Hutchinson, London.

Buchanan, C. D. (1957) *Mixed Blessing: The Motor in Britain*, Leonard Hill, London.

Buck, N. and Gordon, I. (1986) The beneficiaries of employment growth: an analysis of the experience of disadvantaged groups in expanding labour markets, in V. Hausner (ed.) op. cit., Vol. II.

Buck, N., Gordon, I. and Young, K. with Ermisch, J. and Mills, L. (1986) *The London Employment Problem*, Oxford University Press.

Buder, S. (1990) *Visionaries and Planners: The Garden City Movement and the Modern Community*, Oxford University Press, New York.

Building (1991) Supplement: Canary Wharf – A Landmark in Construction, *Building*, October.

Built Environment (1984) Special Number: Women and the Built Environment, *Built Environment*, Vol. 10, no. 1.

Built Environment (1990) Special Number: Women and the Designed Environment, *Built Environment*, Vol. 16, no. 4.

Bullock, N. (1987) Plans for post-war housing in the UK: the case for mixed development and the flat, *Planning Perspectives*, Vol. 2, no. 1, pp. 71–98.

Bulmer, M. (1978a) Change, policy and planning since 1918, in M. Bulmer (ed.) op. cit.

Bulmer, M. (ed.) (1978b) *Mining and Social Change: Durham County in the Twentieth Century*, Croom Helm, London.

Burnett, J. (1978) *A Social History of Housing*, David & Charles, Newton Abbot.

Burns, D. (1990) The decentralization of local authority planning, in J. Montgomery and A. Thornley (eds.) op. cit.

Burns, W. (1963) *New Towns for Old: The Technique of Urban Renewal*, Leonard Hill, London.

Burns, W. (1967) *Newcastle: A Study in Replanning at Newcastle-upon-Tyne*, Leonard Hill, London.

Burton, P. and O'Toole, M. (1993) Urban Development Corporations: post-Fordism in action or Fordism in retrenchment, in R. Imrie and H. Thomas (eds.) op. cit.

BUTLER, S. M. (1981) *Enterprise Zones: Greenlining the Inner Cities*, Heinemann, London.

BVT (Bournville Village Trust) (1941) *When We Build Again: A Study Based into Conditions of Living and Working in Birmingham*, Allen & Unwin, London.

BVT (1955) *The Bournville Village Trust 1900–1955*, BVT, Birmingham.

BYRNE, D. (1989) *Beyond the Inner City*, Open University Press, Milton Keynes.

Cabinet Office (1988) *Action for Cities*, Cabinet Office, London.

CADBURY, G. JNR (1915) *Town Planning with Special Reference to the Birmingham Schemes*, Longmans Green, London.

CALDER, A. (1971) *The People's War: Britain 1939–45*, Panther, London.

CARTER, H. and LEWIS, C. R. (1990) *An Urban Geography of England and Wales in the Nineteenth Century*, Arnold, London.

CARTER, N. and WATTS, C. (1984) *The Cambridge Science Park*, Planning and Development Case Study 4, Royal Institution of Chartered Surveyors, London.

CASSELL, M. (1991) *Long Lease! The Story of Slough Estates 1920–1991*, Pencorp, London.

CASTELLS, M. (1977) *The Urban Question*, Arnold, London.

CEC (Commission of the European Communities) (1990a) *Green Paper on the Urban Environment* (EUR 12902 EN), Commission of the European Communities, Brussels.

CEC (1990b) *Green Paper on the Urban Environment: Expert Contributions* (EUR 13145 EN), Commission of the European Communities, Brussels.

CHADWICK, G. (1971) *A Systems View of Planning*, Pergamon, Oxford.

Chamberlain Committee (Committee to Consider and Advise on the Principles to be Followed in Dealing with Unhealthy Areas) (1920) *Interim Report*, HMSO, London.

Chamberlain Committee (1921) *Final Report*, HMSO, London.

CHARLESWORTH, G. (1984) *A History of British Motorways*, Thomas Telford, London.

Chelmsford Committee (1931) *Interim Report of the Departmental Committee on Regional Development* (Cmd 3915), HMSO, London.

CHERRY, G. E. (1972) *Urban Change and Planning: A History of Urban Development in Britain since 1750*, Foulis, Henley.

CHERRY, G. E. (1974a) The Housing, Town Planning Etc Act 1919, *The Planner*, Vol. 60, no. 5, pp. 681–4.

CHERRY, G. E. (1974b) *The Evolution of British Town Planning*, Leonard Hill, Leighton Buzzard.

CHERRY, G. E. (1975a) *Factors in the Origins of Town Planning in Britain: The Example of Birmingham, 1905–1914*, Centre for Urban and Regional Studies Working Paper no. 36, University of Birmingham.

CHERRY, G. E. (1975b) *Environmental Planning 1939–1969, Vol. II, National Parks and Recreation in the Countryside*, HMSO, London.

CHERRY, G. E. (1980) The place of Neville Chamberlain in British town planning, in G. E. Cherry (ed.) *Shaping an Urban World*, Mansell, London.

CHERRY, G. E. (ed.) (1981) *Pioneers in British Planning*, Architectural Press, London.

CHERRY, G. E. (1988) *Cities and Plans: The Shaping of Urban Britain in the Nineteenth and Twentieth Centuries*, Arnold, London.

CHERRY, G. E. and PENNY, L. (1986) *Holford: A Study in Architecture, Planning and Civic Design*, Mansell, London.

CHILDS, D. (1992) *Britain since 1945* (3rd edn), Routledge, London.

CHURCHILL, W. S. (1951) *The Second World War: Vol. IV: The Hinge of Fate*, Cassell, London.

COATES, D. (1980) *Labour in Power? A Study of the Labour Government 1974–1979*, Longman, London.

COATES, K. and SILBURN, R. (1970) *Poverty: The Forgotten Englishmen*, Penguin, Harmondsworth.

COATES, K. and SILBURN, R. (1980) *Beyond the Bulldozer*, Department of Adult Education, University of Nottingham.

COCHRANE, A. (1983) Local economic policies: like trying to drain the ocean with a teaspoon, in J. Anderson, A. Cochrane and R. Hudson (eds.) *Redundant Spaces in Cities and Regions? Studies in Industrial Decline and Social Change*, Academic Press, London.

COCHRANE, A. (1986) Local employment initiatives: towards a new municipal socialism? in P. Lawless and C. Raban (eds.) op. cit.

COCKBURN, C. (1977) *The Local State: Management of People and Cities*, Pluto, London.

COLEMAN, A. (1985) *Utopia on Trial: Vision and Reality in Planned Housing*, Hilary Shipman, London.

COLENUTT, B. (1993) After the urban development corporations: development elites or people-based regeneration? in R. Imrie and H. Thomas (eds.) op. cit.

COLLISON, P. (1963) *The Cutteslowe Walls: A Study in Social Class*, Faber, London.

COOKE, P. N. (1983) *Theories of Planning and Spatial Development*, Hutchinson, London.

COOKE, P. N. (ed.) (1989) *Localities: The Changing Face of Urban Britain*, Unwin Hyman, London.

COONEY, E. W. (1974) High flats in local authority housing in England and Wales since 1945, in A. Sutcliffe (ed.) op. cit.

COWAN, P., FINE, D., IRELAND, J., JORDAN, C., MERCER, D. and SEARS, A. (1969) *The Office: A Facet of Urban Growth*, Heinemann, London.

CREESE, W. L. (1966) *The Search for Environment: The Garden City Before and After*, Massachusetts Institute of Technology Press, Cambridge, MA.

CREWE, I. (1988) Has the electorate become Thatcherite? in R. Skidelsky (ed.) op. cit.

CRITCHLEY, J. (1987) *Heseltine: The Unauthorised Biography*, Andre Deutsch, London.

CROSS, D. T. and BRISTOW, M. R. (eds.) (1983) *English Structure Planning: A Commentary on Procedure and Practice in the Seventies*, Pion, London.

CROSS, J. (1970) The regional decentralisation of British government departments, *Public Administration*, Vol. 48, pp. 423–41.

CROSSMAN, R. (1975) *The Diaries of a Cabinet Minister, Vol. 1: Minister of Housing 1964–66*, Hamish Hamilton, London.

CROSSMAN, R. (1976) *The Diaries of a Cabinet Minister, Vol. 2: Lord President of the Council and Leader of the House of Commons 1966–1968*, Hamish Hamilton, London.

CULLINGWORTH, J. B. (1960) *Housing Needs and Planning Policy*, Routledge & Kegan Paul, London.

CULLINGWORTH, J. B. (1970) *Town and Country Planning in England and Wales* (3rd edn), Allen & Unwin, London.

CULLINGWORTH, J. B. (1975) *Environmental Planning 1939–1969, Vol. I, Reconstruction and Land Use Planning 1939–1947*, HMSO, London.

CULLINGWORTH, J. B. (1976) *Town and Country Planning in Britain* (6th edn), Allen & Unwin, London.

CULLINGWORTH, J. B. (1979) *Environmental Planning 1939–1969, Vol. III, New Towns Policy*, HMSO, London.

CULLINGWORTH, J. B. (1980) *Environmental Planning 1939–1969, Vol. IV, Land Values, Compensation and Betterment*, HMSO, London.

CULLINGWORTH, J. B. (1988) *Town and Country Planning in Britain* (10th edn), Unwin Hyman, London.

CULPIN, E. G. (1913) *The Garden City Movement up to Date*, GCTPA, London.

DALTON, H. (1957) *The Fateful Years: Memoirs 1931–1945*, Muller, London.

DALTON, H. (1962) *High Tide and After: Memoirs 1945–1960*, Muller, London.

DANIEL, W. W. (1968) *Racial Discrimination in England*, Penguin, Harmondsworth.

DANTZIG, G. and SAATY, T. L. (1973) *Compact City: A Plan for a Liveable Urban Environment*, Freeman, San Francisco.

DARLOW, C. (ed.) (1972) *Enclosed Shopping Centres*, Architectural Press, London.

DAUNTON, M. J. (1983) *House and Home in the Victorian City: Working Class Housing 1850–1914*, Arnold, London.

DAUNTON, M. J. (ed.) (1984) *Councillors and Tenants: Local Authority Housing in English Cities 1919–1939*, Leicester University Press.

DAVIES, J. G. (1972) *The Evangelistic Bureaucrat: A Study of a Planning Exercise in Newcastle-upon-Tyne*, Tavistock, London.

DAVIES, R. L. and CHAMPION, A. G. (1982) *The Future of the City Centre*, Institute of British Geographers Special Publication no. 14, Academic Press, London.

DAWSON, J. A. (1991) Market services in the United Kingdom, in R. J. Johnston and V. Gardiner (eds.) *The Changing Geography of the United Kingdom* (2nd edn), Routledge, London.

DAY, M. G. (1981) The contribution of Sir Raymond Unwin (1863–1940) and R. Barry Parker (1867–1947) to the development of site-planning theory and practice c.1890–1918, in A. Sutcliffe (ed.) op. cit.

DEA (Department of Economic Affairs) (1965a) *The National Plan* (Cmnd 2764), HMSO, London.

DEA (1965b) *The West Midlands: A Regional Study*, HMSO, London.

DEA (1965c) *The North West: A Regional Study*, HMSO, London.

DEAKIN, D. (ed.) (1989) *Wythenshawe: The Story of a Garden City*, Phillimore, Chichester.

DEAKIN, N. (1987) *The Politics of Welfare*, Methuen, London.

DENBY, E. (1938) *Europe Rehoused*, Allen & Unwin, London.

DENMAN, D. R. (1980) *Land in a Free Society*, Centre for Policy Studies, London.

DENNIS, N. (1970) *People and Planning: The Sociology of Housing in Sunderland*, Faber, London.

DENNIS, N. (1972) *Public Participation and Planners' Blight*, Faber, London.

DENNIS, R. (1984) *English Industrial Cities of the Nineteenth Century*, Cambridge University Press.

DENNIS, R. D. (1980) The decline of manufacturing employment in London 1966–74, in A. Evans and D. Eversley (eds.) *The Inner City*, Heinemann, London.

DE SOISSONS, M. (1988) *Welwyn Garden City: A Town Designed for Healthy Living*, Publications for Business, Cambridge.

DIAMOND, D. R. (1991) The City, the 'Big Bang' and office development, in K. Hoggart and D. R. Green (eds.) *London: A New Metropolitan Geography*, Arnold, London.

DITRD (Department of Industry, Trade and Regional Development) (1963) *The North East: A Plan for Development and Growth* (Cmnd 2206), HMSO, London.

Dix, G. (1981) Patrick Abercrombie 1879–1957, in G. E. Cherry (ed.) op. cit.

DOE (Department of Environment) (1972) *Report of Working Party on Local Authority/Private Enterprise Partnership Schemes*, HMSO, London.

DOE (1977a) *Inner London: Policies for Dispersal and Balance: Final Report of the Lambeth Inner Area Study*, HMSO, London.

DOE (1977b) *Change or Decay: Final Report of the Liverpool Inner Area Study*, HMSO, London.

DOE (1977c) *Unequal City: Final Report of the Birmingham Inner Area Study*, HMSO, London.

DOE (1977d) *Policy for the Inner Cities* (Cmnd 6845), HMSO, London.

DOE (1978) *Strategic Plan for the South East: Review Government Statement*, HMSO, London.

DOE (1983) *Streamlining the Cities: Government Proposals for Reorganizing Local Government in Greater London and the Metropolitan Counties* (Cmnd 9063), HMSO, London.

DOE (1986) *An Evaluation of Industrial and Commercial Improvement Areas*, HMSO, London.

DOE (1987a) *An Evaluation of Derelict Land Grant Schemes*, HMSO, London.

DOE (1987b) *An Evaluation of the Enterprise Zone Experiment*, HMSO, London.

DOE (1988) *An Evaluation of the Urban Development Grant Programme*, HMSO, London.

DOE (1989) *Environmental Assessment: A Guide to the Procedures*, HMSO, London.

DOE (1991) *Simplified Planning Zones: Progress and Procedures*, HMSO, London.

DOE (1992a) *The Effects of Major Out of Town Retail Development*, HMSO, London.

DOE (1992b) *The Use of Planning Agreements*, HMSO, London.

DOE (1992c) *The Relationship between House Prices and Land Supply*, HMSO, London.

DOE (1993) *The Effectiveness of Green Belts*, HMSO, London.

DOE, SO (Scottish Office) and WO (Welsh Office) (1974) *Land* (Cmnd 5730), HMSO, London.

DOE and WO (1986) *The Future of Development Plans*, Consultation Paper, HMSO, London.

DOE and WO (1989) *The Future of Development Plans* (Cm 569), HMSO, London.

DOE *et al.* (1990) *This Common Inheritance: Britain's Environmental Strategy* (Cm 1200), HMSO, London.

DOE *et al.* (1991) *This Common Inheritance: The First Year Report: Britain's Environmental Strategy* (Cm 1655), HMSO, London.

DOE *et al.* (1992) *This Common Inheritance: The Second Year Report: Britain's Environmental Strategy* (Cm 2068), HMSO, London.

Dobry Report (1975) *Review of the Development Control System: Final Report*, HMSO, London.

Donnison, D. and Middleton, A. (eds.) (1987) *Regenerating the Inner City: Glasgow's Experience*, Routledge, London.

Donnison, D. and Soto, P. (1980) *The Good City: A Study of Urban Development and Policy in Britain*, Heinemann, London.

DTEDC (Distributive Trades Economic Development Council) (1968) *The Cowley Shopping Centre*, National Economic Development Office, HMSO, London.

DTEDC (1971) *The Future Pattern of Shopping*, National Economic Development Office, HMSO, London.

DTI (Department of Trade and Industry) (1984) *Regional Industrial Development* (Cmnd 9111), HMSO, London.

Dudley Report (1944) *The Design of Dwellings: Report of the Sub-Committee of the Central Housing Advisory Committee*, HMSO, London.

DUNLEAVY, P. (1981) *The Politics of Mass Housing in Britain 1945–75: A Study of Corporate Power and Professional Influence in the Welfare State*, Clarendon Press, Oxford.

DUNLEAVY, P., GAMBLE, A., HOLLIDAY, I. and PEELE, G. (eds.) (1993) *Developments in British Politics 4*, Macmillan, Basingstoke.

DUNNING, J. H. (1963) *Economic Planning and Town Expansion: A Case Study of Basingstoke*, Workers Educational Association, Southampton.

DURANT, R. (1939) *Watling: A Survey of Life on a New Estate*, King, London.

DYOS, H. J. and WOLFF, M. (eds.) (1973) *The Victorian City: Images and Realities*, 2 vols., Routledge & Kegan Paul, London.

EDWARDS, A. M. (1981) *The Design of Suburbia: A Critical Study in Environmental History*, Pembridge, London.

EDWARDS, J. and BATLEY, R. (1978) *The Politics of Positive Discrimination*, Tavistock, London.

ELKIN, S. H. (1974) *Politics and Land Use Planning: The London Experience*, Cambridge University Press.

ELKIN, T., MCLAREN, D. and HILLMAN, M. (1991) *Reviving the City: Towards Sustainable Urban Development*, Policy Studies Institute/Friends of the Earth, London.

ELSON, M. J. (1985) Containment in Hertfordshire – changing attitudes to land release for new employment generating development, in S. Barrett and P. Healey (eds.) *Land Policy: Problems and Alternatives*, Gower, Farnborough.

ELSON, M. J. (1986) *Green Belts: Conflict Mediation in the Urban Fringe*, Heinemann, London.

ELSON, M. J. (1993) *Old Symbols and New Realities – The Context for Green Belt Policies in the UK*, Oxford Brookes University Inaugural Professorial Lecture.

ENGLISH, J., MADIGAN, R. and NORMAN, P. (1976) *Slum Clearance: The Social and Administrative Context in England and Wales*, Croom Helm, London.

ERMISCH, J. and MACLENNAN, D. (1986) Housing policies, markets and urban economic change, in V. Hausner (ed.) op. cit., Vol. 2.

ESHER, L. (1983) *A Broken Wave: The Rebuilding of England 1940–1980*, Harmondsworth, Penguin.

Estates Gazette, issues as cited in text.

EVANS, A. (1988) *No Room! No Room! The Costs of the British Town and Country Planning System.* Occasional Paper no. 79, Institute of Economic Affairs, London.

EVENSON, N. (1979) *Paris: A Century of Change 1878–1978*, Yale University Press, New Haven.

FERRIS, J. (1972) *Participation in Urban Planning – The Barnsbury Case: A Study of Environmental Improvement in London*, Bell, London.

FIELDING, T. and HALFORD, S. (1990) *Patterns and Processes of Urban Change in the UK: Reviews of Urban Research*, HMSO, London.

FILLER, R. (1986) *A History of Welwyn Garden City*, Phillimore, Chichester.

FINER, H. (1941) *Municipal Trading: A Study in Public Administration*, Allen & Unwin, London.

FINER, H. (1950) *English Local Government* (4th edn), Methuen, London.
FINNEGAN, R. (1984) Council housing in Leeds 1919–1939: social policy and urban change, in M. J. Daunton (ed.) op. cit.
FINNEMORE, B. (1989) *Houses from the Factory*, Rivers Oram, London.
FISHMAN, R. (1977) *Urban Utopias in the Twentieth Century: Ebenezer Howard, Frank Lloyd Wright, Le Corbusier*, Basic Books, New York.
FOGARTY, M. P. (1945) *The Prospects of the Industrial Areas of Great Britain*, Methuen, London.
FORSHAW, J. H. and ABERCROMBIE, P. (1943) *County of London Plan*, Macmillan, London.
FORSTER, E. M. (1946) The challenge of our time, in E. M. Forster (1965) *Two Cheers for Democracy*, Penguin, Harmondsworth.
FOTHERGILL, S. and GUDGIN, G. (1982) *Unequal Growth: Urban and Regional Employment Change in the UK*, Heinemann, London.
FOTHERGILL, S., KITSON M. and MONK, S. (1982) *The Impact of the New and Expanded Town Programmes on Industrial Location in Britain 1960–78*. Working Paper no. 3, Industrial Location Research Project, Department of Applied Economics, University of Cambridge.
FOULSHAM, J. (1990) Women's needs and planning – a critical evaluation of recent local authority practice, in J. Montgomery and A. Thornley (eds.) op. cit.
Franks Committee (1957) *Report of the Committee on Administrative Tribunals and Enquiries* (Cmnd 218), HMSO, London.
FRASER, D. (1979) *Power and Authority in the Victorian City*, Blackwell, Oxford.
FRASER, D. (ed.) (1982) *Municipal Reform and the Industrial City*, Leicester University Press.
FRIEND, J. K. and JESSOP, J. N. (1969) *Local Government and Strategic Choice: an Operational Approach to the Processes of Public Planning*, Tavistock, London.
FRIEND, J. K., POWER, J. M. and YEWLETT, C. J. L. (1974) *Public Planning: the Intercorporate Dimension*, Tavistock, London.
GALBRAITH, J. K. (1958) *The Affluent Society*, Hamish Hamilton, London.
GALBRAITH, J. K. (1992) *The Culture of Contentment*, Sinclair Stevenson, London.
GALLEY, K. (1973) Newcastle-upon-Tyne, in J. C. Holliday (ed.) op. cit.
GAMBLE, A. (1988) *The Free Economy and the Strong State*, Macmillan, London.
GANZ, G. (1977) *Government and Industry: The Provision of Financial Assistance to Industry and its Control*, Professional Books, Abingdon.
GARSIDE, P. L. (1988) 'Unhealthy areas': town planning, eugenics and the slums, 1890–1945, *Planning Perspectives*, Vol. 3, no. 1, pp. 24–46.
GARSIDE, P. L. and HEBBERT, M. (eds.) (1989) *British Regionalism 1900–2000*, Mansell, London.
GASKELL, S. M. (1981) 'The suburb salubrious': town planning in practice, in A. Sutcliffe (ed.) op. cit.
GEDDES, P. (1968, orig. 1915) *Cities in Evolution: An Introduction to the Town Planning Movement and the Study of Civics*, Benn, London.
GEORGE, H. (1911, orig. 1880) *Poverty and Progress: An Inquiry into the Cause of Industrial Depression, and Increase of Want with Increase of Wealth: The Remedy*, Dent, London.
GIBBERD, F., HYDE HARVEY, B., WHITE, L. and others (1980) *Harlow: The Story of a New Town*, Publications for Companies, Stevenage.
GIBBON, G. and BELL, R. W. (1939) *History of the London County Council 1889–1939*, Macmillan, London.

GIBSON, M. S. and LANGSTAFF, M. J. (1982) *An Introduction to Urban Renewal*, Hutchinson, London.

GITTINS, D. (1982) *Fair Sex: Family Size and Structure 1900–1939*, Hutchinson, London.

GLASS, R. (ed.) (1948) *The Social Background of a Plan*, Routledge & Kegan Paul, London.

GLASSON, J. (1978) *An Introduction to Regional Planning*, Hutchinson, London.

GOLD, J. R. and GOLD, M. M. (1990) 'A place of delightful prospects': promotional imagery and the selling of suburbia, in L. Zonn (ed.) *Place Images in Media: Portrayal, Experience and Meaning*, Rowman & Littlefield, Savage, Maryland, USA.

GOLD, J. R. and GOLD, M. M. (1994) 'Home at last': building societies, home ownership and the rhetoric of English suburban promotion in the interwar years, in J. R. Gold and S. V. Ward (eds.) *Place Promotion: The Use of Publicity and Marketing to Sell Towns and Regions*, Wiley, Chichester.

GOLD, J. R. and WARD, S. V. (1994) 'We're going to get it right this time': cinematic representations of urban planning and the British new towns 1939–51, in S. C. Aitken and L. Zonn (eds.) *Place, Power, Situation and Spectacle: A Geography of Film*, Rowman & Littlefield, Savage.

GOOBEY, A. R. (1992) *Bricks and Mortals – The Dreams of the 80s and the Nightmare of the 90s: The Inside Story of the Property World*, Century, London.

GORDON, G. (ed.) (1986) *Regional Cities in the UK 1890–1980*, Harper & Row, London.

GOULD, P. (1988) *Early Green Politics: Back to Nature, Back to the Land and Socialism in Britain 1880–1900*, Harvester, Brighton.

GRAVES, R. and HODGE, A. (1971) *The Long Weekend: A Social History of Great Britain 1918–1939*, Penguin, Harmondsworth.

GREED, C. H. (1993) *Introducing Town Planning*, Longman, Harlow.

GREGORY, D. J. (1970) *Green Belts and Development Control: A Case Study in the West Midlands.* Occasional Paper no. 12, Centre for Urban and Regional Studies, University of Birmingham.

GREGORY, T. (1973) Coventry, in J. C. Holliday (ed.) op. cit.

GREY, G. and AMOOQUAYE, E. (1990) A new agenda for race and planning? in J. Montgomery and A. Thornley (eds.) op. cit.

HAGUE, C. (1984) *The Development of British Planning Thought: A Critical Perspective*, Hutchinson, London.

HAGUE, C. (1990) Scotland: back to the future for planning? in J. Montgomery and A. Thornley (eds.) op. cit.

Hailsham, Lord (1975) *The Door Wherein I Went*, Collins, London.

HALL, D., HEBBERT, M. and LUSSER, H. (1993) The planning background, in A. Blowers (ed.) op. cit.

HALL, P. (1963) *London 2000*, Faber, London.

HALL, P. (1973) Manpower and education, in P. Cowan (ed.) *The Future of Planning*, Faber, London.

HALL, P. (1980) *Great Planning Disasters*, Weidenfeld & Nicholson, London.

HALL, P. (1982) Enterprise Zones: a justification, *International Journal of Urban and Regional Research*, Vol. 6, no. 3, pp. 416–21.

HALL, P. (1988) *Cities of Tomorrow: An Intellectual History of Urban Planning and Design in the Twentieth Century*, Blackwell, Oxford.

HALL, P. (1989) *London 2001*, Unwin Hyman, London.

HALL, P. (1992) *Urban and Regional Planning* (3rd edn), Routledge, London.

HALL, P., GRACEY, H., DREWETT, R. and THOMAS, R. (1973) *The Containment of Urban England*, 2 vols., PEP/Allen & Unwin, London.

HALL, S. and JACQUES, M. (eds.) (1983) *The Politics of Thatcherism*, Lawrence & Wishart, London.

HANNINGTON, W. (1937) *The Problem of the Distressed Areas*, Gollancz, London.

HANNINGTON, W. (1977, orig. 1936) *Unemployed Struggles 1919–1936: My Life and Struggles Amongst the Unemployed*, Lawrence & Wishart, London.

Harborne Tenants Ltd (1907) *Harborne Tenants Ltd: Prospectus 1907–8*, Birmingham.

HARDY, D. (1991a) *From Garden Cities to New Towns: Campaigning for Town and Country Planning, 1899–1946*, Spon, London.

HARDY, D. (1991b) *From New Towns to Green Politics: Campaigning For Town and Country Planning, 1946–1990*, Spon, London.

HARDY, D. and WARD, C. (1984) *Arcadia for All: The Legacy of a Makeshift Landscape*, Mansell, London.

HARLOE, M. (1975) *Swindon: A Town in Transition*, Heinemann, London.

HARRISON, M. (1981) Housing and town planning in Manchester before 1914, in A. Sutcliffe (ed.) op. cit.

HARRISON, M. (1991) Thomas Coglan Horsfall and 'the example of Germany', *Planning Perspectives*, Vol. 6, no. 3, pp. 297–314.

HARVEY, D. (1975) *Social Justice and the City*, Arnold, London.

HARVEY, D. (1982) *The Limits to Capital*, Blackwell, Oxford.

HASEGAWA, J. (1992) *Replanning the Blitzed City Centre*, Open University Press, Buckingham.

HASS-KLAU, C. (1990) *The Pedestrian and City Traffic*, Belhaven, London.

HAUSNER, V. (ed.) (1986) *Critical Issues In Urban Economic Development*, 2 vols., Clarendon, Oxford.

HAWTREE, M. (1981) The emergence of the town planning profession, in A. Sutcliffe (ed.) op. cit.

HDCCS (Housing Development Committee of the Corporation of Sheffield) (1962) *Ten Years of Housing in Sheffield 1953–1963*, City Architect's Department, Sheffield.

HDL (Housing Department, Liverpool) (1937) *City of Liverpool Housing 1937*, Housing Committee, Liverpool.

HEADICAR, P. and BIXBY, R. (1992) *Concrete and Tyres: Local Development Effects of Major Roads: M40 Case Study*, Council for the Protection of Rural England, London.

HEALEY, D. (1990) *The Time of My Life*, Penguin, Harmondsworth.

HEALEY, P. (1983) *Local Plans in British Land Use Planning*, Pergamon, Oxford.

HEALEY, P., DAVIS, J., WOOD, M. and ELSON, M. J. (1982) *Wokingham: The Implementation of Strategic Planning Policy in a Growth Area in the South East*, School of Town Planning, Oxford Polytechnic.

HEALEY, P., MCNAMARA, P., ELSON, M. and DOAK, A. (1989) *Land Use Planning and the Mediation of Urban Change: The British Planning System in Practice*, Cambridge University Press.

HEALEY, P. and NABARRO, R. (eds.) (1990) *Land and Property Development in a Changing Context*, Gower, Aldershot.

HEALEY, P., DAVOUDI, S., O'TOOLE, M., TAVSANOGLU, S. and USHER, D. (eds.) (1992) *Rebuilding the City: Property-Led Urban Regeneration*, Spon, London.

Heap, D. (1961) Green belts and open spaces: the English scene today, *Journal of Planning and Property Law*, January, pp. 16–24.
Heap, D. (1991) *An Outline of Planning Law*, Sweet & Maxwell, London.
Hebbert, M. (1981) Frederic Osborn 1885–1978, in G. E. Cherry (ed.) op. cit.
Hebbert, M. (1983) The daring experiment – social scientists and land use planning in 1940s Britain, *Environment and Planning B*, Vol. 10, pp. 3–17.
Hebbert, M. (1992) The British garden city: metamorphosis, in S. V. Ward (ed.) op. cit.
Heim, C. E. (1990) The Treasury as developer-capitalist? British new town building in the 1950s, *Journal of Economic History*, Vol. L, no. 4, pp. 903–924.
Hennessy, P. (1993) *Never Again: Britain 1945–51*, Vintage, London.
Hennock, E. P. (1973) *Fit and Proper Persons: Ideal and Reality in Nineteenth Century Urban Government*, Arnold, London.
Herbert Commission (Royal Commission on Local Government in Greater London) (1960) *Report* (Cmnd 1164), HMSO, London.
Herington, J. (1984) *The Outer City*, Harper & Row, London.
Herington, J. (1989) *Planning Processes: An Introduction for Geographers*, Cambridge University Press.
Heseltine, M. (1979) Secretary of State's address, *Report of Proceedings of the Town and Country Planning Summer School 1979*, Royal Town Planning Institute, London, pp. 25–9.
Heseltine, M. (1987) *Where There's a Will*, Hutchinson, London.
Hewison, R. (1987) *The Heritage Industry: Britain in a Climate of Decline*, Methuen, London.
Hill, D. M. (1970) *Participating in Local Affairs*, Penguin, Harmondsworth.
Hill, O. (1883) *Homes of the London Poor*, Macmillan, London.
Hill, R. (1986) Urban transport: from technical process to social policy, in P. Lawless and C. Raban (eds.) op. cit.
Hobbs, P. (1992) The economic determinants of post-war British planning, *Progress in Planning*, Vol. 38, no. 3, Pergamon, Oxford.
Hobhouse Committee (1947) *Report of the National Parks Committee (England and Wales)* (Cmd 7121), HMSO, London.
Hole, W. V., Adderson, I. M. and Pountney, M. T. (1979) *Washington New Town: The Early Years*, HMSO, London.
Holland, S. (1976) *The Regional Problem*, Macmillan, London.
Holley, S. (1983) *Washington: Quicker by Quango – The History of Washington New Town*, Publications for Companies, Stevenage.
Holliday, J. C. (ed.) (1973) *City Centre Redevelopment: A Study of British City Centre Planning and Case Studies of Five English City Centres*, Charles Knight, London.
Holmes, M. (1985) *The Labour Government, 1974–1979, Political Aims and Economic Reality*, Macmillan, Basingstoke.
Home, R. (1989) *Planning Use Classes: A Guide to the Use Classes Orders*, Blackwell Scientific Publications, Oxford.
Hornby, W. (1958) *Factories and Plant*, HMSO/Longman, London.
Horsey, M. (1988) Multi-storey council housing in Britain: introduction and spread, *Planning Perspectives*, Vol. 3, no. 2, pp. 167–96.
Horsfall, T. C. (1904) *The Improvement of the Dwellings and Surroundings of the People: The Example of Germany*, Manchester University Press.
House of Commons Select Committee on Employment (1989) *Third Report: The*

Employment Effects of Urban Development Corporations, HC 327 I and II, HMSO, London.

HOSSAIN, S. (1990) Race and planning – the Camden experience, in J. Montgomery and A. Thornley (eds.) op. cit.

HOWARD, E. (1898) *To-morrow: A Peaceful Path to Real Reform*, Swan Sonnenschein, London.

HOWARD, E. (1902) *Garden Cities of To-morrow*, Swan Sonnenschein, London.

HUBBARD, E. and SHIPPOBOTTOM, M. (1988) *A Guide to Port Sunlight Village*, Liverpool University Press.

HUDSON, R. (1989) *Wrecking a Region*, Pion, London.

HUGHES, M. R. (ed.) (1971) *The Letters of Lewis Mumford and Frederic J. Osborn: A Transatlantic Dialogue*, Adams & Dart, Bath.

Hunt Committee (1969) *The Intermediate Areas: Report of Departmental Committee* (Cmnd 3998), HMSO, London.

HYDER, J. (1913) *The Case for Land Nationalisation*, Simpkin, Marshall, Hamilton, Kent, London.

IMRIE, R. and THOMAS, H. (1993a) Urban policy and the urban development corporations, in R. Imrie and H. Thomas (eds.) op. cit.

IMRIE, R. and THOMAS, H. (eds.) (1993b) *British Urban Policy and the Urban Development Corporations*, Paul Chapman, London.

JACKSON, A. A. (1991) *Semi-Detached London: Suburban Development, Life and Transport 1900–1939* (2nd edn), Wild Swan, Didcot.

JACOBS, J. (1964) *The Death and Life of Great American Cities: The Failure of Town Planning*, Penguin, Harmondsworth.

JAY, D. (1980) *Change and Fortune: A Political Record*, Hutchinson, London.

JEFFERYS, J. B. (1954) *Retail Trading in Britain 1850–1950*, Economic and Social Studies no. 13, National Institute of Economic and Social Research, Cambridge University Press.

JENKIN, P. (1984) Secretary of State's address [to the RTPI Summer School, September 1983]: the Rt Hon. Patrick Jenkin, MP, *The Planner*, Vol. 70. no. 2, pp. 15–17.

JENKINS, S. (1975) *Landlords to London: The Story of a Capital and its Growth*, Constable, London.

JEVONS, R. and MADGE, J. (1946) *Housing Estates: A Study of Bristol Corporation Policy and Practice between the Wars*, Arrowsmith, Bristol.

JOHNSON, J. H. (ed.) (1974) *Suburban Growth: Geographical Processes at the Edge of the Western City*, Wiley, London.

JOHNSON, J. H. and POOLEY, C. G. (eds.) (1982) *The Structure of Nineteenth Century Cities*, Croom Helm, London.

JOHNSON-MARSHALL, P. (1966) *Rebuilding Cities*, Edinburgh University Press.

JONES, C. and PATRICK, J. (1992) The merchant city as an example of housing-led urban regeneration, in P. Healey *et al.* (eds.) op. cit.

JONES, G. S. (1971) *Outcast London: A Study of the Relationship between Classes in Victorian Society*, Oxford University Press.

JOWELL, J. (1983) Structure plans and social engineering, in *Structure Plans and Local Plans – Planning in Crisis*, Occasional paper of the *Journal of Planning and Environment Law*, Sweet & Maxwell, London.

KEYNES, J. M. (1973, orig. 1936) *The Collected Writings of John Maynard Keynes: Vol. 7: The General Theory of Employment, Interest and Money*, Macmillan/Royal Economic Society, London.

KING, A. D. (1982) 'Town planning': A note on the origins and use of the term, *Planning History Bulletin*, Vol. 4, no. 2, pp. 15–17.

KNEVITT, C. (1975) Macclesfield: the self-help GIA, *Architects Journal*, Vol. 162, pp. 995–1002.

KOHAN, C. M. (1952) *Works and Buildings*, HMSO/Longman, London.

KROPOTKIN, P. (1974, orig. 1899) *Fields, Factories and Workshops Tomorrow*, Allen & Unwin, London.

LARSSON, L. O. (1984) Metropolis architecture, in A. Sutcliffe (ed.) *Metropolis 1890–1940*, Mansell, London.

LAW, C. M. (1981) *British Regional Development since World War I*, Methuen, London.

LAW, C. M. (1992) Property-led urban regeneration in inner Manchester, in P. Healey *et al.* (eds.) op. cit.

LAW, C. M., in association with GRIME, E. K., GRUNDY, C. L., SENIOR, M. L. and TUPPEN, J. N. (1988) *The Uncertain Future of the Urban Core*, Routledge, London.

LAWLESS, P. (1986) *The Evolution of Spatial Policy*, Pion, London.

LAWLESS, P. (1989) *Britain's Inner Cities* (2nd edn), Paul Chapman, London.

LAWLESS, P. (1991) Urban policy in the Thatcher decade: English inner city policy 1979–1990, *Environment and Planning C: Government and Policy*, Vol. 9, pp. 15–30.

LAWLESS, P. and BROWN, F. (1986) *Urban Growth and Change in Britain: An Introduction*, Harper & Row, London.

LAWLESS, P. and RABAN, C. (eds.) (1986) *The Contemporary British City*, Harper & Row, London.

LAWTON, R. and POOLEY, C. G. (1986) Liverpool and Merseyside, in G. Gordon (ed.) op. cit.

LCC (London County Council) (1937) *London Housing*, LCC, London.

LCC (1961) *The Planning of a New Town*, LCC, London.

LE CORBUSIER (1922) Une ville contemporaire, in Le Corbusier and P. Jeanneret (eds.) (1964) *Oeuvre Complète de 1910–1929*, Les Editions d'Architecture, Zurich.

LE CORBUSIER (1925) Plan Voisin (de Paris), in Le Corbusier and P. Jeanneret (eds.) (1964) *Oeuvre Complète de 1929–1934*, Les Editions d'Architecture, Zurich.

LE CORBUSIER, trans. by ETCHELLS, F. (1946, orig. 1927) *Towards a New Architecture*, Architectural Press, London (originally published in French in 1923 as *Vers Une Architecture*, Editions Cres, Paris).

LEE, J. M. (1963) *Social Leaders and Public Persons*, Clarendon, Oxford.

Leitch Committee (1977) *Report of the Advisory Committee on Trunk Road Assessment*, HMSO, London.

LEWIS, J. and TOWNSEND, A. (eds.) (1989) *The North–South Divide: Regional Change in Britain in the 1980s*, Paul Chapman, London.

LEWIS, N. (1992) *Inner City Regeneration: The Demise of Regional and Local Government*, Open University Press, Buckingham.

LGB (Local Government Board) (1919) *Manual on the preparation of State-aided Housing Schemes*, HMSO, London.

LITTLE, J., PEAKE, L. and RICHARDSON, P. (eds.) (1988) *Women in Cities: Gender and the Urban Environment*, Macmillan, Basingstoke.

LIVINGSTONE, K. (1987) *If Voting Changed Anything They'd Abolish It*, Collins, London.

LLEC (Liberal Land Enquiry Committee) (1914) *The Land, Vol. 2: Urban*, Hodder & Stoughton, London.

LOCK, D. (1989) Second honeymoon in the marriage of town and country? *Town and Country Planning*, Vol. 58, no. 6, pp. 174–5.

LOCK, M. (1946) *The Middlesbrough Survey and Plan*, Middlesbrough Corporation.

LOEBL, H. (1988) *Government Factories and the Origins of British Regional Policy 1934–1948*, Avebury, Aldershot.

LONG, J. R. (1962) *The Wythall Inquiry*, Estates Gazette, London.

LONGLEY, P., BATTY, M., SHEPHERD, J. and SADLER, G. (1992) Do green belts change the shape of urban areas? A preliminary survey of the settlement geography of south east England, *Regional Studies*, Vol. 26, no. 5, pp. 437–52.

LONIE, A. A. and BEGG, H. M. (1979) Comment: further evidence on the quest for an effective regional policy, *Regional Studies*, Vol. 13, pp. 497–500.

LUTYENS, E. and ABERCROMBIE, L. P. (1945) *A Plan for the City and County of Kingston-upon-Hull*, Brown, Hull.

LYDDON, D. (1987) The development plan: vision or vacuum? in B. Robson (ed.) op. cit.

MCAUSLAN, P. (1980) *Ideologies of Planning Law*, Pergamon, Oxford.

MCCALLUM, J. (1979) The development of British regional policy, in D. M. Maclennan, and J. B. Parr (eds.) *Regional Policy: Past Experience and New Directions*, Martin Robertson, Oxford.

MCCORMICK, J. (1993) *Environmental Politics*, in P. Dunleavy *et al.* (eds.) pp. 267–83.

MCCRONE, G. (1969) *Regional Policy in Britain*, Allen & Unwin, London.

MCDOUGALL, G. (1979) The state, capital and land: the history of town planning revisited, *International Journal of Urban and Regional Research*, Vol. 3, no. 3, pp. 361–80.

M'GONIGLE, G. C. M. and KIRBY, J. (1936) *Poverty and Public Health*, Gollancz, London.

MCKAY, D. H. and COX, A. W. (1979) *The Politics of Urban Change*, Croom Helm, Beckenham.

MACKAY, R. R. (1992) 1992 and relations with the EEC, in P. Townroe and R. Martin (eds.) op. cit.

MACKINTOSH, M. and WAINWRIGHT, H. (1987) *A Taste of Power*, Verso, London.

MCLOUGHLIN, J. B. (1969) *Urban and Regional Planning: A Systems Approach*, Faber, London.

MACMILLAN, H. (1969) *Tides of Fortune 1945–1955*, Macmillan, London.

MANDELKER, D. R. (1962) *Green Belts and Urban Growth: English Town and Country Planning*, University of Wisconsin Press, Madison.

MANZONI, H. J. B. (1939) *The Building of 50,000 Municipal Houses*, City of Birmingham.

Marley Committee (1935) *Report of the Departmental Committee on Garden Cities and Satellite Towns*, HMSO, London.

MARMOT, A. F. (1982) The legacy of Le Corbusier and high rise housing, *Built Environment*, Vol. 7, no. 2, pp. 82–95.

MARRIOTT, O. (1969) *The Property Boom*, Pan, London.

MARSH, D. L. (1965) *The Changing Social Structure of England and Wales 1871–1961*, Routledge & Kegan Paul, London.

MARSHALL, R. J. and MASSER, I. (1982) British planning methodology: three historical perspectives, *Built Environment*, Vol. 7, no. 2, pp. 121–9.

MARTIN, R. L. (1988) The political economy of Britain's north–south divide, *Transactions of the Institute of British Geographers*, New Series, Vol. 13, pp. 389–418.

MARTINS, M. R. (1986) *An Organizational Approach to Regional Planning*, Gower, Aldershot.

MARWICK, A. (1964) Middle opinion in the thirties: planning, progress and political agreement, *English Historical Review*, Vol. LXXIX, no. 311, pp. 285–98.

MARWICK, A. (1967) *The Deluge: British Society and the First World War*, Penguin, Harmondsworth.

MARWICK, A. (1970) *Britain in the Century of Total War: War, Peace and Social Change 1900–1967*, Penguin, Harmondsworth.

MASON, T. and TIRATSOO, N. (1990) People, politics and planning: the reconstruction of Coventry's city centre, 1940–53, in J. Diefendorf (ed.) *Rebuilding Europe's Blitzed Cities*, Macmillan, Basingstoke.

MASSEY, D. (1989) Regional planning 1909–1939, in P. L. Garside and M. Hebbert (eds.) op. cit.

MASSEY, D. B. (1979) In what sense a regional problem? *Regional Studies*, Vol. 13, pp. 233–43.

MASSEY, D. B. (1982) Enterprise Zones: a political issue, *International Journal of Urban and Regional Research*, Vol. 6, no. 3, pp. 429–34.

MASSEY, D. B. (1984) *Spatial Divisions of Labour: Social Structures and the Geography of Production*, Macmillan, Basingstoke.

MASTERMAN, C. F. G. (1909) *The Condition of England*, Methuen, London.

Maud Report (1967) *The Management of Local Government*, HMSO, London.

MAWSON, J. and MILLER, D. (1986) Interventionist approaches in local employment and economic development: the experience of Labour local authorities, in V. Hausner (ed.) op. cit., Vol. 1, pp. 145–99.

MEEGAN, R. (1989) Paradise postponed: the growth and decline of Merseyside's outer estates, in P. N. Cooke (ed.) op. cit.

MELLER, H. (1990) *Patrick Geddes: Social Evolutionist and City Planner*, Routledge, London.

MELLING, J. (ed.) (1980) *Housing, Social Policy and the State*, Croom Helm, London.

MERRETT, S. (1979) *State Housing in Britain*, Routledge & Kegan Paul, London.

MESS, H. A. (1928) *Industrial Tyneside: A Social Survey*, Benn, London.

MH (Ministry of Health) (1921) *Report of the South Wales Regional Survey Committee*, HMSO, London.

MH & MW (Ministry of Health and Ministry of Works) (1944) *Housing Manual*, HMSO, London.

MHLG (Ministry of Housing and Local Government) (1951) *Town and Country Planning 1943–1951 Progress Report* (Cmd 8204), HMSO, London.

MHLG (1952a) *Town and Country Planning Act, 1947: Amendment of Financial Provisions* (Cmd 8699), HMSO, London.

MHLG (1952b) *The Density of Residential Areas*, HMSO, London.

MHLG (1955) *Green Belts* (Circular no. 42/55), HMSO, London.

MHLG (1957) *Green Belts* (Circular no. 50/57), HMSO, London.

MHLG (1958) *Flats and Houses 1958*, HMSO, London.

MHLG (1962) *The Green Belts*, HMSO, London.

MHLG (1963) *London – Employment: Housing: Land* (Cmnd 1952), HMSO, London.

MHLG (1964) *The South East Study 1961–1981*, HMSO, London.

MHLG (1966) *The Deeplish Study: Improvement Possibilities in a District of Rochdale*, HMSO, London.

MHLG (1968) *Report of the Inquiry into the Collapse of Flats at Ronan Point, Canning Town*, HMSO, London.

MHLG (1970a) *Living in a Slum: A Study of St Mary's, Oldham*, HMSO, London.

MHLG (1970b) *Moving out of a Slum: A Study of People Moving from St Mary's, Oldham*, HMSO, London.

MHLG and MT (Ministry of Transport) (1962) *Town Centres: Approach to Renewal*, HMSO, London.

MHLG and WO (Welsh Office) (1967) *Town and Country Planning* (Cmnd 3333), HMSO, London.

MHLG and WO (1968) *Old Houses into New Homes* (Cmnd 3602), HMSO, London.

MIDDLETON, M. (1991) *Cities in Transition: The Regeneration of Britain's Inner Cities*, Michael Joseph, London.

MILIUTIN, N. A. (1974) *Sotsgorod: The Problem of Building Socialist Cities*, Massachusetts Institute of Technology Press, Cambridge.

MILLER, M. (1989) *Letchworth: The First Garden City*, Phillimore, Chichester.

MILLER, M. (1992) *Raymond Unwin: Garden Cities and Town Planning*, Leicester University Press.

MILLER, M. and GRAY, A. S. (1992) *Hampstead Garden Suburb*, Phillimore, Chichester.

MILLS, L. and YOUNG, K. (1986) Local authorities and economic development: a preliminary analysis, in V. Hausner (ed.) op. cit., Vol. I.

MINETT, M. J. (1974) The Housing, Town Planning, Etc 1909, *The Planner*, Vol. 60, no. 5, pp. 676–80.

Minister without Portfolio *et al.* (1985) *Lifting the Burden* (Cmnd 9571), HMSO, London.

MITCHELL, E. (1967) *The Plan that Pleased*, Town and Country Planning Association, London.

MLNR (Ministry of Land and Natural Resources) (1965) *The Land Commission* (Cmnd 2771), HMSO, London.

MONTGOMERY, J. (1990) Counter-revolution: out-of-town shopping and the future of town centres, in J. Montgomery and A. Thornley (eds.) op. cit.

MONTGOMERY, J. and THORNLEY, A. (eds.) (1990) *Radical Planning Initiatives: New Directions for Urban Planning in the 1990s*, Gower, Aldershot.

MOOR, N. (1979) The contribution and influence of office developers and their companies on the location and growth of office activities, in P. W. Daniels (ed.) *Spatial Patterns of Office Growth and Location*, Wiley, Chichester.

MOORE, B., RHODES, J. and TYLER, P. (1986) *The Effects of Government Regional Economic Policy*, Department of Trade and Industry, HMSO, London.

MORGAN, K. O. (1984) *Labour in Power 1945–51*, Clarendon, Oxford.

MORGAN, K. O. (1992) *The People's Peace: British History 1945–1990*, Oxford University Press.

MORTIMER, J. (1990) *Titmuss Regained*, Penguin, Harmondsworth.

MOWAT, C. L. (1956) *Britain between the Wars 1918–1940*, Methuen, London.

MT (Ministry of Transport) (1963) *Traffic in Towns: A Study of the Long Term Problems of Traffic in Urban Areas: Reports of the Steering Group and Working Group appointed by the Minister of Transport*, HMSO, London.

MTCP (Ministry of Town and Country Planning) (1945) *National Parks in England and Wales* (Dower Report) (Cmd 6628), HMSO, London.

MTCP (Ministry of Town and Country Planning) (1947) *Town and Country Planning Act, 1947: Explanatory Memorandum* (Cmd 7006), HMSO, London.

MTCP (1948) *Town and Country Planning Act, 1947: Explanatory Memorandum, Part II – Notes on Sections*, HMSO, London.

MTCP and SO (Scottish Office) (1944) *The Control of Land Use* (Cmd 6537), HMSO, London.

MUCHNICK, D. H. (1970) *Urban Renewal in Liverpool: A Study of the Politics of Redevelopment*. Occasional Papers on Social Administration no. 33, Bell, London.

MUNTON, R. (1983) *London's Green Belt: Containment in Practice*, Allen & Unwin, London.

MURRAY, M. (1992) *The Politics and Pragmatism of Urban Containment: Belfast since 1940*, Avebury, Aldershot.

MUTHESIUS, S. (1982) *The English Terraced House*, Yale University Press, New Haven, CT.

NAIRN, I (ed.) (1955) Outrage, Special Number, *Architectural Review*, Vol. 117, no. 702, pp. 363–460.

NCDP (National Community Development Project) (1976) *Whatever Happened to Council Housing?* CDP Information and Intelligence Unit, London.

NEALE, C. (1984) *South Woodham Ferrers – The Essex Design Guide in Practice*, Royal Institution of Chartered Surveyors Planning and Development Case Study 3, Surveyors Publications, London.

NEDC (National Economic Development Council) (1963) *Conditions Favourable for Faster Growth*, HMSO, London.

NEPC (Northern Regional Economic Planning Council) (1966) *Challenge of the Changing North*, HMSO, London.

NETTLEFOLD, J. S. (1908) *Practical Housing*, Garden City Press, Letchworth.

NETTLEFOLD, J. S. (1914) *Practical Town Planning*, St Catherine Press, London.

New Townsmen [HOWARD, E., OSBORN, F. J., PURDOM, C. B. and TAYLOR, W. G.] (1918) *New Towns after the War*, Dent, London.

NICOLSON, H. (1967) *Diaries and Letters 1939–1945*, Collins, London.

NICHOLSON, M. (1970) *The Environmental Revolution*, Penguin, Harmondsworth.

NORTHFIELD, Lord (1989) Private sector development of new country towns, *Town and Country Planning*, Vol. 58, no. 1, pp. 14–15.

NPD (Newcastle-upon-Tyne Planning Department) (1973) *Planning Progress and Policy 1973: Newcastle-Upon-Tyne*, Planning Department, Newcastle.

NUGENT, N. (1993) *The European Dimension*, in P. Dunleavy *et al.* (eds.), pp. 40–68.

OGILVY, A. A. (1975) *Bracknell and its Migrants: 21 Years of a New Town*, HMSO, London.

OLIVER, P., DAVIS I. and BENTLEY, I. (1981) *Dunroamin: The Suburban Semi and its Enemies*, Barrie & Jenkins, London.

OPCS (Office of Population Censuses and Surveys) (1992) *1991 Census: Preliminary Report*, HMSO, London.

ORLANS, H. (1952) *Stevenage: A Sociological Study of a New Town*, Routledge & Kegan Paul, London.

ORWELL, G. (1962, orig. 1938) *Homage to Catalonia*, Penguin, Harmondsworth.

ORWELL, G. (1970) *The Collected Essays, Journalism and Letters of George Orwell, Vol. 1: An Age Like This*, Penguin, Harmondsworth.

OSBORN, F. J. (1959) *Can Man Plan? and Other Verses*, Harrap, London.

OSBORN, F. J. and WHITTICK, A. (1977) *The New Towns: Their Origins, Achievements and Progress*, Leonard Hill, London.

PAG (Planning Advisory Group) (1965) *The Future of Development Plans*, HMSO, London.

Pahl, R. E. (1970) *Whose City?* Longman, London.

Parker-Morris Committee (1961) *Homes for Today and Tomorrow: Report of the Committee Appointed by the Central Housing Advisory Committee to Consider the Standards of Design and Equipment Applicable to Family Dwellings and Other Residential Accommodation whether Provided by Public Authorities or Private Enterprise*, HMSO, London.

Parkinson, M., Foley, B. and Judd, D. (eds.) (1988) *Regenerating the Inner Cities: The UK Crisis and the US Experience*, Manchester University Press.

Parsons, D. W. (1986) *The Political Economy of British Regional Policy*, Croom Helm, Beckenham.

Patton, K. (1978) The foundations of Peterlee New Town, in M. Bulmer, (ed.) op. cit.

Pearce, D. C. (1989) The Yorkshire and Humberside Economic Planning Council 1965–1979, in P. L. Garside and M. Hebbert, (eds.) op. cit.

Pearson, L (1988) *The Architectural and Social History of Co-operative Living*, Macmillan, London.

Peden, G. C. (1991) *British Economic Policy: Lloyd George to Margaret Thatcher* (2nd edn), Philip Allan, Hemel Hempstead.

PEP (Political and Economic Planning) (1934) *Housing England*, PEP, London.

PEP (Political and Economic Planning) (1939) *Report on the Location of Industry*, PEP, London.

Perry, C. A. (1939) *Housing for the Machine Age*, Russell Sage Foundation, New York.

Phillips, A. (1993) *The Best in Science, Office and Business Park Design*, Batsford, London.

Pickup, L., Stokes, G., Meadowcroft, S., Goodwin, P., Tyson, B. and Kenny, F. (1991) *Bus Deregulation in Metropolitan Areas*, Avebury, Aldershot.

Pilcher Committee (1975) *Commercial Property Development: First Report of the Advisory Committee*, HMSO, London.

Pinto-Duschinsky, M. (1970) Bread and circuses? The Conservatives in office 1951–1964, in V. Bogdanor and R. Skidelsky (eds.) op. cit.

Pitfield, D. E. (1978) The quest for an effective regional policy 1934–7, *Regional Studies*, Vol. 12, pp. 429–43.

Plowden, S. (1972) *Towns Against Traffic*, Andre Deutsch, London.

Plowden, W. (1973) *The Motor Car and Politics in Britain*, Penguin, Harmondsworth.

Porritt, J. and Winner, D. (1988) *The Coming of the Greens*, Fontana, London.

Potter, S. (1990) Britain's development corporations, *Town and Country Planning*, Vol. 59, no. 11, pp. 294–8.

PPG2 (1988) *Planning Policy Guidance 2: Green Belts*, Department of Environment/HMSO, London.

PPG9 (1988) *Planning Policy Guidance 9: Regional Guidance for the South East*, Department of the Environment/HMSO, London.

PPG12 (1988) *Planning Policy Guidance 12: Local Plans*, Department of the Environment and Welsh Office/HMSO, London.

PPG15 (1990) *Planning Policy Guidance 15: Regional Planning Guidance, Structure Plans and the Content of Development Plans*, Department of the Environment/HMSO, London.

Property Advisory Group (1981) *Planning Gain*, HMSO, London.

Punter, J. (1985) *A History of Aesthetic Control 2: The Control of Development in England and Wales 1947–1985*, Department of Land Management, University of Reading.

Punter, J. (1991) A microcosm of design control in post-war Britain: a case study of office development in central Bristol 1940–90, *Planning Perspectives*, Vol. 6, no. 3, pp 315–47.

Purdom, C. B. (1925) *The Building of Satellite Towns: A Contribution to the Study of Town Development and Regional Planning*, Dent, London.

Ratcliffe, J. R. (1976) *Land Policy: An Exploration of the Nature of Land in Society*, Hutchinson, London.

Ravetz, A. (1974a) *Model Estate: Planned Housing at Quarry Hill*, Croom Helm, London.

Ravetz, A. (1974b) From working class tenement to modern flat: local authorities and multi-storey housing between the wars, in A. Sutcliffe (ed.) op. cit.

Ravetz, A. (1980) *Remaking Cities: Contradictions of the Recent Urban Environment*, Croom Helm, London.

Ravetz, A. (1986) *The Governance of Space: Town Planning in Modern Society*, Faber, London.

Read, D. (1972) *Edwardian England 1901–15*, Harrap, London.

Reade, E. (1987) *British Town and Country Planning*, Open University Press, Buckingham.

Redcliffe-Maud Commission (Royal Commission on Local Government in England) (1969a) *Local Government Reform: Summary* (Cmnd 4039), HMSO, London.

Redcliffe-Maud Commission (1969b) *Report* (Cmnd 4040, I–II), 3 vols., HMSO, London.

Redfern, P. (1982) Profile of our cities, *Population Trends*, no. 30, pp. 21–32.

Reid, M. (1982) *The Secondary Banking Crisis*, Macmillan, London.

Reith Committee (1946a) *Interim Report of the New Towns Committee* (Cmd 6759), HMSO, London.

Reith Committee (1946b) *Second Interim Report of the New Towns Committee* (Cmd 6794), HMSO, London.

Reith Committee (1946c) *Final Report of the New Towns Committee* (Cmd 6876), HMSO, London.

Reith, J. C. W. (1949) *Into the Wind*, Hodder & Stoughton, London.

Rentoul, J., with Ratford, J. (1989) *Me and Mine: The Triumph of the New Individualism?* Unwin Hyman, London.

Reynolds, J. (1983) *The Great Paternalist: Titus Salt and the Growth of Nineteenth Century Bradford*, Temple Smith, London.

Reynolds, J. P. (1952) Thomas Coglan Horsfall and the town planning movement in England, *Town Planning Review*, Vol. XXII, no. 1, pp. 52–66.

Richardson, H. W. and Aldcroft, D. H. (1968) *Building in the British Economy between the Wars*, Allen & Unwin, London.

Riddell, P. (1989) *The Thatcher Decade: How Britain has Changed during the 1980s*, Blackwell, Oxford.

Riden, P. (1988) *Rebuilding a Valley: A History of Cwmbran Development Corporation*, Cwmbran Development Corporation.

Ridley, N. (1987) Address by Nicholas Ridley, Secretary of State for the Environment [to the Town and Country Planning Summer School 1986], *The Planner*, Vol. 73, no. 2, pp. 39–41.

Ridley, N. (1992) *'My Style of Government': The Thatcher Years*, Fontana, London.

Roberts, J. (1981) *Pedestrian Precincts in Britain*, Transport and Environment Studies, TEST, London.

Roberts, M. (1991) *Living in a Man-Made World: Gender Assumptions in Modern Housing Design*, Routledge, London.

Roberts, P. and Whitney, D. (1993) The new partnership: interagency co-operation and urban policy in Leeds, in R. Imrie and H. Thomas (eds.) op. cit.

Robson, B. (ed.) (1987) *Managing the City: The Aims and Impacts of Urban Policy*, Croom Helm, Beckenham.

Robson, B. (1988) *Those Inner Cities: Reconciling the Economic and Social Aims of Urban Policy*, Clarendon Press, Oxford.

Robson, W. A. (1939) *The Government and Misgovernment of London*, Allen & Unwin, London.

Rothman, B. (1982) *The 1932 Kinder Trespass*, Willow, Altrincham.

Rowntree, B. S. (1901) *Poverty: A Study of Town Life*, Macmillan, London.

Russell, B. (1981) *Building Systems, Industrialization and Architecture*, Wiley, London.

Rydin, Y. (1986) *Housing Land Policy*, Gower, Aldershot.

Saunders, P. (1979) *Urban Politics: A Sociological Interpretation*, Hutchinson, London.

Saville, J. (1988) *The Labour Movement in Britain: A Commentary*, Faber, London.

Scarman Report (1981) *The Brixton Disorders 10–12 April 1981* (Cmnd 8427), HMSO, London.

Schaffer, F. (1970) *The New Town Story*, Paladin, London.

Schaffer, F. (1974) The Town and Country Planning Act 1947, *The Planner*, Vol. 60, no. 5, pp. 690–5.

Schuster Committee (1950) *Report of the Committee on the Qualifications of Planners* (Cmd 8059), HMSO, London.

Scott Committee (1942) *Report of the Committee on Land Utilisation in Rural Areas* (Cmd 6378), HMSO, London.

SDD (Scottish Development Department) (1963) *Central Scotland: A Programme for Development and Growth* (Cmnd 2288), HMSO, Edinburgh.

Seeley, I. H. (1974) *Planned Expansion of Country Towns*, Godwin, London.

SEEPC (South East Economic Planning Council) (1967) *A Strategy for the South East*, HMSO, London.

SEJPT (South East Joint Planning Team) (1970) *Strategic Plan for the South East*, HMSO, London.

SEJPT (1976) *Strategy for the South East: 1976 Review*, HMSO, London.

Sewel, J., Twine, F. and Williams, N. (1984) The sale of council houses: some empirical evidence, *Urban Studies*, Vol. 21, pp. 439–50.

Shanks, M. (1977) *Planning and Politics: the British Experience 1960–76*, Allen & Unwin, London.

Sharp, E. G. (1980) The London County Council green belt scheme – a note on some primary sources, *Planning History Bulletin*, Vol. 2, no. 2, pp. 12–16.

Sharp, E. (1969) *Ministry of Housing and Local Government*, Allen & Unwin, London.

Sharp, T. (1932) *Town and Countryside: Some Aspects of Urban and Rural Development*, Oxford University Press.

Sharp, T. (1940) *Town Planning*, Penguin, Harmondsworth.

Sharp, T. (1957) Planning now, *Journal of the Town Planning Institute*, Vol. XLIII, no. 5, pp. 133–41.

SHARP, T. (1966) Planning planning, *Journal of the Town Planning Institute*, Vol. LII, no. 6, pp. 209–15.

SHEAIL, J. (1981) *Rural Conservation in Interwar Britain*, Clarendon Press, Oxford.

SHORT, J. R., FLEMING, S. and WITT, S. (1986) *Housebuilding, Planning and Community Action*, Routledge & Kegan Paul, London.

SHORT, J. R., WITT, S. and FLEMING, S. (1987) Conflict and compromise in the built environment: housebuilding in central Berkshire, *Transactions of the Institute of British Geographers*, New Series, Vol. 12, pp. 29–47.

SILKIN, J. (1987) *Changing Battlefields: The Challenge to the Labour Party*, Hamish Hamilton, London.

SIMPSON, M. (1985) *Thomas Adams and the Modern Planning Movement: Britain, Canada and the United States*, Mansell, London.

SINCLAIR, G. (1992) *The Lost Land: Land Use Change in England 1945–1990*, Council for the Protection of Rural England, London.

SISSONS, M. and FRENCH, P. (eds.) (1964) *Age of Austerity 1945–1951*, Penguin, Harmondsworth.

SITTE, C. (trans. COLLINS, G. R. and COLLINS, C. C.) (1965) *City Planning according to Artistic Principles*, Phaidon, London.

Skeffington Committee (1969) *People and Planning: Report of the Committee on Public Participation in Planning*, HMSO, London.

SKIDELSKY, R. (1967) *Politicians and the Slump: The Labour Government of 1929–1931*, Macmillan, London.

SKIDELSKY, R. (ed.) (1988) *Thatcherism*, Chatto & Windus, London.

SKILLETER, K. J. (1993) The role of public utility societies in early British town planning and housing reform, 1901–36, *Planning Perspectives*, Vol. 8, no. 2, pp. 125–65.

SKINNER, D. N. (1976) *A Situation Report on Green Belts in Scotland*, Countryside Commission for Scotland, Perth.

SMIGIELSKI, K. (1960) London: a metropolis in disintegration, *Journal of the Town Planning Institute*, Vol. XLVI, no. 8, pp. 208–10.

SMITH, G. C. and FORD, R. G. (1990) Social change in outer city British public housing, *Habitat International*, Vol. 14, no. 4, pp. 89–96.

SMITH, R. (1979) *East Kilbride: The Biography of a New Town 1947–1973*, HMSO, Edinburgh.

SMITH, R. and FARMER, E. (1985) Housing, population and decentralisation, in R. Smith and U. Wannop (eds.) *Strategic Planning in Action: The Impact of the Clyde Valley Plan 1946–1982*, Gower, Aldershot.

SMITH, R. and WHYSALL, P. (1990) The Addison Act and the local authority response: housing policy formulation and implementation in Nottingham 1917–1922, *Town Planning Review*, Vol. 61, no. 2, pp. 185–208.

SMITH, R., WHYSALL, P. and BEVRIN, C. (1986) Local authority inertia in housing improvement 1890–1914, *Town Planning Review*, Vol. 57, no. 4, pp. 404–24.

SMITH, T. D. (1970) *An Autobiography*, Oriel, Newcastle.

SPCL (Stockley Park Consortium Ltd) (c.1992) *Operating Your Business at Stockley Park, Heathrow*, SPCL, London.

STAMP, L. D. (1962) *The Land of Britain: Its Use and Misuse*, Longman, London.

STANSFIELD, K. (1981) Thomas Sharp 1901–1978, in G. E. Cherry (ed.) op. cit.

STARKIE, D. A. (1982) *The Motorway Age: Roads and Traffic Policies in Post-War Britain*, Pergamon, Oxford.

STEIN, C. S. (1958) *Toward New Towns for America* (2nd edn), Liverpool University Press.

STEPHENSON, G. (ed. DE MARCO, C.) (1992) *On a Human Scale: A Life in City Design*, Fremantle Arts Centre/Liverpool University Press.

STILGOE, H. E. (1910) Town planning in the light of the Housing, Town Planning Etc Act 1909, *Proceedings of the Institution of Municipal and County Engineers*, Vol. XXXVII, pp. 11–45.

STJTPC (South Tyneside Joint Town Planning Committee) (1928) *The South Tyneside Regional Town Planning Scheme*, STJTPC, Gateshead.

SUTCLIFFE, A. (1974a) A century of flats in Birmingham 1873–1973, in A. Sutcliffe (ed.) op. cit.

SUTCLIFFE, A. (ed.) (1974b) *Multi-Storey Living: The British Working Class Experience*, Croom Helm, London.

SUTCLIFFE, A. (1981a) *Towards the Planned City: Germany, Britain, the United States and France 1780–1914*, Basil Blackwell, Oxford.

SUTCLIFFE, A. (ed.) (1981b) *British Town Planning: The Formative Years*, Leicester University Press.

SUTCLIFFE, A. (1982) The growth of public intervention in the British urban environment during the nineteenth century: a structural approach, in J. H. Johnson and C. G. Pooley (eds.) op. cit.

SUTCLIFFE, A. (1986) The 'midlands metropolis': Birmingham 1890–1980, in G. Gordon (ed.) op. cit.

SUTCLIFFE, A. (1988) Britain's first town planning act: a review of the 1909 achievement, *Town Planning Review*, Vol. 59, no. 3, pp. 289–303.

SUTCLIFFE, A. and SMITH, R. (1974) *Birmingham 1939–1970*, Oxford University Press.

SWENARTON, M. (1981) *Homes Fit for Heroes: The Politics and Architecture of Early State Housing in Britain*, Heinemann, London.

TARN, J. N. (1973) *Five Per Cent Philanthropy: An Account of Housing in Urban Areas Between 1840 and 1914*, Cambridge University Press.

TCPSS (Town and Country Planning Summer School) (1949) *Report of the Summer School*, TCPSS, St Andrews.

TEBBIT, N. (1988) *Upwardly Mobile*, Weidenfeld & Nicolson, London.

THATCHER, M. (1989) *Speeches to the Conservative Party Conference 1975–1988*, Conservative Political Centre, London.

THOMAS, D. (1970) *London's Green Belt*, Faber, London.

THOMAS H. and HEALEY, P. (1991) *Dilemmas of Planning Practice: Ethics, Legitimacy and the Validation of Knowledge*, Gower, Aldershot.

THOMPSON, R. (1990) An achievable alternative for planning, in J. Montgomery and A. Thornley (eds.) op. cit.

THORNLEY, A. (1991) *Urban Planning under Thatcherism: The Challenge of the Market*, Routledge, London.

TOWNROE, P. and MARTIN, R. (eds.) (1992) *Regional Development in the 1990s: The British Isles in the 1990s*, Jessica Kingsley/Regional Studies Association, London.

TPI (Town Planning Institute) (1956) *Report on Planning in the London Region*, TPI, London.

TRIPP, H. A. (1942) *Town Planning and Road Traffic*, Arnold, London.

TUCKETT, I. (1990) Coin street: there *is* another way . . . , in J. Montgomery and A. Thornley (eds.) op. cit.

Tudor Walters Committee (1918) *Report of the Committee to Consider Questions of Building Construction in Connection with the Provision of Dwellings for the Working Classes in England and Wales, and Scotland* (Cd 9191), HMSO, London.

TYME, J. (1978) *Motorways versus Democracy: Public Inquiries into Road Proposals and their Political Consequences*, Macmillan, London.

UNDERWOOD, J. (1980) *Planners in Search of a Role: A Participant Observation in a London Borough.* Occasional Paper no. 6, School of Advanced Urban Studies, University of Bristol.

UNWIN, R. (1909) *Town Planning in Practice: An Introduction to the Art of Designing Cities and Suburbs*, Fisher Unwin, London.

UNWIN, R. (1912) *Nothing Gained by Overcrowding: How the Garden City Type of Development May Benefit Both Owner and Occupier*, GCTPA, London.

Uthwatt Committee (1941) *Interim Report of the Expert Committee on Compensation and Betterment* (Cmd 6291), HMSO, London.

Uthwatt Committee (1942) *Final Report of the Expert Committee on Compensation and Betterment* (Cmd 6386), HMSO, London.

WANNOP, U. (1984) The evolution and roles of the Scottish Development Agency, *Town Planning Review*, Vol. 55, pp. 313–21.

WARD, C. (1973) *Anarchism in Action*, Allen & Unwin, London.

WARD, C. (1976) *Housing: An Anarchist Approach*, Freedom Press, London.

WARD, S. V. (1974) The Town and Country Planning Act 1932, *The Planner*, Vol. 60, no. 5, pp. 685–9.

WARD, S. V. (1975) *Planning, Politics and Social Change 1939–1945*, Department of Town Planning Working Paper no. 1, Polytechnic of the South Bank, London.

WARD, S. V. (1984) List Q: a missing link in interwar public investment, *Public Administration*, Vol. 62, pp. 348–58.

WARD, S. V. (1986) Planmaking versus implementation: the example of List Q and the depressed areas 1922–1939, *Planning Perspectives*, Vol. 1, no. 1, pp. 3–26.

WARD, S. V. (1988) *The Geography of Interwar Britain: The State and Uneven Development*, Routledge, London.

WARD, S. V. (1990) Local industrial promotion and development policies 1899–1940, *Local Economy*, Vol. 5, no. 2, pp. 100–19.

WARD, S. V. (1991) The fairy ring, *Planning History*, Vol. 13, no. 1, pp. 29–31.

WARD, S. V. (1992a) The garden city introduced, in S. V. Ward (ed.) op. cit.

WARD, S. V. (ed.) (1992b) *The Garden City: Past, Present and Future*, Spon, London.

WATES, N. (1976) *The Battle for Tolmers Square*, Routledge & Kegan Paul, London.

WATES, N. and KNEVITT, C. (1987) *Community Architecture*, Penguin, Harmondsworth.

WATSON, J. P. and ABERCROMBIE, L. P. (1943) *A Plan for Plymouth*, Underhill, Plymouth.

WEIGHTMAN, G. and HUMPHRIES, S. (1984) *The Making of Modern London 1914–1939*, Sidgwick & Jackson, London.

Wheatley Commission (Royal Commission on Local Government in Scotland) (1969) *Report* (Cmnd 4150 I–II), 2 vols., HMSO, Edinburgh.

WHITEHAND, J. W. R. (1992) *The Making of the Urban Landscape*, Institute of British Geographers Special Publication no. 26, Blackwell, Oxford.

WHITEHOUSE, P. (1964) *Partners in Property: A History and Analysis of Institutional Finance for Property Development*, Birn Shaw, London.

WILKINSON, E. (1939) *The Town that was Murdered*, Gollancz, London.

WILLIAMS, N. (1939) *Population Problems of New Estates with Special Reference to Norris Green*, Liverpool University Press.

WILLIAMS-ELLIS, C. (1975, orig. 1928) *England and the Octopus*, Blackie, Glasgow.

WILSON, H. (1974) *The Labour Government 1964–70: A Personal Record*, Penguin, Harmondsworth.

WINTER, J., COOMBES, T. and FARTHING, S. (1993) Satisfaction with space around the home on large private sector estates: lessons from surveys in southern England and south Wales 1985–89, *Town Planning Review*, Vol. 64, no. 1, pp. 65–88.

WOHL, A. S. (1974) *The Eternal Slum: Housing and Social Policy in Victorian London*, Arnold, London.

WOHL, A. S. (1983) *Endangered Lives: Public Health in Victorian London*, Dent, London.

Woolton, Earl of (1959) *Memoirs*, Cassell, London.

WRIGHT, J. F. (1979) *Britain in the Age of Economic Management: An Economic History since 1939*, Oxford University Press.

WRIGHT, M. (1982) *Lord Leverhulme's Unknown Venture*, Hutchinson Benham, London.

YELLING, J. A. (1986) *Slums and Slum Clearance in Victorian London*, Allen & Unwin, London.

YELLING, J. A. (1992) *Slums and Redevelopment: Policy and Practice in England, 1918–45*, UCL Press, London.

YOUNG, H. (1989) *One of Us: A Biography of Margaret Thatcher*, Macmillan, London.

YOUNG, K. and GARSIDE, P. L. (1982) *Metropolitan London: Politics and Urban Change 1837–1981*, Arnold, London.

YOUNG, M. and WILLMOTT, P. (1957) *Family and Kinship in East London*, Routledge & Kegan Paul, London.

YOUNG, T. (1934) *Becontree and Dagenham*, Pilgrim Trust and Becontree Social Survey Committee, London.

INDEX

Abercrombie, Patrick 70, 72, 79, 87, 95, 103, 163, 266–7
Adams, Thomas 34, 36, 54
Addison, Christopher 41–3, 52
affluent society 117–24, 152–3, 177, 191, 201
Aldridge, Henry R. 23
Amos, Jim 205
architects (and planning): before 1914 26, 29, 34, 36, 1914–1939 50, 53, 63, 69, 78, after 1939 95, 104, 111, 126, 155-8, 216
Arndale Centres 150–1, 243
Attlee, Clement 100–1, 120, 194

back-to-back housing 11, 16, 60–1
Barlow Commission 78–9, 80, 86–7, 260
Barnett, Henrietta 29
Becontree-Dagenham, London 49, 53
Bellamy, Edward 24
betterment 3, before 1947 33, 46, 88, 90–1, 1947–1974 108, 114, 124, 126, 130, 138, after 1974 210, 222, 252
Beveridge Report 92
birth rate 118, 120
Board of Trade: before 1952 73, 79, 81, 100, 110, 1952–74 130, 169-70, 172, 175-6, 181
bombing 82–5, 88–9, 94
Bor, Walter 136, 157
Bournville 21–2, 24, 26, 29, 56
Brooke, Henry 164–5
Brown, George 122, 129–30, 150, 171–2, 181, 196
Brundtland Report 218, 220
Buchanan, Colin 148–9
building licences 83, 99–100, 110, 114, 143
building societies 44–5, 62, 262
Bull Ring Centre, Birmingham 150
Burnham, Daniel 37
Burns, John 33
Burns, Wilfred 136, 153, 155
by-law housing 16–17, 21–2, 29, 35, 37

Cadbury, George 21–3, 26, 29, 51. George Jnr 31
Callaghan, James 192, 203
Canary Wharf, London 239–41, 272
central areas: before 1952 13, 16, 37, 61–3, 95–6, 1952–74 143–52, 158, 186, after 1974 224, 235, 238–9, 242–4, planning impacts 260–2, 264, 276
Central Land Board 108, 114, 130
Centre for Policy Studies 205, 219
Chamberlain, Joseph 14, 27
Chamberlain, Neville 42–3, 52, 65, 73, 76, 78–9, 86
Channel Tunnel 195, 248
Chelmsford Committee 71–2
Cherry, Gordon 2, 36
Churchill, Winston S. 86, 92, 100, 116, 119, 124, 226
CIAM – see Congres Internationaux Architecture Moderne
City Grant 232
City Beautiful 37, 62, 241
City Challenge 234
civic design 36–7, 96
Civic Amenities Act 1967 135–6
Civic Trust 135
Clyde Valley Plan 1946 102–3, 164, 254, 266
co-partnership societies 21, 23–4, 28–9, 33
Coin Street, London 216–8
Commission for the New Towns 177, 249
Community Land Act 1975 196–7, 208
Community Development Projects 197
compact city 218, 256–7
compensation 3, before 1947 33, 46–7, 56, 88, 90–1, after 1947 107–9, 114, 124–6, 138–9, 161, 374–5, 272
comprehensive development areas 94, 109–10, 133, 143–7, 149, 152
Congres Internationaux Architecture Moderne (CIAM) 63, 65, 69
conservation areas 135–6, 211, 220–1, 266

Consortium Developments 251–4, 256
containment 2–3, 8–9, origins 39, 56, 58–60, 97–8, policies 1955–74 127, 142, 155, 161–8, 186, policies after 1974 224, 254–8, impacts 260, 264–7, 269, 274
Control of Pollution Act 1974 203
Control of Land Use 1944 94–5
Control of Office and Industrial Development Act 1965 129–30, 171
corporate planning 204–5
Council for the Preservation/Protection of Rural England (CPRE) 56, 256, 267
County of London Plan 1943 95–8, 106
crime 191, 227
critical theory 204
Crosland, Anthony 192–3, 196, 210, 249
Crossman, Richard 129, 131–3, 152, 165–7, 172, 181
'culture of contentment' 191, 207

Dalton, Hugh 76, 100
decentralization (and planning) 2–3, 8–9, origins 24, 39, 51–6, 58, 71, policies to 1974 87, 97, 102–6, 142, 155, 162, 166, 176–86, policies since 1974: 224, 238, 243, 248–54, 258; impacts 261, 266–70
deindustrialization 189–90
Dennison, S. R. 90
depressed areas 71–6, 77–8, 81, 87, 99–100, 113, 168–9, 244
deregulation 214, 222, 239, 244, 252–4, 256, 273
Derelict Land Grant (DLG) 230, 237
development charge 108, 114–5, 130
development control 47–8, 109–10, 127, 140, 208, 210, 214
development land 47, 108, 114, 116, 126, 130–1, 165–6, 195–6, 250–2, 271–2, 278
Development Areas 100, 169, 173–4, 246
Development Corporation, New Town 7, 102, 105, 113, 230
Development Plans 109, 111, 113, 126–8, 131, 137, 162–4, 171, 212–4, 219–20, 271
Distribution of Industry Act 1945 100, 110, 113
Dobry Report 210
Docklands, London 218, 224, 228, 230, 235–42, 248
Dower, John 92, 112
Drancy-la-Muette, Paris 64–5, 69, 157
'dual market' (in land) 125
Dudley Report 96–7

Economic Affairs, Department of (DEA) 121–2, 129–31, 171–2, 180–1, 186
economic planning: before 1974 77, 120–1, 126, 131, 171–3 after 1974 189, 191–5, 245–8, 260

Eden, Anthony 119, 123
Edgbaston, Birmingham 15, 32
employment policy 99–100, 190
engineers (and planning) 34, 36–7, 47, 65, 136, 155
Englsih Heritage 211
Enterprise Zones 208–9, 230–3
Environment, Department of the (DOE): before 1979 137–8, 172, 195, 197–9, 203, after 1979 210–1, 213, 221, 229, 252, 258
Environmental Protection Act 1990 220
environmental pollution 201, 203
environmentalism 201–3, 278
European (Economic) Community 194, 207, 247
European Regional Development Fund (ERDF) 247–8
Euston Centre, London 144–5, 151
Eversley David 172
Expanded Towns 112–3, 171, 182, 184–6, 249; impacts 263, 266–8, 272, 274

Federation of British Industries 79, 84, 114
'five per cent philanthropy' 17
Five Year Programme (of slum clearance) 66
Forshaw, J. H. 95
Forster, E. M. 103
Friends of the Earth 203, 257

Galbraith, J. K. 118, 191, 207
Galley, Ken 205
garden cities 5, before 1914: 23–9, 41, 51–5, 63, 65, after 1939 72, 103, 105, 113, 157, 217–8
Garden City (and Town Planning) Association 2, 24, 28, 33, 56, 86
garden festivals 231, 236
garden suburbs 28–9, 32, 34, 37, 41–3, 45, 51
Geddes, Patrick 29–30, 36, 109
General Development Order (GDO) 210, 213, 219
General Improvement Area (GIA) 160–1
gentrification 161, 264
George, Henry 20
German influences: before 1939 27–8, 30–1, 36–7, 52, 62-3, 65, 77–8, after 1939 85, 96, 236, 247
Gibson, Donald 95–6, 143
Glasgow Eastern Area Renewal (GEAR) 225, 246
Greater London Development Plan 138
Greater London Regional Planning Committee (GLRPC) 53, 58, 71
Greater London Council (GLC) 139, 212, 217–8, 234, 247
Greater London Plan 1944 95, 97–9, 102–3, 106, 112, 266

green belts: origins 54–60, 71, 97–8, after 1955 155, 163–8, 170, 249–50, 253, 254–8, impacts 264, 268, 273
green movement 214, 216–9, 256, 278
Greenwood, Anthony 132-3
Greenwood Arthur 43, 46–7, 54
growth point 180, 182
growth pole idea 5, 171, 173

Hackney, Rod 216
Hailsham, Lord 170–1, 180, 228
Hall, Peter 2, 136, 208–9, 222, 271
Hammerson 151
Hampstead Garden Suburb, London 29, 41
Harborne Garden Suburb, Birmingham 29, 32
Harmsworth, Alfred 26, 32
Haydock Park, Lancashire 186
Health, Ministry of 44, 53, 54, 72, 73, 88–9
Heath, Edward 119, 122, 138-9, 152, 169–70, 174, 192–6, 226
heritage 211, 221, 237, 243
Heseltine, Michael 205, 207, 209–11, 221, 252, and inner cities 227–9, 232, 235–6
Hichens, Lionel 76
high rise housing 155–60, 206
Hill, Octavia 20, 29
Hillington Trading Estate, Glasgow 75
Hnchingbrooke, Viscount 103
Hobhouse Committee 112
Holford, William 92
homelessness 191, 275
Hong Kong 209
Horsfall, Thomas Coghlan 27–8, 31, 36
Housebuilders Association 91
Housing Act 1969 160–1
Housing and Local Government, Ministry of: and planning system 115, 124, 129, 138, and strategic policies 146–7, 149, 155, 162, 164–5, 170–2, 180, 185–6
Housing and Planning Act 1986 46, 212–3, 221
housing reform 2, 20–4, 31, 36, 41, 52
Housing, Town Planning Etc. Act 1909 10, 32–5, 37, 39
Housing, Town Planning Etc. Act 1919 41–2, 44
Housing of the Working Classes Act 1890 17–8
Howard, Ebenezer 23–8, 31, 38, 105, 113, 216–7, 252
Hyams, Harry 152

Idris, Thomas 26
immigration 119, 262
industrial planning 72, 75–6, 221
Industrial Development Certificiate (IDC) 100, 110, 129, 245
Industrial Improvement Areas 198, 200, 225, 229
inner cities 152–61, 197–200, 224–43, 257, 262–4, 270
Inner Urban Areas Act 1978 197–200
Institute of Economic Affairs 205
Intermediate Areas 174, 246

Jarrow 73–5, 77
Jenkin, Patrick 207, 255–6
Joseph, Keith 131, 155, 165

Keynes, John Maynard 71–2, 77, Keynesianism 99, 117, 120, 168, 188–9, 192, 223
Kropotkin, Peter 24

Labour, Ministry of 73
Laing 149–50
Land Commission 129–30, 138, 196
Land Compensation Act 1973 138–9
Land Nationalisation Society 20, 23–4, 26
land nationalization 91, 196
Land and Natural Resources, Ministry of 129–30
Land and Planning, proposed Ministry of 129
land reform 2, 19–20, 24, 26, 279
Land Restoration League 20
'land saving' 161–3
land taxation 20, 33, 195–6
Land Utilisation Survey 49, 161
Le Corbusier 63, 65, 144, 157
Leitch Committee 203
Letchworth Garden City: before 1914 24, 26–9, 31, 36, 1914–1939 41, 52–3, 56, 75
Lever, William H. 21–2, 26
Levy, Joe 144–5, 151
Lifting the Burden 212, 214, 219-21
Lightmoor, Telford 217
linear city 65, 183
List Q 72
Livingstone, Ken 212, 217–8
Lloyd George, David 42, 92
Local Government Board 33, 36, 41
Local Government and Planning, Ministry of 114
Local Government, Planning and Land Act 1980 208
local government reform 136, 139–40, 166–7, 171, 201, 204, 222
local plans: origins 131–3, 135, 140, 201, pressures to change 198, 208, 212, 214–5, 221
Location of Offices Bureau 130, 244–5
Lock, Max 111
London Passenger Transport Board (LPTB) 44, 49

London County Council (LCC): before 1914 14–5, 17, 28–9, 33, 1914–52 49, 53, 58–60, 65, 69, 95, 101, 1952-65 126, 136, 144–5, 152, 155, 176, 179, 265
Lower Earley, Berkshire 251–2
Lutyens, Edwin L. 29, 78, 95

MacDonald, Ramsay 43
Macmillan, Harold 76, 163, 176, 189, 226
Major, John 192, 221–2, 229, 234
Marley Committee 54
MARS Group 65, 183
Marshall, Alfred 24
Maudling, Reginald 160, 169
May, Ernst 52–3
Milton Keynes 181, 183
mixed development 69, 96–7, 155
modernism 62–5, 67–9, 143, 149, 152–60
Morris, William 24, 26, 31, 37
Morrison, Herbert 58–60, 102
Morrison W. S. 92–3, 95
Mortimer, John 207, 253
Mosley, Oswald 71, 77
motor vehicles: 1914–1939 39, 54–5, 61–2, 71, after 1939 118, 146, 148–9, 183, 190, 203, and urban change 261, 265–6, 269, 276
motorways 125, 139, 170, 248, 251
municipal housing: before 1914 16–19, 22–3, 28–9, 32–3, interwar 42, 48–51, 66–9, and urban change 262

National Enterprise Board 192, 245
National Economic Development Council 121
National Housing Reform Council 23, 27–8, 32
National Housing and Town Planning Council 32
National Plan 122, 171–2
nationalization 121, 192
Nazis 77–8
neighbourhood unit 54, 96–7, 102, 104–5, 178, 184, 265
Nettlefold, John S, 26–9, 31–4, 51, 196
Neville, Ralph 26
New Earswick, York 29, 41
New Towns 3, 5, 7, origins 52, 98, 102–7, 112–3, 1952–74 142, 146, 156, 165–6, 171, 176–86, after 1974 210, 217, 224, 248–54, impacts 263, 266–7, 272, 274–5, 277
Nicolson, Harold 85
nimby 207–8, 256–8
north-south divide 244

Office Development Permit (ODP) 129, 171–2, 239, 243, 245
Orwell, George 50, 66
Osborn, Frederic J. 52, 77, 93, 102, 157, 161
Our Common Future 218
outer cities 265–9, 270
owner-occupation 4, 44–5, 49–50, 130, 249, and urban change 262, 273–4
Oxford Preservation Trust 56

Paris 51, 63–5
Parker, Barry 26, 28, 29, 34, 53–4, 58
Parker-Morris standards 155
partnership: (with developers) 127, 143–6, 151–2, 158, 196, 235–7, 242, (under 1978 Inner Urban Areas Act) 198, 224–5, 229
Patten, Chris 290, 254
Perroux, Francois 171
Perry, Clarence 54, 102, 105
Pick, Frank 44, 49
Pilgrim suicide 124–5
place marketing 232–4, 248
Planning Advisory Group (PAG) 131–3, 137, 140, 148, 201
planning aid 216, 277
Planning and Compensation Act 1991 221–2
planning gain 127, 210–1, 222, 232, 251–2
planning obligations 222
Political and Economic Planning (PEP) 77, 79, 260
Port Sunlight 21, 24, 26, 28, 37
'positive planning' 195, 207
Poulson, John 160
poverty 20, 29, 72, 78, 92, 119, 191, and inner cities 197, 226–7, 247
precinct 96, 99, 143, 148–51, 177
private tenants 274–5
private housebuilding 41, 43–8, 102, 124, 249–54, 256
privatization 193–4, 249
Property Advisory Group 210, 213
property owning democracy 123–4
Public Health Act 1875 16
public interest, the 4, 15, 110–12, 206–8
public involvement 111–12, 133, 216
public participation 133–5, 202
public sector tenants 194, 274
public transport: before 1939 13, 15, 16, 43–5, 49, after 1945 183, 243–4, 261, 265, 275
Purdom, C. B. 52

Quarry Hill, Leeds 69

racism 119, 191
Radburn layout 54–5, 96, 162, 178
rationing 80–1, 106, 117, 120, 144, 146, 176
Ravenseft 143, 177
red-lining 262–4
Redcliffe-Maud Commission 139, 166

redevelopment(and planning) 2–3, 6–7, 9, 1914–1939 39, 43, 46, 60, 62, 65–9, 76, 1939–52, 83, 87, 91, 94, 97–8, 106, 108, 1952–74 131, 139, 143–61, 166, 179, 186–7, since 1974 197, 204–6, 224, 235, 241–2, 249, impacts 260–3, 274, 277
Regional Economic Planning Councils (REOCs) 171–4, 247
regional planning: pre 1952 53, 58, 69–73, 1952–1974 121, 129, 131, 137–8, 166, 169–74, since 1974 188, 224, 244–5, 247–8, 250–2, 258
regional policy: before 1952 73–6, 99–100, 110, 1952–74 126, 150, 168–71, 173–6, 186, since 1974 220, 224, 234, 244–7, 258, impacts 269–71
regional problem 71–2, 245–6, 269–70
regional shopping centres 239
Reigh, Lord 87–9, 92, 94–7, 102–3, 115
Rent and Mortgage Interest (War Restrictions) Act 1915 41
ribbon development 48
Ridley, Nicholas 207, 214, 219, 232, 252–4, 256
riots 119, 191, 226–8, 235
Robertson, John R. 26
Ronan Point, London 158–9, 206
Roosevelt, Franklin D. 54
Rowntree, Joseph 29
Rowntree, Seebohm 29, 30

Salt, Titus 21
Sandys, Duncan 135, 163–5
satellite towns 5, before 1939 52–4, 58, 63, 70–1, 72, 75, since 1939 179, 265
Schuster Committee 111
Scott Committee 89–93, 112, 161
Scottish Development Agency (SDA) 245
self-containment 53, 105, 265–7
SERPLAN 247
Sharp, Evelyn 129, 156
Sharp, Thomas 86, 92, 127, 131
Shaw, George Bernard 23
Shore, Peter 196, 210, 224, 228–9, 232, 235, 248–9
Sieff, Israel 76
Silkin, Lewis 101–3, 106–8, 111–2
Silkin, John 195–6
Simplified Planning Zones (SPZ) 212–3
Sitte, Camillo 36–7
Skeffington Committee 133–5
Slough Trading Estate 53, 57
slums: before 1939 12, 28, 32, 61, 66–7, after 1939 119, 152–5, 158, 176, 262
Smigielski, Konrad 136, 146
Smith, T. Dan 136, 160, 170, 172
Soria y Mata 65
South East Study 171, 180, 266
South Woodham Ferrers, Essex 251
South Wales Regional Survey 69
Special Areas 73–6, 79, 100
Special Development Areas 174, 246
Speke, Liverpool 53, 75, 179
Stamp, L. Dudley 49, 90, 161
Stein, Clarence 54–5
Stephenson, Gordon 92
Stevenage New Town 103–5, 177, 183
Stewart, Malcolm 76, 79
Stilgoe, Henry R. 34
strategic planning 9, 87, 106, 161, 166, 168, retreat from 203, 208, 212, 224, 248, 257–8
'streets in the sky' 156–8
structure plans: origins 132–3, 137–40, downgrading 198, 201, 208, 211–2, 214, 247, 252, reprieved 220–2, and green belts 254
sub-regional planning studies 133, 137, 166, 172
suburbanization: inter-war 39–53, 77, 87, post-war 162–3, 179, and planning impacts 267–8
suburbs 2, 5, before 1939 13, 28, 33–5, 37, after 1945 185, 218, 254, and planning impacts 263–4
surveyors (and planning) 34–7, 114
sustainable development 218, 278–9
systems planning 137, 203

Team Valley Trading Estate, Gateshead 75
Tebbit, Norman 227
'technological revolution' 121–2, 127, 131, 247
tenements 11–12, 16–18, 20
Thatcher, Margaret and Thatcherism 188, 191–5, 205–23, 225–36, 240, 242, 247–57, 279
This Common Inheritance 220–2, 248
Town and Country Planning, Ministry of 92–3, 95–6, 104, 106–8
Town and Country Planning Act 1932 46–8, 67, 94, 103, 109
Town and Country Planning Act 1944 94, 108
Town and Country Planning Act 1947: main features 106, 108–14, modifications 116, 124–6, 130–7, 140, 143, 146, 161
Town and Country Planning Acts 1953, 1954 and 1959 124–5
Town and Country Planning Act 1962 and 1963 126
Town and Country Planning Act 1968 131–3
Town and Country Planning Act 1971 138, 210
Town and Country Planning Act 1972 138

Town and Country Planning Acts 1990 220–1
Town and Country Planning Association 86, 102, 157, 161, 203, 216–7
Town and Country Planning (Interim Development) Act 1943 93–4
Town Development Act 1952 112–3, 116, 163, 176–9
Town Planning Institute 2, 10, 36, 85, 111, 136
town planning schemes 33-6, 41–2, 44, 47–8, 69–70, 94, 109
Traffic in Towns Report 146–9
Trafford Park Industrial Estate, Manchester 75, 230, 236
tramways 13–4, 16, 22, 28, supertrams 237, 243
Transport, Ministry of 78, 138, 146–8, 170
Treforest trading Estate, Pontypridd 75
Tripp, H. Alker 96, 148
Tudor Walters Report 41, 43, 51

unemployment: before 1952 31, 71–8, 81, 92, 99, and post-war affluence 117, 119, 168–9, 174, after 1974 188, 190, 192, 194, 197, 226–7, 242, 244–5, 247, and planning impacts 265, 269
Unhealthy Areas Commitee 52, 58, 65
unitary development plans 212
Unwin, Raymond: before 1914 26–9, 34, 37–8, 1914–39 41, 43, 47, 53–4, 58–60, 71
Urban Development/Regeneration Grants (UDG/URG) 232
Urban Programme 225, 237
Urban Development Corporations (UDCs) 208–10, 218, 226, 229–31, 234–7, 239–43, 249
US influences: before 1974 30, 54–7, 89, 118, 149, since 1974 191, 194, 227, 232, 261
Use Classes Order 1987 213
Uthwatt Committee 88–96, 101, 106, 108, 115, 130, 195

Victoria Park, Manchester 15
Vienna 63–5, 66, 69
Villeurbanne, Lyons 65

Waldegrave, William 256
Walker, Peter 167
Wallace, Alfred Russel 20
Watson, J. Paton 95
Welsh Development Agency (WDA) 245
Welwyn Garden City 52–3, 56, 75, 103
Wheatley Act 42–3, 66
Wilson, Harold 119, 121–2, 129, 131, 138, 174, 189, 191–2
women 31–2, 40, 117–8, 218, 275–6
Works, Ministry of: and Buildings 83, 87, and Planning 89
Wrights, Henry 54–5
Wythenshawe, Manchester 53–4, 58, 70, 75, 179

Young, Michael and Willmott, Peter 160

zoning 2, 4, before 1939 28, 33, 36, 47, 63, after 1939 105, 109, 213–4, impacts 263, 275